AF594976

Applications of Remote Sensing in Coastal Areas

Applications of Remote Sensing in Coastal Areas

Special Issue Editors

Konstantinos Topouzelis
Apostolos Papakonstantinou
Suman Singha
XiaoMing Li
Dimitris Poursanidis

MDPI • Basel • Beijing • Wuhan • Barcelona • Belgrade • Manchester • Tokyo • Cluj • Tianjin

Special Issue Editors
Konstantinos Topouzelis
University of the Aegean
Greece

Apostolos Papakonstantinou
University of the Aegean
Greece

Suman Singha
German Aerospace Center (DLR)
Germany

XiaoMing Li
Key Laboratory of Digital Earth Science
China

Dimitris Poursanidis
Foundation for Research and Technology
Greece

Editorial Office
MDPI
St. Alban-Anlage 66
4052 Basel, Switzerland

This is a reprint of articles from the Special Issue published online in the open access journal *Remote Sensing* (ISSN 2072-4292) (available at: https://www.mdpi.com/journal/remotesensing/special_issues/Coastal_Areas).

For citation purposes, cite each article independently as indicated on the article page online and as indicated below:

LastName, A.A.; LastName, B.B.; LastName, C.C. Article Title. *Journal Name* **Year**, *Article Number*, Page Range.

ISBN 978-3-03928-658-4 (Hbk)
ISBN 978-3-03928-659-1 (PDF)

Contents

About the Special Issue Editors

Konstantinos Topouzelis is Assistant Professor at the Department of Marine Sciences, University of the Aegean, where he leads the Marine Remote Sensing Group. He holds a bachelor's degree in Environmental Studies (University of the Aegean, Department of Environment), a master's degree in Remote Sensing (University of Dundee, Department of Applied Physics & Mechanical Engineering) and a Ph.D. in pollution monitoring from space (National Technical University of Athens, School of Rural and Surveying Engineering). His main research interest is on the analysis of remote sensing datasets, including satellite and aerial images, for marine and coastal applications. His expertise includes automatic detection of oceanographic phenomena, object-based image analysis, image processing algorithms, and coastal mapping.

Apostolos Papakonstantinou is currently Senior Postdoc Researcher in the Marine Remote Sensing Group, conducting his second postdoctoral research position funded by the Greek State Scholarships Foundation on the detection and mapping of plastic litter in coastal zones using unmanned aerial systems and deep learning techniques. His previous postdoc that was also funded by the Greek State Scholarships Foundation was on the potential use of unmanned aerial systems for the geovisualization and mapping of coastal environments. Additionally, it is a Contracted Lecturer at the Department of Civil Engineering and Geomatics at Cyprus University of Technology. He holds a bachelor's degree in Physics (University of Ioannina, Department of Physics), an M.Sc. in Cultural Informatics and Communication (Department of Cultural Informatics and Communication, University of the Aegean), and a Ph.D. in Geoinformatics, with specialization in Cartography and Geovisualization (University of the Aegean, Department of Geography). His current research focuses on the use of UAS in 2D and 3D mapping of spatiotemporal phenomena in the coastal environment. His expertise includes UAS data acquisition and image processing for 2–3D geovisualization applications, computational methods, and algorithms in automated cartographic visualization, scale issues in cartographic representation, and coastal mapping.

Suman Singha received his M.Tech. degree in remote sensing from the Indian Institute of Technology Roorkee, India, in 2009, and M.Sc. degree in remote sensing and Ph.D. degree in microwave remote sensing from the University of Hull, Hull, U.K., in 2010 and 2014, respectively. He was a Visiting Scientist with the European Maritime Safety Agency, Lisbon, Portugal, in 2012. Since 2013, he has been Research Scientist with the Remote Sensing Technology Institute, German Aerospace Center, Bremen, Germany. His research interests include the application of artificial intelligence and machine vision approaches in classifying traditional and polarimetric synthetic aperture radar images with an emphasis on sea ice classification and oil spill detection.

XiaoMing Li Professor at the Aerospace Information Research Institute, Chinese Academy of Science, where he leads the Marine Remote Sensing team. He obtained his Ph.D. degrees from the University of Hamburg, Germany, and Ocean University of China, in the subjects of Geophysics and Ocean Remote Sensing, respectively. His main research is in SAR oceanography, focusing on the retrieval of marine-meteorology parameters of ocean surface wind, wave, and current, and observations of oceanic and atmospheric dynamic processes.

Dimitris Poursanidis is Research Fellow at the Remote Sensing lab of the Foundation for Research and Technology, Hellas. He holds a Ph.D. in Marine Ecology and an M.Sc. in Coastal Zone Management from the University of the Aegean, Dept. of Marine Science, Greece. He has been a research visitor in ITC (University of Twente's Faculty of Geo-Information Science and Earth Observation) for six months working on Ecological Modeling and Remote Sensing, whilst he has experience working on research and development projects in both the academic and industry sector. His scientific interests are concerned with (a) the collection and use of high-quality field data using scientific diving and modern tools and methods for data analysis and conservation decisions, (b) the use of Earth Observation and remote sensing data for the mapping and monitoring of coastal habitats (seagrass, coral reefs, wetlands), (c) understanding of the limits of EO in the estimation of satellite bathymetry and seascape mapping, (d) the application of ecological modeling using EO data in support of conservation decisions and protected areas management, and (e) the use of the state-of-the-art tools and methods for ecological monitoring.

Editorial

Editorial on Special Issue "Applications of Remote Sensing in Coastal Areas"

Konstantinos Topouzelis [1,*], Apostolos Papakonstantinou [1], Suman Singha [2], XiaoMing Li [3] and Dimitris Poursanidis [4]

1 Department of Marine Sciences, University of the Aegean, University Hill, 81100 Lesvos, Greece; apapak@aegean.gr

2 German Aerospace Center (DLR), Remote Sensing Technology Institute, Am Fallturm 9, 28359 Bremen, Germany; suman.singha@dlr.de

3 Key Laboratory of Digital Earth Science, Institute of Remote Sensing and Digital Earth, Chinese Academy of Sciences, Beijing 100094, China; lixm@radi.ac.cn

4 Foundation for Research and Technology, Institute of Applied and Computational Mathematics, Remote Sensing Lab, N. Plastira 100, 70013 Heraklion, Crete, Greece; dpoursanidis@iacm.forth.gr

* Correspondence: topouzelis@marine.aegean.gr

Received: 11 March 2020; Accepted: 16 March 2020; Published: 18 March 2020

Coastal areas are remarkable regions with high spatiotemporal variability. Many domains are affected by their physical and biological processes, from tourism to biodiversity and productivity. Coastal ecosystems perform several critical ecosystem services and functions such as water oxygenation and nutrient provision, seafloor and beach stabilization (as sediment is controlled and trapped within the rhizomes of the seagrass meadows), carbon burial, and areas for nursery and refuge of several commercial and endemic species. Knowledge of the spatial distribution of marine habitats is prerequisite information for the conservation and sustainable use of marine resources. This Special Issue contains 14 papers in several fields of coastal remote sensing and reveals the potential of remote sensing in integrated coastal management.

Cliff coasts are dynamic environments that can retreat very quickly. However, short-term changes and factors contributing to cliff coast erosion have not received as much attention as dune coasts. Terefenko et al. [1] conducted work at three soft-cliff systems in the southern Baltic Sea; these have been monitored with the use of terrestrial laser scanner technology over a period of almost two years to generate a time series of 13 topographic surveys. Digital elevation models constructed for those surveys allowed the extraction of several geomorphological indicators describing coastal dynamics. Combined with observational and modeled datasets on hydrological and meteorological conditions, descriptive and statistical analyses were performed to evaluate cliff coast erosion. A new statistical model of short-term cliff erosion was developed by using a non-parametric Bayesian network approach. The results revealed the complexity and diversity of the physical processes influencing both beach and cliff erosion. Wind, waves, sea levels, and precipitation were shown to have different impacts on each part of the coastal profile. At each level, different indicators were useful for describing the conditional dependency between storm conditions and erosion. These results are an important step toward a predictive model of cliff erosion.

Poursanidis et al. [2], for first time, exploit the capabilities of PlanetScope Cubesats for the calculation of coastal bathymetry. High spatial and temporal resolution satellite remote sensing estimates are the silver bullet for monitoring of coastal marine areas globally. From 2000, when the first commercial satellite platforms appeared offering high spatial resolution data, mapping of coastal habitats and extraction of bathymetric information have been possible at local scales. Since then, several platforms have offered such data, although not at high temporal resolution, making the selection of suitable images challenging, especially in areas with high cloud coverage. PlanetScope

CubeSats appear to cover this gap by providing their relevant imagery. The current study is the first that examines the suitability of them for calculating satellite-derived bathymetry. The availability of daily data allows the selection of the most qualitatively suitable images within the desired timeframe. The application of an empirical method of spaceborne bathymetry estimation provides promising results, with depth errors that fit to the requirements of the International Hydrographic Organization at the Category Zone of Confidence for the inclusion of these data in navigation maps. While this is a pilot study in a small area, more studies in areas with diverse water types are required for solid conclusions on the requirements and limitations of such approaches in coastal bathymetry estimations.

Coastal dunes provide the hinterland with natural protection from marine dynamics. The specialized plant species that constitute dune vegetation communities are descriptive of the dune evolution status, which in turn reveals the ongoing coastal dynamics. De Giglio et al. [3] work towards demonstrating the applicability of a low-cost unmanned aerial system for the classification of dune vegetation, in order to determine the level of detail achievable for the identification of vegetation communities and define the best-performing classification method for the dune environment according to pixel-based and object-based approaches. These goals were pursued by studying the North Adriatic coastal dunes of Casal Borsetti (Ravenna, Italy). Four classification algorithms were applied to three-band orthoimages (red, green, and near-infrared). All classification maps were validated through ground truthing, and comparisons were performed for the three statistical methods, based on the k coefficient and on correctly or incorrectly classified pixel proportions of two maps. All classifications recognized the five vegetation classes considered, and high spatial resolution maps were produced (0.15 m). For both pixel-based and object-based methods, the support vector machine algorithm demonstrated a better accuracy for class recognition. The comparison revealed that an object approach is the better technique, although the required level of detail determines the final decision.

Coastal areas harbor the most threatened ecosystems on Earth, and cost-effective ways to monitor and protect them are urgently needed, but they represent a challenge for habitat mapping and multitemporal observations. The availability of open access remotely sensed data with increasing spatial and spectral resolution is promising in this context. Thus, in a sector of the Mediterranean coast (Lazio region, Italy), Marzialetti et al. [4] tested the strength of a phenology-based vegetation mapping approach and statistically compared results with previous studies, making use of open source products across all the processing chain. We identified five accurate land cover classes in three hierarchical levels, with good values of agreement with previous studies for the first and the second hierarchical levels. The implemented procedure resulted as being effective for mapping a highly fragmented coastal dune system. This is encouraging to take advantage of Earth observations through remote sensing technology in an open source perspective, even at the fine scale of highly fragmented sand dunes landscapes.

One of the most important linear features on the Earth's surface is coastline; thus, the detection and monitoring of dynamic coastlines through time and space is critical for tracking changes in vulnerable coastal zones and managing increasingly threatened water resources. In their study, Bishop-Taylor et al. [5] evaluated a method for mapping waterlines at subpixel accuracy from satellite remote sensing data, combining a synthetic landscape approach with high-resolution WorldView-2 satellite imagery. Their method reproduced, with confidence, both absolute waterline positions and relative shape at a resolution that exceeds that of whole-pixel thresholding methods in environments without extreme contrast between water and land. Their subpixel waterline extraction method is available as an open source tool and has low computational overhead; thus, it is suitable for continental-scale or full time-depth analyses aimed at accurately mapping and monitoring dynamic waterlines.

Sha et al. [6] revealed asymmetric oceanic thermal responses corresponding to an island wind wake and proposed associated mechanisms with ocean heat advection terms. Using multisensor remote sensing observations including advanced synthetic aperture radar (ASAR), the work investigated the sub-mesoscale features of the local wind wake. Then, by combining the satellite observations and model results, the ocean heat advection terms were reconstructed and compared with the air–sea heat flux. The results highlighted the contribution of the wind-driven heat advection in the regional ocean thermal dynamic process.

Unmanned aerial systems (UAS) are used in an increasing number of applications, especially for data acquisition, to map spatiotemporal changes. Papakonstantinou et al. [7] used UAS true-color RGB (tc-RGB) and multispectral high-resolution orthomosaics by applying object-based image analysis (OBIA) to map marine habitats. Furthermore, they examined the usefulness of bathymetry and the roughness of information derived from the echo sounder as training data to the UAS-OBIA methodology, applying three different scenarios using k-nearest neighbor (k-NN), fuzzy rules, and their combination as classifiers. High-resolution classification maps produced from the methodology followed, providing valuable information regarding the current state of the habitat species and enabling in-depth analysis of change detections caused by anthropogenic interventions and other factors.

Sea-level rise, storm surges, and many other ocean dynamics affect coast vulnerability and lead to hazards such as beach erosion, sedimentation, or inundation. By using 40 years of satellite observations from multisensors, Waqas et al. [8] analyzed spatial and temporal variations of barrier islands (BIs) along the Indus Delta region of the Pakistan coast. They concluded that approximately 75% of these BIs are vulnerable to the ocean controlling factors. Therefore, coastal protection and management along the Indus Delta should be adopted to defend against the erosive action of the ocean.

On the other hand, reclaimed lands or islands may also suffer the problem of ground subsidence in addition to the erosion by the ocean. Liu et al. [9] showed that the maximum annual subsidence rate of the new airport constructed on the reclaimed land of the Xiamen City reached −130 mm/year between 2015 and 2016, based on the interferometry of Sentinel-1 SAR data.

To map fine variations of the coastal zone, Almeida et al. [10] proposed a method of deriving high-resolution (2.0 m) DEM from the Pleiades satellite data with a pixel size of 0.7 m, which showed a good agreement with ground truth measurements by GPS.

Besides spaceborne optical and radar sensors, which are widely used for coastal monitoring, other instruments exhibit their advantages for some applications. Ma et al. [11] proposed a method of detecting photon signals of Lidar, by which they estimated surface profiles of different surface types in coastal zones.

Song et al. [12] proposed a semi-global subpixel shoreline localization method based on Landsat 8 Operational Land Imager (OLI) data. Authors selected the port of Caofeidian and the Xiamen coastal area as the study areas and utilized higher-resolution shoreline extracted from GF-2, which is capable of acquiring optical images with a spatial resolution better than 1 m. The proposed methodology utilizes global spectral information and shoreline morphological features coupled with local water index homogeneity features to determine the artificial shoreline with an RMSE of less than 5 m.

Cao et al. [13] analyzed 40 features extracted via polarimetric decomposition in the full-polarimetric (FP), simulated compact-polarimetric (CP), and dual-polarimetric (DP) SAR modes for ship detection using the Euclidean distance and mutual information. Authors found that that the features in CP SAR were better than those of FP or DP SAR in general. The study also proposed a CP SAR based feature, named 'phase factor'. In the framework of the study, the authors concluded the 'phase factor' based detector performed better than other traditional ship detection techniques in low, medium, and high sea states.

Su et al. [14] proposed sea ice information indexes using medium resolution Sentinel-3 Ocean and Land Color Instrument (OLCI) images and validated the index performance using Sentinel-2 MultiSpectral Instrument (MSI) images with higher spatial resolution. The study evaluated the proposed Enhanced Normalized Difference Sea Ice Information Index based on 4 OLCI bands (B12, B16, B20, and B21). Authors demonstrated that the proposed index effectively detected sea ice information in the Bohai Sea and suppressed most background information compared to other established methods.

Author Contributions: The guest editors contributed equally to all aspects of this editorial. All guest editors have read and agreed to the published version of the manuscript.

Acknowledgments: The guest editors would like to thank the authors who contributed to this Special Issue and to the reviewers who dedicated their time for providing the authors with valuable and constructive recommendations.

Conflicts of Interest: The guest editors declare no conflicts of interest.

References

1. Terefenko, P.; Paprotny, D.; Giza, A.; Morales-Nápoles, O.; Kubicki, A.; Walczakiewicz, S. Monitoring cliff erosion with LiDAR surveys and bayesian network-based data analysis. *Remote Sens.* **2019**, *11*, 843. [CrossRef]
2. Poursanidis, D.; Traganos, D.; Chrysoulakis, N.; Reinartz, P. Cubesats allow high spatiotemporal estimates of satellite-derived bathymetry. *Remote Sens.* **2019**, *11*, 1299. [CrossRef]
3. De Giglio, M.; Greggio, N.; Goffo, F.; Merloni, N.; Dubbini, M.; Barbarella, M. Comparison of pixel- and object-based classification methods of unmanned aerial vehicle data applied to coastal dune vegetation communities: Casal borsetti case study. *Remote Sens.* **2019**, *11*, 1416. [CrossRef]
4. Marzialetti, F.; Giulio, S.; Malavasi, M.; Sperandii, M.G.; Acosta, A.T.R.; Carranza, M.L. Capturing coastal dune natural vegetation types using a phenology-based mapping approach: The potential of Sentinel-2. *Remote Sens.* **2019**, *11*, 1506. [CrossRef]
5. Bishop-Taylor, R.; Sagar, S.; Lymburner, L.; Alam, I.; Sixsmith, J. Sub-pixel waterline extraction: Characterising accuracy and sensitivity to indices and spectra. *Remote Sens.* **2019**, *11*, 2984. [CrossRef]
6. Sha, J.; Li, X.M.; Chen, X.; Zhang, T. Satellite observations of wind wake and associated oceanic thermal responses: A case study of Hainan Island wind wake. *Remote Sens.* **2019**, *11*, 3036. [CrossRef]
7. Papakonstantinou, A.; Stamati, C.; Topouzelis, K. Comparison of True-Color and Multispectral Unmanned Aerial Systems Imagery for Marine Habitat Mapping Using Object-Based Image Analysis. *Remote Sens.* **2020**, *12*, 554. [CrossRef]
8. Waqas, M.; Nazeer, M.; Shahzad, M.I.; Zia, I. Spatial and temporal variability of open-ocean barrier islands along the Indus Delta region. *Remote Sens.* **2019**, *11*, 437. [CrossRef]
9. Liu, X.; Zhao, C.; Zhang, Q.; Yang, C.; Zhang, J. Characterizing and monitoring ground settlement of marine reclamation land of Xiamen New Airport, China with Sentinel-1 SAR Datasets. *Remote Sens.* **2019**, *11*, 585. [CrossRef]
10. Almeida, L.P.; Almar, R.; Bergsma, E.W.J.; Berthier, E.; Baptista, P.; Garel, E.; Dada, O.A.; Alves, B. Deriving high spatial-resolution coastal topography from sub-meter satellite stereo imagery. *Remote Sens.* **2019**, *11*, 590. [CrossRef]
11. Ma, Y.; Zhang, W.; Sun, J.; Li, G.; Wang, X.H.; Li, S.; Xu, N. Photon-counting lidar: An adaptive signal detection method for different land cover types in coastal areas. *Remote Sens.* **2019**, *11*, 471. [CrossRef]
12. Song, Y.; Liu, F.; Ling, F.; Yue, L. Automatic semi-global artificial shoreline subpixel localization algorithm for Landsat imagery. *Remote Sens.* **2019**, *11*, 1779. [CrossRef]
13. Cao, C.; Zhang, J.; Meng, J.; Zhang, X.; Mao, X. Analysis of ship detection performance with full-, compact- and dual-polarimetric SAR. *Remote Sens.* **2019**, *11*, 2160. [CrossRef]
14. Su, H.; Ji, B.; Wang, Y. Sea Ice extent detection in the Bohai Sea Using Sentinel-3 OLCI Data. *Remote Sens.* **2019**, *11*, 2436. [CrossRef]

Article

Monitoring Cliff Erosion with LiDAR Surveys and Bayesian Network-based Data Analysis

Paweł Terefenko [1,*], Dominik Paprotny [2,3], Andrzej Giza [1], Oswaldo Morales-Nápoles [2], Adam Kubicki [4] and Szymon Walczakiewicz [1]

1 Institute of Marine and Coastal Sciences, Faculty of Geosciences, University of Szczecin, 70-453 Szczecin, Poland; andrzej.giza@usz.edu.pl (A.G.); szymon.walczakiewicz@usz.edu.pl (S.W.)

2 Department of Hydraulic Engineering, Faculty of Civil Engineering and Geosciences, Delft University of Technology, 2628 CN Delft, The Netherlands; paprotny@gfz-potsdam.de (D.P.); O.MoralesNapoles@tudelft.nl (O.M.-N)

3 Section Hydrology, Helmholtz Centre Potsdam, GFZ German Research Centre for Geosciences, 14473 Potsdam, Germany

4 GEO Ingenieurservice Nord-West, 26386 Wilhelmshaven, Germany; kubicki@geogroup.de

* Correspondence: pawel.terefenko@usz.edu.pl; Tel.: +48-9144-42-354

Received: 15 March 2019; Accepted: 4 April 2019; Published: 8 April 2019

Abstract: Cliff coasts are dynamic environments that can retreat very quickly. However, the short-term changes and factors contributing to cliff coast erosion have not received as much attention as dune coasts. In this study, three soft-cliff systems in the southern Baltic Sea were monitored with the use of terrestrial laser scanner technology over a period of almost two years to generate a time series of thirteen topographic surveys. Digital elevation models constructed for those surveys allowed the extraction of several geomorphological indicators describing coastal dynamics. Combined with observational and modeled datasets on hydrological and meteorological conditions, descriptive and statistical analyses were performed to evaluate cliff coast erosion. A new statistical model of short-term cliff erosion was developed by using a non-parametric Bayesian network approach. The results revealed the complexity and diversity of the physical processes influencing both beach and cliff erosion. Wind, waves, sea levels, and precipitation were shown to have different impacts on each part of the coastal profile. At each level, different indicators were useful for describing the conditional dependency between storm conditions and erosion. These results are an important step toward a predictive model of cliff erosion.

Keywords: cliff coastlines; time-series analysis; terrestrial laser scanner; southern Baltic Sea; non-parametric Bayesian network

1. Introduction

Coastal areas are highly susceptible to changes in hydrometeorological conditions, as they constitute the boundary between land and sea. The geomorphological resilience of a particular segment of coast depends on several variables including storm intensity and topographical properties, because most changes appear during severe storms or as an effect of a series of subsequent storms [1].

Soft cliff coasts experience storms strongly, and they can retreat relatively fast. However, most monitoring systems, analyses, and models have been implemented along dune coasts [2–6], largely because of the technical difficulties in registering the morphological changes on cliff coasts. Despite such difficulties, mainly connected with accessibility of high cliffs, the factors influencing cliff erosion have been investigated through quantitative numerical methods. These approaches have varied from simple correlation matrices [7] to stochastic simulations [8] and from local to continental scales [9]. In recent years, Bayesian networks (BNs) have gained popularity as probabilistic tools for both descriptive and predictive applications [10]. However, the available studies using BNs have only addressed long-term shoreline changes [11–13], of which only Hapke and Plant [11] carried out an analysis limited strictly to cliff coasts. Furthermore, all applications have been based on discrete BNs, which generally dominate coastal hazard analyses [14]. Short-term cliff erosion has not been investigated with BNs in either discrete or continuous mode.

This study aims to propose reproducible solutions for analyzing the relationship between the erosion rate on coastal cliffs and selected variables. For this purpose, obtaining very precise topographic data was paramount [15]. The light detection and ranging (LiDAR) surveys enabled gathering datasets that were used to analyze erosion speed and its relationship to various elements that influence the geosystem of coastal cliff zones. The geomorphological analysis was based on several commonly considered indicators: sediment budgets [16], mean sea level contour [5,17], cliff base line [1,18], and cliff top line [19].

All indicators were monitored on three study sites in the southern Baltic Sea coast for a period of 1.5 years, resulting in a time series of 13 LiDAR datasets. A preliminary descriptive analysis of these results was presented by Terefenko et al. [1], but this preliminary analysis was based on only one test site and on the first five topographic surveys. In the present study, the analysis has been extended in time and space, and an original statistical model of the geomorphological response of a beach and cliff system has been developed using a non-parametric, continuous Bayesian network. This methodology will provide a foundation for creating a probabilistic solution in the prediction of unconsolidated coastal cliffs erosion.

2. Materials and Methods

2.1. Study Sites

The cliff retreat analysis was performed for a non-tidal basin of the Baltic Sea (Figure 1). The Baltic Sea is dominated by winds from southwest and west directions. The prevailing directions in particular seasons are as follows: spring—east and northeast; summer—southwest and northwest; autumn—northwest; winter—north, south, southwest, and northwest. The highest strength of wind (> 6°B) reaches from November to March [20].

In recent decades, the highest absolute amplitude of sea level changes in the study area was recorded during year 1984 (2.79 m), whereas the most extreme storm surge occurred in November 1995 (+1.61 m above mean) [21]. However, extreme value analysis have shown that a 100-year storm surge in the western part of the Polish coast could reach +1.71 m above mean, and a 500-year event would exceed 2 m [22].

The study area covered three 500 m long cliff sites that have different geomorphological configurations. The first two research areas were located in Poland near two popular seaside resorts, Międzyzdroje (Wolin Island, Biała Góra cliffs) and Wicie, representing similar northwestern coastal exposures but with different geomorphological contexts. The third area was located in Germany next to the Bansin resort (Usedom Island, Langer Berg cliffs) and was characterized not only by different exposure (northeastern), but also by a much wider beach protecting the cliffs. Detailed in situ investigations were not performed for any of the analyzed cliff test sites.

The cliff formations selected to represent the effects of marine abrasion have long been subjects of widespread research interest. Moraine hills built of glacial and glaciofluvial deposits, till, and eolian deposition predominate the relief of these areas in which the landscape varies greatly from beaches to its characteristic element: high cliffs. This region is among the stormiest in Europe, experiencing high surges and strong winds [23]. The erosion rate has been frequently debated, as different rates are measured using a variety of techniques, either directly in the field (both with traditional and modern measurement techniques) or by analyzing historical maps [24].

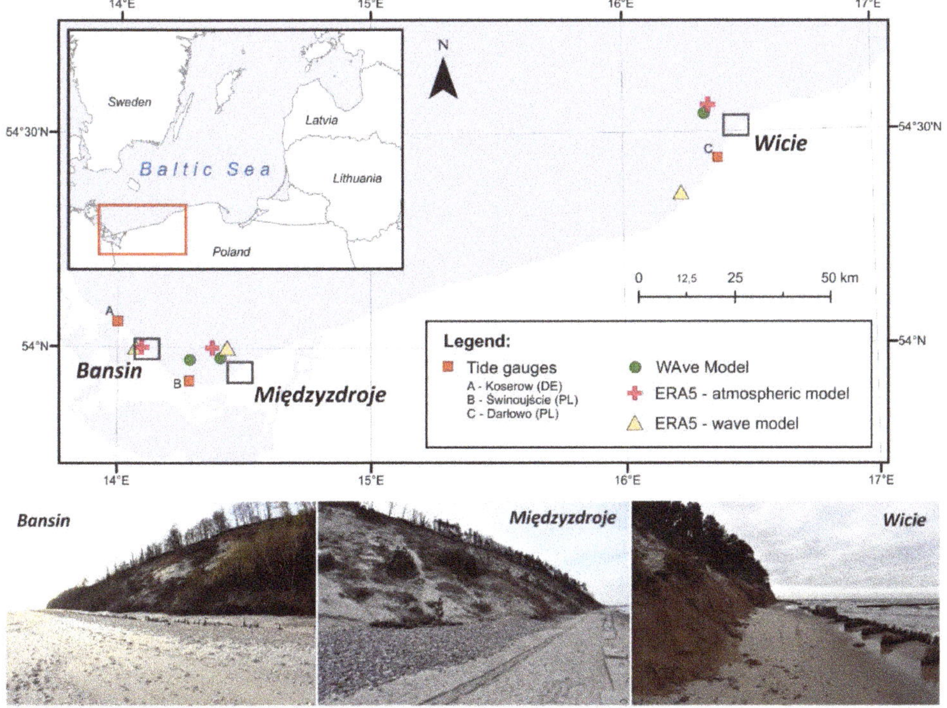

Figure 1. Location map showing the study sites, tide gauges, and grid data points.

The cliff coast of the Usedom Island, with the highest cliff (ca. 58 m) at this section named Streckelsberg, was subjected to largest coastal erosion in the area, endangering the town of Koserow, located partially on its stoss side. Generally Usedam Island cliff has been protected since the end of the 19th Century [25] by a cliff rampart, strengthened by a triple wall, groynes, three wave-breakers, and sand nourishment in modern times. For the study need a shorter, but unprotected by human made structures, an active cliff section of similar height (ca. 54 m) was chosen. This cliff, named Langer Berg, retreated ca. 100 m in 300 years [26] and is leaded by a sandy beach up to 30 meters wide.

As storms, wind, precipitation, and the sun contribute to the cliffs' erosion the Wolin cliffs (ca. 90 m high in heights parts and ca. 57 m high in investigation site), the cliffs retreat approximately 80 cm per year, although the exact erosion rate is a subject that has been discussed for years [1,16,24]. The front of the high cliffs is protected by a series of flat concrete blocks, reaching up on average up to several meters, mostly covered by mix of sand and gravels beach, dogged deep into the sand and uncovered occasionally by strong storms [1].

The Wicie study site represents a slightly different geomorphological context. The beach in front of the cliffs is covered by mix of sand and gravels similarly to Międzyzdroje test site, but its width varies from less than 1 m to up to 20 m, depending of the analyzed section. The cliff face itself is much lower in highest sections, reaching only 11 m. The investigated area is protected by a series of manmade groins. No detailed geomorphological or geological investigations have been performed on this section of the Polish coast.

2.2. Data

The data used in this study covered a survey timeline from November 2016 to June 2018. Thirty-nine topographic surveys (thirteen for each study site) were conducted with terrestrial laser scanner (TLS) technology. The significant advantage of TLS data collection compared to traditional techniques or airborne laser scanning is related to time limitations. Coasts are extremely dynamic environments. To track cliff changes and identify the processes of its modifications, data must be collected frequently over consistent time intervals [27]. Data collection using classic field methods is a long and laborious process, which in the case of numerous and extensive research areas may not provide the required results. The implemented laser-based survey technique allowed for rapid and accurate collection of large amounts of topographic data. During the last decade, TLS has been successfully applied to topographic surveys and to the monitoring of coastal processes [28–30]. In this study, highly accurate measurements of coastal changes were performed with the use of Riegl VZ-400 equipment. Each of the 500 m long test sites were scanned from 10 spots, acquiring 90 to 100 points per square meter, with an estimated vertical accuracy of more than 5 mm. A list of all surveys and the resulting analytical periods is included in Supplementary Information 1 (Table S1).

The hydrometeorological data used in this study combined both observational and modeled datasets (Table 1). Wave parameters from the high-resolution operational WAve Model (WAM) were validated for the Baltic Sea in the framework of the Hindcast of dynamic processes of the ocean and coastal areas of Europe (HIPOCAS) project [31]. One minor gap lasting 6–7 h for two WAM points corresponding to the Bansin and Międzyzdroje cliffs was filled by interpolation. Three larger gaps in the wave and wind parameters for all locations, lasting a total of 36 days (within December 2016, June 2017, and February 2018), were filled using the fifth major global climate reanalysis dataset produced by the European Center for Medium-Range Weather Forecasts (ERA5) [32]. As the resolution of the ERA5 reanalysis model, which represents wave conditions further from the coast, is far coarser than the WAM data, the ERA5 values were corrected by a constant factor for each location, variable, and data gap. The constant factor was computed by dividing the average WAM values for the available days within each month during which a gap occurred by the average ERA5 reanalysis values.

Table 1. Sources of hydromet variables of interest by study area and so eorological data. Locations of tide gauges and grid data points are shown in Figure 1.

Variable	Source	Provider	Resolution
Wave parameters	WAM wave model hindcast	Interdisciplinary Centre for Mathematical and Computational Modelling of Warsaw University (ICM)	hourly, 1/12°
Wave parameters	ERA5 wave reanalysis	European Center for Medium-Range Weather Forecasts (ECMWF)	hourly, 0.36°
Sea level	Observations at Koserow, Świnoujście and Darłowo	German Federal Institute of Hydrology (BfG), Institute of Meteorology and Water Management (IMGW)	hourly, at tide gauges
Temperature, precipitation	ERA5 atmosphere reanalysis	European Center for Medium-Range Weather Forecasts (ECMWF)	hourly, 0.28°

Information on water levels was derived from tide gauges located at the shortest distance from each case study site through personal communication with the institutions responsible for the gauge upkeep. Finally, hourly precipitation and temperature data were collected from the ERA5 reanalysis model.

2.3. Geomorphological Indicators

Depending on the study objectives, five major geomorphological indicators were extracted from the LiDAR-derived digital elevation models (DEMs), namely shoreline retreat, beach volume balance, cliff foot retreat, cliff volume balance and cliff top retreat (Figure 2.). Because part of the topographic measurements were realized directly after storms while the water level was still quite high, some limitations in the high-resolution dataset availability caused the shoreline retreat indicator to be extracted as a 1 m contour above mean sea level (MSL) instead of at zero MSL.

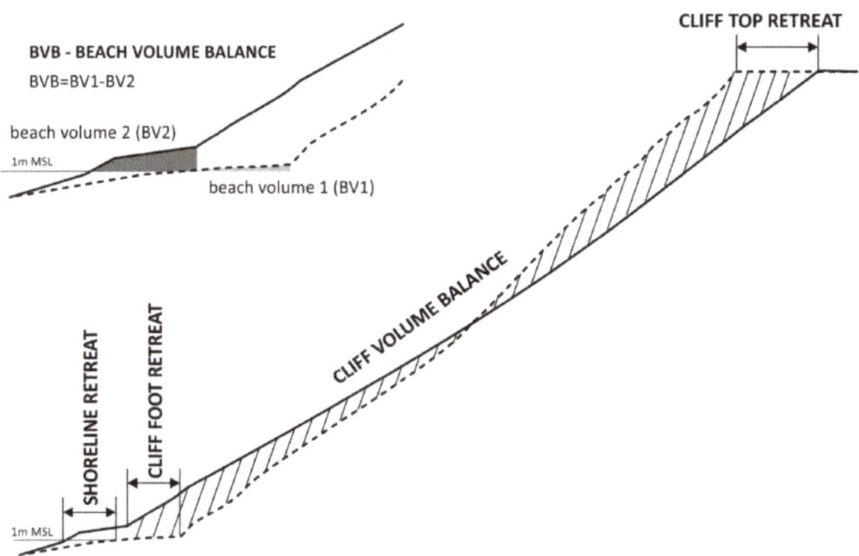

Figure 2. Scheme of measuring procedure of major geomorphological indicators extracted from the LiDAR-derived digital elevation models.

Extracting the shoreline contour from DEM was rather straightforward; however, acquiring the cliff base line was more challenging and required some deliberation. Because the purpose of this study was to create reproducible solutions for analyzing the relationship between the erosion rates on coastal cliffs, a comparable procedure for extracting cliff base line was needed. While several studies realized on cliffs analyzed volumetric changes, to our knowledge, all assumed manual delineation of the cliff baseline, relying mainly on aerial photographs, topographic maps, or in situ surveys [1,18,33,34]. Some attempts of advanced automatic delineation were performed on the cliff bases of generalized coastal shoreline vectors by approximating the distance between shoreline and the cliff top [19]. In our study, a simplified methodology was implemented that considered a rapid change in altitude (higher than 0.5 m for a distance of 1 m). This procedure appeared to be a sufficient solution, because the delineation of the cliff base line can be a subject of interpretation, even by operators during field surveys. Moreover, as presented by Palaseanu-Lovejoy et al. [19], the manual digitization of geomorphological breaklines on DEMs not only has lower precision but also lacks reproducibility. The assumed simplified procedure was fully reproducible and comparable for all test sites and was a sufficient indicator that was independent of human skill.

Mapping the cliff top line and its migration over time is one of the most common methodologies for investigating cliff recession [24]. Traditionally obtained during field surveys or based on hand-digitized procedures [35], the cliff top line can also be extracted automatically [19]. Due to TLS limitations mainly related to data shortages on parts of the cliff edge densely overgrown by vegetation, the highest available point existing on two successive topographic surveys was assumed as the cliff top line for the analyzed time period.

Finally, to explore how the beach–cliff system changed between each LiDAR survey, line indicator migration as well as volumetric changes was analyzed. The results were separately determined for beach and cliff areas between the lines in 50 m wide sections. Similarly, for the needs of Bayesian network analysis, all line indicators were marked on profiles using the same 50 m spacing as the volumetric measurements.

2.4. Bayesian Networks

Bayesian networks, also known as Bayesian belief nets, are graphical, probabilistic models [36] that have a wide range of applications in the environmental sciences, particularly in coastal zone problems [10,14]. The main advantage of BNs is the ability to model complex processes and, at least for models with a small number of nodes, the explicit representation of uncertainty and intuitive interpretation. BNs can be discrete or continuous, depending on the type of data available. In this study, a continuous BN was applied as it better suits the data collected (for discussion on pros and cons of various BN types, we refer to Hanea et al. [37]).

In general, a BN consists of a directed acyclic graph with associated conditional probability distributions [38,39]. The graph consists of "nodes" and "arcs" in which the nodes represent random variables connected by arcs, which represent the dependencies between variables. Arcs have a defined direction: the node on the upper end is known as the "parent" node, and the node on the lower (receiving) end is the "child" node. Each variable is conditionally independent of all predecessors given its parents: if one conditionalizes the parent node and there is no arc connecting the child node with any of the predecessors of the parent node (directly or through another parent node), the conditional distribution of the child node does not change if the predecessors of the parent node are conditionalized. The joint probability density f(x_1,x_2, … ,x_n) for a given node is therefore written as

$$f(x_1, x_2, \ldots, x_n) = \prod_{i=1}^{n} f\left(x_i \middle| x_{pa(i)}\right) \quad (1)$$

where pa(i) is the set of parent nodes of X_i. One possibility of BNs is to update the probability distribution of child nodes given the new evidence at parent nodes. Two elements are needed to quantify a BN: the marginal distribution for each node and a dependency model for each arc. In this study, we used non-parametric margins, which were the same as the empirical distribution of data collected for this study. The dependencies were represented by normal (Gaussian) copulas. Basically, a copula is a joint distribution on the unit hypercube with uniform (0,1) margins. While there are many types of copulas (we refer to Joe [39] for detailed descriptions), the assumption of a normal copula is a limitation of the available computer code [38]—though most dependencies between variables used here did not indicate tail dependence—a property that can be represented as either normal, Frank, or Plackett copulas. A goodness-of-fit test for copulas proposed by Genest et al. [40] indicates that several copula types are, on average, similarly suitable for the analysis (Frank, Plackett, t, Gumbel, Gaussian), while others much less (Clayton and Joe copulas). A normal copula was parameterized using Spearman's rank correlation coefficient; hence, in all cases, the results refer to this measure of correlation. For the detailed procedure of obtaining conditional probabilities from a non-parametric continuous BN with a normal copula, we followed the procedure of Hanea et al. [37]. The algorithms from that study were implemented in the Uninet software used to build our model.

The configuration of nodes and arcs is researcher dependent. Yet a good BN incorporates existing knowledge of the process in question, in this case the factors influencing the cliff erosion and the physical processes in action. For this study, a total of 41 variables were tested while preparing the BN. The full list of variables and their descriptions is available in the SI1 file. Five erosion indicators (Section 2.3) and two further geomorphological indicators, namely beach width (i.e., between shoreline and cliff foot) and cliff slope (i.e., above cliff foot), were used as variables. The following rules were used to design the BN model in this study:

1. Cliff erosion indicators were connected with each other, starting from the shoreline retreat indicator and moving toward the cliff top.
2. In every case, the cliff erosion indicator was used as the first parent node when other parent nodes were added.
3. Meteorological, hydrological, and morphological variables were added starting from the shoreline retreat (Shore) node and moving toward the cliff top.

4. Each variable was connected only to one node containing a cliff erosion indicator.
5. Meteorological and hydrological variables were not given any parent nodes and were not connected with each other.
6. The first meteorological or hydrological variable to be connected with a cliff erosion indictor was the variable with the highest unconditional correlation within the model. The unconditional correlation matrix is shown in Supplementary Information 2.
7. Further meteorological or hydrological variables were selected based on the conditional correlation with cliff erosion indicators.
8. Only parent nodes with (conditional) correlations higher than 0.1 were included in the model, except for the parents of cliff top retreat (Top), where only the correlation with the cliff volume balance (Cliff) exceeded this threshold.

The meteorological and hydrological factors, such as significant wave height, wave direction, mean wave period, peak wave period, water level, wind speed, temperature, and precipitation, were used in several configurations where applicable: mean (total), maximum (minimum) values between measurement campaigns, mean value during storm surges, and the 95th (5th) percentile during the period between measurement campaigns. Synthetic indicators of storm conditions were also investigated, including storm energy [41], accumulated excess energy [42], and wave power [43]. For the purposes of this study, a storm surge was defined as a water level of at least 0.45 m above mean sea level (545 cm Normal Null); this value was selected on the basis of (unconditional) correlations between erosion and hydrological variables. Moreover, if after a storm, the water level fell below this level for less than 6 h before the next storm, the whole series was considered to be one storm surge. The value of the upper and lower percentile in some indicators was similarly selected to maximize (unconditional) correlations across multiple variables.

3. Results

3.1. Hydrological Conditions during the Period of Study

Many storms reached the coast during the measurement period. Using the definition of storm surge described in Section 2.4 (based on sea levels of at least 0.45 m above mean sea level), a total of 61 storms affected the cliffs in Bansin, compared to 43 in Międzyzdroje and 62 in Wicie. The distribution of surges was highly uneven, as shown in Figure 3. The most intense period lasted from late November 2016 to mid-January 2017. Around 10 surges were distinguished during that period, with water levels exceeding 1.4 m above average at all locations on 4–5 January 2017. This water level corresponded to an event with a return period of 15–20 years [44]. The maximum water level of 1.55 m was observed at the Koserow tide gauge close to the Bansin cliffs. Conversely, the waves reached their maximum height throughout late 2016, culminating on 7 December 2016.

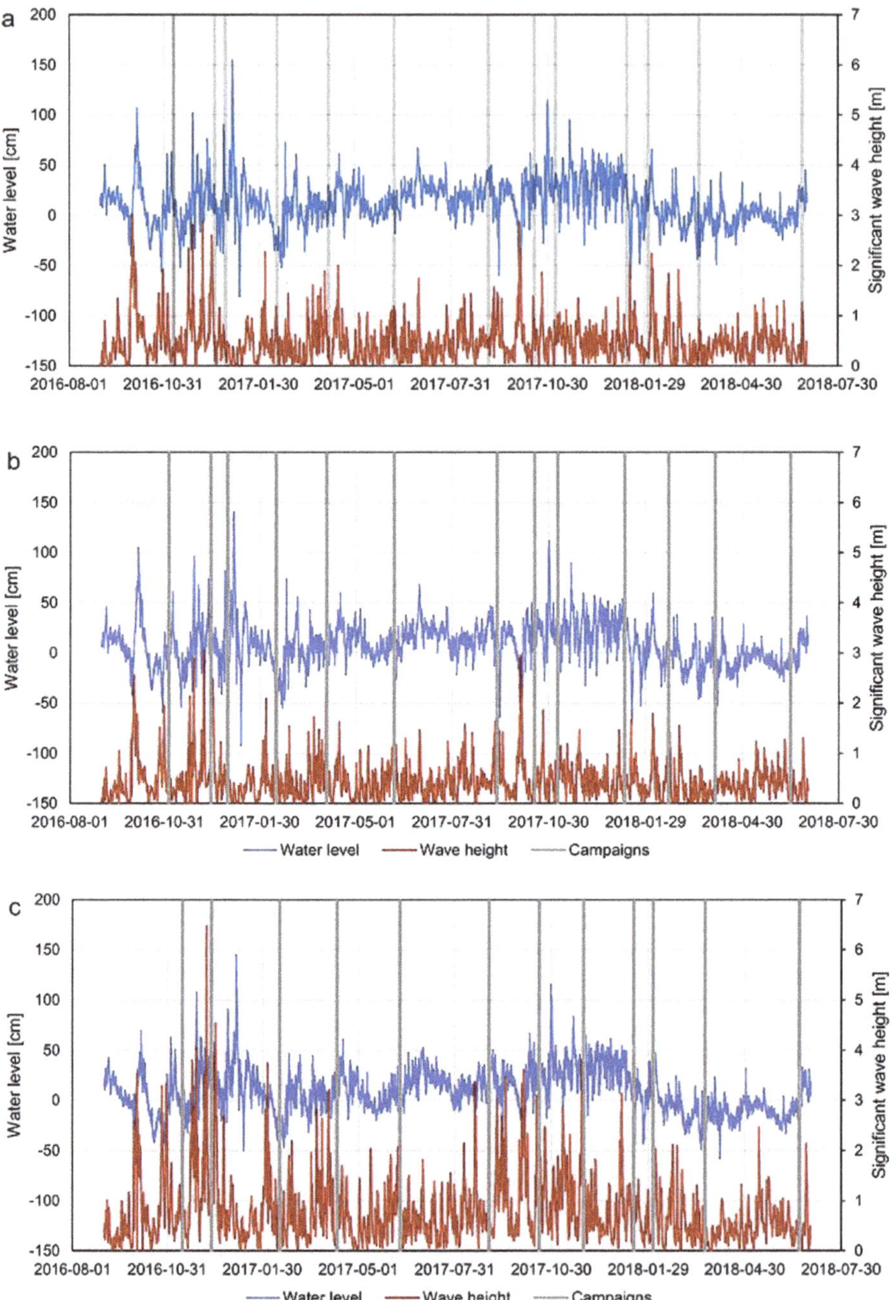

Figure 3. Water levels, significant wave height, and timing of the LiDAR measurement campaigns at (**a**) Bansin, (**b**) Międzyzdroje, and (**c**) Wicie. Water levels obtained from the tide gauge measurements, and wave heights obtained from the WAM model.

Another period of stormy weather lasted from mid-October 2017 to early January 2018, during which around 20 surges affected the coast. However, neither the water levels nor the wave heights were as extreme as those during the 2016–2017 storm season. The most intense storm in the 2017–2018 storm season occurred around 29–30 October 2017 during which the water levels slightly exceeded 1 m above mean in all study areas. Considering the stricter definition of storm surge presented by Wiśniewski and Wolski [44], i.e., the exceedance of water levels of 0.6 m above mean, the first half of the study period had three times more storms than the long-term average of about four per year, including a very unusual occurrence of a storm surge in June; the second half of the study period was close to an average year.

3.2. Descriptive Analysis of Cliff Erosion

During the monitoring period, the sediment budget was definitely negative with a total loss of 49,330 m3. Erosion was most significant on the cliffs (over 58,000 m^3), while a positive budget was observed on the beaches, with a value slightly exceeding 9000 m^3. This positive balance shows that not all of the cliffs' material was swept into the sea, but some of it remained on the beaches.

Erosion and sedimentation were unevenly distributed in time and space (Figure 4). At the beginning of the 2016–2017 storm season, erosion was principally visible on the beach (over 80% of total erosion volume in Bansin and Międzyzdroje). As the successive lowering of beach proceeded, the proportions changed, and the cliff erosion started to dominate, reaching over 85% of the total erosion volume. Due to the very narrow beach, the Wicie area suffered cliff-dominated erosion of more than 90% of the total loss in this coast section. In fact, the sediment budget was obviously negative both for the beach and cliff during the winter season. The maximum negative volume of eroded material measured between the third and fourth topographic campaigns was also the highest during the monitoring period. Erosion volume on the beach varied at different test sites, reaching from 627 to 2191 and 2566 m^3 for Międzyzdroje, Bansin, and Wicie beaches, respectively. However, the first group of severe storms affected the cliff face much stronger than the beach, exceeding the maximum volumes of 6000, 12,000, and 18,000 m^3 for Międzyzdroje, Bansin, and Wicie cliffs, respectively. Notwithstanding the clear erosion dominance across the whole study area during the 2016–2017 storm season, the retreat of the cliff top was relatively small compared to changes of the 1 m contour line and the cliff base line. While the cliff top retreated by a maximum of 11 m in Wicie, the average change on all areas was less than 1 m, and the median was only 0.03 m. The maximum changes of shoreline and cliff base lines were similar, reaching around 11 m. However, the average change of shoreline and cliff base lines of 2.5 and 1.3 m, respectively, as well as medians of 1.7 and 0.15 m, respectively, suggested more even distribution.

The period between storm seasons contained higher variability in both the time and space distributions, even though the total volumes were much lower. Furthermore, the compilation of the next five surveys revealed both accumulation and erosional patterns with a rather modest positive overall sediment budget (1800 m^3). Before the 2017 winter season approached, the dominant processes were much weaker, but cliff erosion still occurred along with the overall recovery of beach height and length. The volume values between surveys fluctuated from −2870 to 9280 m^3 and −3520 to 3683 m^3, respectively, for beach and cliff. However, the negative values for the beach and the positive for the cliffs were a consequence of landslide processes that pushed the cliff base line in the seaward direction rather than significant erosion or deposition episodes.

The second period of stormy weather as well as the following spring season (2017–2018) revealed strong similarities to the corresponding earlier periods. This observation was supported by a comparison of data from the last four topographic surveys. Erosion was still principally visible on the cliffs, though the water levels and wave heights were not as extreme as those during the 2016–2017 storm season. The much weaker waves were not able to clean all the debris, and in some of the investigated areas, the cliff base line migrated seawards, and the volume values presented an inverse pattern to what was observed during the first storm season. The after-storm period was again characterized by beach recovery processes.

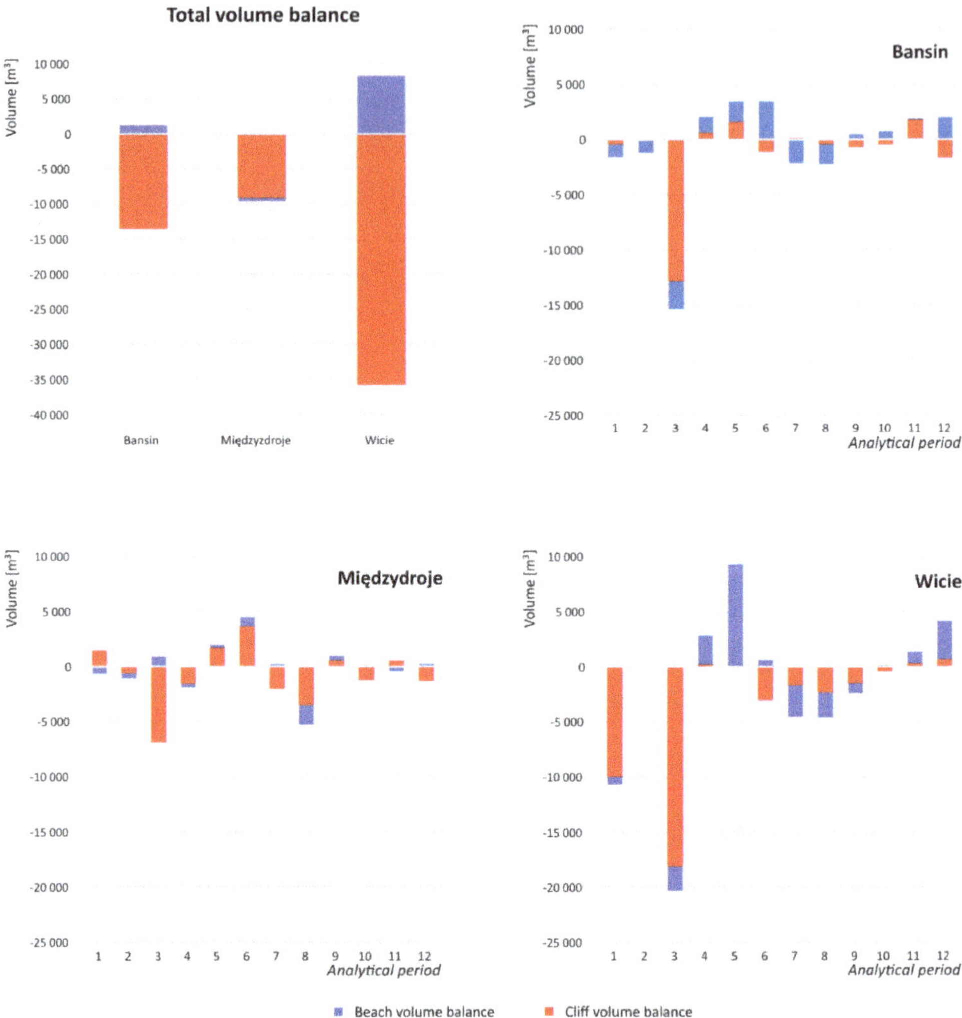

Figure 4. Distribution of sediment changes in time (between LiDAR campaigns) and space (different test sites).

3.3. Statistical Analysis of Cliff Erosion

The statistical analysis was performed using the BN presented in Figure 5. The final model, constructed following the procedure explained in Section 2.4., included five cliff erosion indicators explained by two morphological factors, eleven hydrological factors, and two meteorological factors. The morphological factors were additionally explained by two hydrological factors and one meteorological factor.

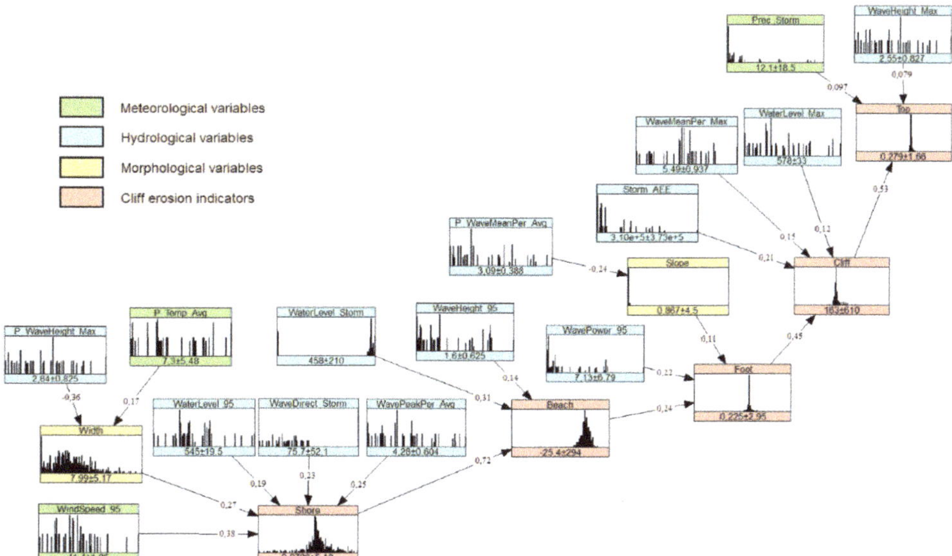

Figure 5. Proposed Bayesian network for cliff coast erosion. The ordering of parent variables is clockwise, starting from the leftmost node. The numbers below the histograms indicate the average and standard deviation, and the numbers on the arcs are Spearman's (conditional) rank correlations. See the SI1 file for full explanations of variables. The letter "P" before the name of some variables indicates that the values are for the preceding period, rather than for the period during which the erosion occurred.

The shoreline is the most dynamic component of the coastline; therefore, its changes (Shore) have the highest number of explanatory variables. The highest correlation was observed with the 95th percentile of wind speed (WindSpeed_95), which gave a slightly higher correlation than the wave height indicators. A likely explanation for this relationship is that wind is more dynamic than offshore waves containing significant inertia and hence is a better predictor of the small wind-driven waves that contribute to shoreline retreat. The second factor influencing shoreline retreat was the width of the beach (Width) before the occurrence of erosion. Wider beaches have more material to be eroded, resulting in larger shoreline retreat. The beach width was influenced by both the maximum wave height (P_WaveHeight_Max), which resulted in shorter beaches, and the average temperature (P_Temp_Avg), which is an indicator of the time of the year, as beaches tend to be shorter during the autumn and winter storm season than during the warmer spring or summer. Other factors contributing to shoreline retreat were the 95th percentile of water levels (WaterLevel_95), average wave direction during storm surges (WaveDirect_Storm), and average wave peak period (WavePeakPer_Avg), all of which resulted in higher and longer waves attacking the shoreline, resulting in erosion.

Beach volume balance (Beach) was highly correlated (0.72) with shoreline retreat, which incorporated the influence of several factors. The average water level during storms (WaterLevel_Storm) further contributed to beach erosion, as higher baseline sea levels allowed waves to reach further onto the beach, while the 95th percentile of significant wave height (WaveHeight_95) indicated the importance of high waves in beach erosion.

Cliff foot retreat (Foot) showed a relatively low correlation (0.24) with beach volume balance, as more complex mechanisms were observed: material from cliff erosion could be deposited on the beach, which would result in a weak dependency between beach and cliff erosion. However, some of the waves eroding the beach still cut into the cliff. Specifically, waves that were both particularly high and long contributed to cliff foot retreat, as revealed by the wave power (WavePower_95) indicator, which was proportional to the product of significant wave height and the mean wave period. Additionally, the cliff was more prone to erosion if more vertical than inclined, as shown by the cliff slope (Slope) variable. The cliff slope showed the highest correlation with the average mean wave period in the preceding period (P_WaveMeanPer_Avg), where stormy periods resulted in lower cliff slopes due to erosion.

The cliff volume balance (Cliff) depended primarily on waves undercutting the cliff, resulting in the eventual collapse of the cliff. Erosion was further increased by very high waves, as shown by the accumulated excess energy (Storm_AEE) indicator. The accumulated excess energy indicator represented the energy of waves above a 2 m threshold (including sea level), which was close to the average elevation of cliff foots in the study area; hence, this indicator counted only the waves that actually eroded the cliff. Two other variables correlated with the cliff volume balance were the maximum mean wave period (WaveMeanPer_Max), which indicated the occurrence of very long waves, and the maximum water level (WaterLevel_Max), as the high baseline sea level increased the number of waves that could reach the cliff.

Finally, erosion of the cliff could also result in retreat of the cliff top (Top). This erosion indicator was the least dynamic and depended mostly on factors already included in previous erosion indicators. Some correlation existed with the total precipitation recorded during storms (Prec_Storm), as rainfall could weaken the structure of the cliff, making it more susceptible to collapse. Other factors showed only a small conditional correlation; the largest was for the maximum wave height (WaveHeight_Max), which indicated the occurrence of extreme waves having the biggest impact on the cliff.

The model was validated by analyzing the correlation between predicted and observed changes in the variables of interest (Table 2). This was carried out for different choices of input sample, thus analyzing how transferable is the model between locations. The small sample size resulted in a non-negligible variation of results between different model runs; therefore, the results shown are averages of 100 model runs per each variant of location or sample source. A split-sample validation (using half of the data as input sample, and the other half to run the model) showed only marginally lower performance than using the same data for both purposes. Of the three study sites, data from the Bansin cliff is the most transferable. For individual variables, the highest correlation between modeled and observed data is for beach volume balance, followed by shoreline retreat and beach width (correlations of 0.4-0.6). Correlations for cliff foot and volume balance are in the 0.3–0.4 range, and lower for the cliff top, which was the least dynamic part of the cliff in the timeframe of the study.

Table 2. Validation results for variables of interest by study area and source of sample for the model. Values indicate Spearman's rank correlation.

Study Area	Source of Data	Variable						
		Shore	Beach	Foot	Cliff	Top	Width	Slope
All	All	0.50	0.60	0.36	0.31	0.19	0.40	0.24
	All (split-sample)	0.48	0.59	0.35	0.30	0.18	0.37	0.24
	Bansin	0.50	0.59	0.34	0.30	0.17	0.33	0.23
	Międzyzdroje	0.49	0.62	0.34	0.25	0.18	0.36	−0.11
	Wicie	0.47	0.59	0.34	0.31	0.17	0.37	0.13
Bansin	All	0.60	0.74	0.32	0.25	0.01	0.26	0.19
	Międzyzdroje + Wicie	0.59	0.71	0.32	0.19	0.01	0.20	0.19
Międzyzdroje	All	0.41	0.29	0.46	0.10	−0.16	0.42	−0.02
	Bansin + Wicie	0.41	0.28	0.46	0.07	−0.16	0.40	−0.02
Wicie	All	0.50	0.72	0.22	0.50	0.45	0.15	0.01
	Bansin + Międzyzdroje	0.47	0.70	0.26	0.47	0.31	0.12	0.02

4. Discussion

The tracking of cliff changes requires very detailed topographic data to be acquired repeatedly in time, not only for revealing patterns of coastal behavior [18,45] but also for providing better understanding of the relations between processes and indicators. As the comparison of two datasets provided only a cumulative result for coastal analysis [15,16], multiple measurements enabled the analysis of both isolated events and storm series on erosion, as well as the processes for cliff modifications.

In this study, we demonstrated that changes to coastal cliffs are very complex, and physical processes that influenced both beach and cliff may be responsible for erosion processes. Our results confirmed the impact of sea activity as well as enabled evaluation of the effects of unfavorable weather conditions to coastal cliffs [18,24]. In fact, the cliff coast develops as a result of numerous overlapping processes. While storm surges undercut and destabilize cliff faces [46], waves are mainly responsible for temporal shoreline changes with correlation to temperature, which acts as a season indicator. Consequently, the beach is successively eroded and lowered, resulting in the occurrence of favorable hydrometeorological conditions for cliff erosion. These conditions are not directly linked to the highest waves, but to longest waves during maximum water levels. While high-magnitude events advance cliff face erosion, when these events weaken, part of the transported debris is lost, which starts the process of beach recovery [47,48]. Finally, the most powerful events were not able to directly influence the cliff top line. As presented by Kostrzewski et al. [24], changes of the cliff top line are linked with precipitation factors, especially during storm events.

In this study, as suggested by Andrews et al. [27], numerous topographic "snapshots" realized more than several times during a year were analyzed with increasingly popular Bayesian networks. This analysis enabled an understanding of the complex changes of coastal systems from the event scale to seasonal variations. The BN model presented here is the first BN application for analyzing short-term cliff erosion and therefore is not comparable with the few existing models due to the different spatial or temporal scales and model designs. Some similarities could be found; however, as certain common factors were identified to contribute to erosion, such as the cliff/beach slope, sea level, and wave height. On the other hand, recurring variables were not included in this study, such as the tidal range and geology/geomorphology of the coast. Tides have negligible amplitude along the coast in question. The qualitative properties of the cliffs were not included due to the similarity of the study sites. Moreover, inclusion of the geomorphology would necessitate the use of a discrete or hybrid BN, which would require a very different model set up in the context of our relatively small sample size.

In this study, the model was used for data analysis without making predictions. The inclusion of prediction capability in our model would require validation based on another cliff erosion dataset. For instance, the annual cliff top erosion since 1985 for multiple sections of the Wolin Island cliffs [24] could be used for this purpose. However, such an analysis is limited by the availability of hydrometeorological data. Existing reanalyses (ERA5, ERA-Interim) have much lower resolution than the WAM model used here; therefore, the wave conditions indicated in those reanalyses differ substantially from those in WAM: they show much bigger wave heights. Moreover, tests with an operational BN model have shown that such models are too sensitive, given the amount of data available. Therefore, more LiDAR scanning campaigns performed would be needed to improve the performance of the model, especially for the less dynamic upper parts of the cliff. The model than could be reworked using ERA5 as the input hydrometeorological dataset, which planned to be extended back to 1950 [32]. Moreover, the assumption of a normal copula for modeling the dependencies would need to be validated before the model could be used for prediction [49], and the graph would need to be further investigated to better represent the joint distribution [50]. The SI1 file presents an example of a modified BN with many additional arcs between the hydrometeorological variables, as those are the most highly correlated, and such connections are relevant for properly representing the joint distribution.

5. Conclusions

1. Our study demonstrates the advantages of using Bayesian network for analysis of surface morphological changes on cliff coasts even on relatively short analyzed shore segments. Despite the site-specific geomorphological settings for different test areas, the implementation of the proposed Bayesian network model enabled the determination of relationships between the erosion rates and selected factors. The proposed model explained the general behavior of the cliff coast with respect to different hydrometeorological conditions, indicating variables most relevant at each segment along the profile. Validation of the model showed good performance along the beach and cliff foot, but weaker in predicting cliff mass balance or cliff top recession.
2. Our study proves that high temporal resolution in TLS surveys enables the analysis of correlations between the influence of several factors (wave height, length and period, water level, storm energy, precipitation, etc.) and the geomorphological response of coast during isolated storm events, as well as with cumulative effects for season-long analysis. In general, a presentation of short and mid-term analyses expands possibilities in coastal morphological studies. Although we have seen a rapid increase of TLS usage in recent years, most of these have focused on a small quantity of realized surveys or long-term analysis.
3. The automatic extraction of all geomorphological indicators from DEMs enabled reproducible and comparable cliff recession analysis. However, caution should be taken when interpreting the beach recovery, because some erosion and deposition processes may be masked by an automatic delineation of the cliff base line.

Supplementary Materials: The following are available online at http://www.mdpi.com/2072-4292/11/7/843/s1: Table S1.1. Analytical periods used in the study according to dates of surveys by test site, Table S1.2. Candidate variables for the Bayesian Network model and their description, Description S1. Equations of synthetic indicators, Figure S1.1. A Bayesian Network for cliff erosion witb additional arc between hydrometeorological variables, Figure S1.2. D-calibration score for the unsaturated BN from the paper, Figure S1.3. D-calibration score for the saturated BN from this supplement. The following are available online at http://www.mdpi.com/2072-4292/11/7/843/s2: Table S2.1. Spearman's rank correlations.

Author Contributions: Conceptualization, methodology, formal analysis and writing original draft: P.T. and D.P.; writing–review and editing: P.T., D.P. and A.G.; data curation: P.T., A.G.; resources and visualization: A.G. and S.W.; validation D.P.; supervision: O.M-N. and A.K; funding acquisition: P.T.

Funding: This research was funded by the National Science Centre, Poland, grant number UMO-2015/17/D/ST10/02191 "Coastal cliffs under retreat imposed by different forcing processes in multiple timescales-CLIFFREAT".

Conflicts of Interest: The authors declare no conflict of interest.

References

1. Terefenko, P.; Giza, A.; Paprotny, D.; Kubicki, A.; Winowski, M. Cliff retreat induced by series of storms at Międzyzdroje (Poland). *J. Coastal Res.* **2018**, *85*, 181–185. [CrossRef]
2. Furmańczyk, K.; Andrzejewski, P.; Benedyczak, R.; Bugajny, N.; Cieszyński, .; Dudzińska-Nowak, J.; Giza, A.; Paprotny, D.; Terefenko, P.; Zawiślak, T. Recording of selected effects and hazards caused by current and expected storm events in the Baltic Sea coastal zone. *J. Coastal Res.* **2014**, *70*, 338–342. [CrossRef]
3. Paprotny, D.; Andrzejewski, P.; Terefenko, P.; Furmańczyk, K. Application of Empirical Wave Run-Up Formulas to the Polish Baltic Sea Coast. *PLoS ONE* **2014**, *9*, e105437. [CrossRef] [PubMed]
4. Bugajny, N.; Furmańczyk, K. Comparison of Short-Term Changes Caused by Storms along Natural and Protected Sections of the Dziwnow Spit, Southern Baltic Coast. *J. Coastal Res.* **2017**, *33*, 775–785. [CrossRef]
5. Deng, J.; Harff, J.; Zhang, W.; Schneider, R.; Dudzińska-Nowak, J.; Giza, A.; Terefenko, P.; Furmańczyk, K. The Dynamic Equilibrium Shore Model for the Reconstruction and Future Projection of Coastal Morphodynamics. In *Coastline Changes of the Baltic Sea from South to East*; Harff, J., Furmańczyk, K., VonStorch, H., Eds.; Springer: Cham, Switzerland, 2017; pp. 87–106.
6. Uścinowicz, G.; Szarafin, T. Short-term prognosis of development of barrier-type coasts (Southern Baltic Sea). *Ocean Coast. Manag.* **2018**, *165*, 258–267. [CrossRef]

7. Prémaillon, M.; Regard, V.; Dewez, T.J.B.; Auda, Y. GlobR2C2 (Global Recession Rates of Coastal Cliffs): A global relational database to investigate coastal rocky cliff erosion rate variations. *Earth Surf. Dyn.* **2018**, *6*, 651–668. [CrossRef]
8. Hall, J.W.; Meadowcroft, I.C.; Lee, E.M.; van Gelder, P.H.A.J.M. Stochastic simulation of episodic soft coastal cliff recession. *Coast. Eng.* **2002**, *46*, 159–174. [CrossRef]
9. Le Cozannet, G.; Garcin, M.; Yates, M.; Idier, D.; Meyssignac, B. Approaches to evaluate the recent impacts of sea-level rise on shoreline changes. *Earth-Sci. Rev.* **2014**, *138*, 47–60. [CrossRef]
10. Beuzen, T.; Splinter, K.D.; Marshall, L.A.; Turner, I.L.; Harley, M.D.; Palmsten, M.L. Bayesian Networks in coastal engineering: 5 Distinguishing descriptive and predictive applications. *Coast. Eng.* **2018**, *135*, 16–30. [CrossRef]
11. Hapke, C.; Plant, N. Predicting coastal cliff erosion using a Bayesian probabilistic model. *Mar. Geol.* **2010**, *278*, 140–149. [CrossRef]
12. Gutierrez, B.T.; Plant, N.G.; Thieler, E.R. A Bayesian network to predict coastal vulnerability to sea level rise. *J. Geophys. Res.* **2011**, *116*, F02009. [CrossRef]
13. Yates, M.L.; Le Cozannet, G. Brief communication "Evaluating European Coastal Evolution using Bayesian Networks". *Nat. Hazards Earth Syst. Sci.* **2012**, *12*, 1173–1177. [CrossRef]
14. Jäger, W.S.; Christie, E.K.; Hanea, A.M.; den Heijer, C.; Spencer, T. A Bayesian network approach for coastal risk analysis and decision making. *Coast. Eng.* **2018**, *134*, 48–61. [CrossRef]
15. Le Mauff, B.; Juigner, M.; Ba, A.; Robin, M.; Launeau, P.; Fattal, P. Coastal monitoring solutions of the geomorphological response of beach-dune systems using multi-temporal LiDAR datasets (Vendee coast, France). *Geomorphology* **2018**, *304*, 121–140. [CrossRef]
16. Kolander, R.; Morche, D.; Bimböse, M. Quantification of moraine cliff erosion on Wolin Island (Baltic Sea, northwest Poland. *Baltica* **2016**, *26*, 37–44. [CrossRef]
17. Nunes, M.; Ferreira, O.; Loureiro, C.; Baily, B. Beach and cliff retreat induced by storm groups at Forte Novo, Algarve (Portugal). *J. Coastal Res.* **2011**, *64*, 795–799.
18. Warrick, J.A.; Ritchie, A.C.; Adelman, G.; Adelman, K.; Limber, P.W. New techniques to measure cliff change form historical oblique aerial photographs and structure-for-motion photogrammetry. *J. Coastal Res.* **2017**, *33*, 39–55. [CrossRef]
19. Palaseanu-Lovejoy, M.; Danielson, J.; Thatcher, C.; Foxgrover, A.; Barnard, P.; Brock, J.; Young, A. Automatic Delineation of Seacliff Limits using Lidar-derived High-resolution DEMs in Southern California. *J. Coastal Res.* **2016**, *76*, 162–173. [CrossRef]
20. Rokiciński, K. Geograficzna i hydrometeorologiczna charakterystyka Morza Bałtyckiego jako obszaru prowadzenia działań asymetrycznych. *Zeszyty Naukowe Akad. Marynarki Wojennej* **2007**, *48*, 65–82.
21. Wolski, T.; Wiśniewski, B.; Giza, A.; Kowalewska-Kalkowska, H.; Boman, H.; Grabbi-Kaiv, S.; Hammarklint, T.; Holfort, J.; Lydeikaitė, Ž. Extreme sea levels at selected stations on the Baltic Sea coast. *Oceanologia* **2014**, *56*, 259–290. [CrossRef]
22. Paprotny, D.; Terefenko, P. New estimates of potential impacts of sea level rise and coastal floods in Poland. *Nat. Hazards* **2017**, *85*, 1249–1277. [CrossRef]
23. Vousdoukas, M.I.; Voukouvalas, E.; Annunziato, A.; Giardino, A.; Feyen, L. Projections of extreme storm surge levels along Europe. *Clim. Dyn.* **2016**, *47*, 3171–3190. [CrossRef]
24. Kostrzewski, A.; Zwoliński, Z.; Winowski, M.; Tylkowski, J.; Samołyk, M. Cliff top recession rate and cliff hazards for the sea coast of Wolin Island (Southern Baltic). *Baltica* **2015**, *28*, 109–120. [CrossRef]
25. Schumacher, W. Coastal dynamics and coastal protection of the Island of Usedom. *Greifswalder Geogr. Arbeiten* **2002**, *27*, 131–134.
26. Schwarzer, K.; Diesing, M.; Larson, M.; Niedermeyer, R.O.; Schumacher, W.; Furmanczyk, K. Coastline evolution at different time scales: Examples from the Pomeranian Bight, southern Baltic Sea. *Mar. Geol.* **2003**, *194*, 79–101. [CrossRef]
27. Andrews, B.P.; Gares, P.A.; Colby, J.B. Techniques for GIS modeling of coastal dunes. *Geomorphology* **2002**, *48*, 289–308. [CrossRef]
28. Vousdoukas, M.; Kirupakaramoorthy, T.; Oumeraci, H.; de la Torre, M.; Wübbold, F.; Wagner, B.; Schimmels, S. The role of combined laser scanning and video techniques in monitoring wave-by-wave swash zone processes. *Coast. Eng.* **2014**, *83*, 150–165. [CrossRef]

29. Almeida, L.P.; Masselink, G.; Russell, P.E.; Davidson, M.A. Observations of gravel beach dynamics during high energy wave conditions using a laser scanner. *Geomorphology* **2015**, *228*, 15–27. [CrossRef]
30. Terefenko, P.; Zelaya Wziątek, D.; Dalyot, S.; Boski, T.; Pinheiro Lima-Filho, F. A High-Precision LiDAR-Based Method for Surveying and Classifying Coastal Notches. *ISPRS Int. J. Geo-Inf.* **2018**, *7*, 295. [CrossRef]
31. Cieślikiewicz, W.; Paplińska-Swerpel, B. A 44-year hindcast of wind wave fields over the Baltic Sea. *J. Coastal Eng.* **2008**, *55*, 894–905. [CrossRef]
32. Hersbach, H.; Dee, D. ERA5 Reanalysis Is in Production. 2016. Available online: https://www.ecmwf.int/en/newsletter/147/news/era5-reanalysis-production (accessed on 23 November 2018).
33. Rosser, N.J.; Brain, M.J.; Petley, D.N.; Lim, M.; Norman, E.C. Coastline retreat via progressive failure of rocky coastal cliffs. *Geology* **2013**, *41*, 939–942. [CrossRef]
34. Johnstone, E.; Raymond, J.; Olsen, J.M.; Driscoll, N. Morphological Expressions of Coastal Cliff Erosion Processes in San Diego County. *J. Coastal Res.* **2016**, *76*, 174–184. [CrossRef]
35. Hapke, C.J.; Reid, D. *National Assessment of Shoreline Change, Part 4: Historical Coastal Cliff Retreat along the California Coast*; USGS Open-File Report 2007-1133; U.S. Department of the Interior, U.S. Geological Survey: Santa Cruz, USA, 2007; 51p.
36. Kurowicka, D.; Cooke, R. *Uncertainty Analysis with High Dimensional Dependence Modelling*; John Wiley & Sons Ltd.: Chichester, UK, 2006.
37. Hanea, A.M.; Kurowicka, D.; Cooke, R.M. Hybrid Method for Quantifying and Analyzing Bayesian Belief Nets. *Qual. Reliab. Eng. Int.* **2006**, *22*, 709–729. [CrossRef]
38. Hanea, A.; Morales Nápoles, O.; Dan Ababei, D. Non-parametric Bayesian networks: Improving theory and reviewing applications. *Reliab. Eng. Sys. Saf.* **2015**, *144*, 265–284. [CrossRef]
39. Joe, H. *Dependence Modeling with Copulas*; Chapman & Hall/CRC: London, UK, 2014.
40. Genest, C.; Rémillard, B.; Beaudoin, D. Goodness-of-fit tests for copulas: A review and a power study. *Insur. Math. Econ.* **2009**, *44*, 199–213. [CrossRef]
41. Furmańczyk, K.K.; Dudzińska-Nowak, J.; Furmańczyk, K.A.; Paplińska-Swerpel, B.; Brzezowska, N. Critical storm thresholds for the generation of significant dune erosion at Dziwnow Spit, Poland. *Geomorphology* **2012**, *143*, 62–68. [CrossRef]
42. Hackney, C.; Darby, S.E.; Leyland, J. Modelling the response of soft cliffs to climate change: A statistical, process-response model using accumulated excess energy. *Geomorphology* **2013**, *187*, 108–121. [CrossRef]
43. Earlie, C.; Masselink, G.; Russell, P. The role of beach morphology on coastal cliff erosion under extreme waves. *Earth Surf. Process. Landf.* **2018**, *43*, 1213–1228. [CrossRef]
44. Wiśniewski, B.; Wolski, T. *Katalogi Wezbrań i Obniżeń Sztormowych Poziomów Morza oraz Ekstremalne Poziomy wód na Polskim Wybrzeżu*; Maritime University of Szczecin: Szczecin, Poland, 2009.
45. Young, A.P. Recent deep-seated coastal landsliding at San Onofre State Beach, California. *Geomorphology* **2015**, *228*, 200–212. [CrossRef]
46. Marques, F.; Matildes, R.; Redweik, P. Statistically based sea cliff instability hazard assessment of Burgau—Lagos coastal section (Algarve, Portugal). *J. Coastal Res.* **2011**, *64*, 927–993.
47. Varnes, D.J. Slope Movement Types and Processes. In *Special Report 176: Landslides: Analysis and Control*; Schuster, R.L., Krizek, R.J., Eds.; Transportation and Road Research Board; National Academy of Science: Washington, DC, USA, 1978; pp. 11–33.
48. Bray, M.J.; Hooke, J.M. Prediction of soft-cliff retreat with accelerating sea-level rise. *J. Coastal Res.* **1997**, *13*, 453–467.
49. Paprotny, D.; Morales-Nápoles, O. Estimating extreme river discharges in Europe through a Bayesian network. *Hydrol. Earth Syst. Sci.* **2017**, *21*, 2615–2636. [CrossRef]
50. Morales Nápoles, O.; Hanea, A.M.; Worm, D.T.H. Experimental results about the assessments of conditional rank correlations by experts: Example with air pollution estimates. In *Safety, Reliability and Risk Analysis: Beyond the Horizon*; CRC Press/Balkema: Leiden, Holland, 2013; pp. 1359–1366.

Technical Note

Cubesats Allow High Spatiotemporal Estimates of Satellite-Derived Bathymetry

Dimitris Poursanidis [1,*], Dimosthenis Traganos [2], Nektarios Chrysoulakis [1] and Peter Reinartz [3]

1 Foundation for Research and Technology—Hellas (FORTH), Institute of Applied and Computational Mathematics, N. Plastira 100, Vassilika Vouton, 70013 Heraklion, Greece; zedd2@iacm.forth.gr

2 German Aerospace Center (DLR), Remote Sensing Technology Institute, Rutherfordstraße 2, 12489 Berlin, Germany; Dimosthenis.Traganos@dlr.de

3 German Aerospace Center (DLR), Earth Observation Center (EOC), 82234 Weßling, Germany; peter.reinartz@dlr.de

* Correspondence: dpoursanidis@iacm.forth.gr; Tel.: +30-2810391774

Received: 31 March 2019; Accepted: 29 May 2019; Published: 31 May 2019

Abstract: High spatial and temporal resolution satellite remote sensing estimates are the silver bullet for monitoring of coastal marine areas globally. From 2000, when the first commercial satellite platforms appeared, offering high spatial resolution data, the mapping of coastal habitats and the extraction of bathymetric information have been possible at local scales. Since then, several platforms have offered such data, although not at high temporal resolution, making the selection of suitable images challenging, especially in areas with high cloud coverage. PlanetScope CubeSats appear to cover this gap by providing their relevant imagery. The current study is the first that examines the suitability of them for the calculation of the Satellite-derived Bathymetry. The availability of daily data allows the selection of the most qualitatively suitable images within the desired timeframe. The application of an empirical method of spaceborne bathymetry estimation provides promising results, with depth errors that fit to the requirements of the International Hydrographic Organization at the Category Zone of Confidence for the inclusion of these data in navigation maps. While this is a pilot study in a small area, more studies in areas with diverse water types are required for solid conclusions on the requirements and limitations of such approaches in coastal bathymetry estimations.

Keywords: satellite-derived bathymetry; hydrography; CubeSats; hypertemporal; zones of confidence; PlanetScope

1. Introduction

Bathymetry is the center of several important biogeophysical processes such as primary production and the development of marine forests and seagrass meadows—influenced by the exponential decrease of light with depth. The spatial variation can also define the topographic properties of the studied seascape, e.g., slope, aspect, rugosity, terrain roughness, and the bathymetric position index [1,2]. The importance of bathymetry as a product and its use in nautical charts [3] under several categories (Zones of Confidence—ZOC) is high in areas of maritime navigation. Traditionally, bathymetry has been estimated by the implementation of hydroacoustic tools and methods like the Single-Beam (SBES) and Multi-Beam Echo Sounders (MBES), Airborne Lidar Bathymetry (ALB), and LIDAR devices, installed on vessels following specially designed sailing lines with a specific geometry [4]. These methods, and especially the MBES and ALB, can provide highly accurate information in multiple scales. However, depending on the extent of the project area, they require a large amount of effort and are costly [5] when compared with newly adopted approaches such as the Satellite-derived bathymetry (SDB).

During the last four decades, many SDB-related studies have emphasized the potential utilization of satellite remote sensing sensors for bathymetric calculations in clear shallow waters in a plethora of spatiotemporal resolutions: From Lyzenga [6,7] in 1978 and 1981 using the first Earth Observation satellite, the optical Landsat Multispectral Scanner (spatial resolution of about 79 m and temporal one of 18 days) to 1983 and the implementation of the first spaceborne synthetic aperture radar satellite, SEASAT (25-m spatial resolution) [8]; and from an inversion of spaceborne altimetry data from Geosat and the ERS-1 (12-km spatial resolution) in 1997 [9] to a global 500-m bathymetry map utilizing Cryosat-2 and Jason-1 data in 2014 [10]. Analytical, semi-analytical and empirical methods have been developed for the estimation of bathymetry up to 30-m depth (Table 1 at reference [1]). The analytical and semi-analytical methods are based on the physics of light transmission in water using different parameters of the atmospheric, water surface, water column, and bottom layers; such parameterization renders these methods more complicated and of greater computational demand to retrieve bathymetry data, but also of higher accuracy than the empirical methods [1]. Lyzenga showed that bathymetry can be estimated over clear shallow water using satellite remote sensing data with a multi-band log linear algorithm. Since then, this method has been utilized in various approaches or with small modifications to derive bathymetry using different spaceborne data [11–15]. Depending on the application and the scale of data needs, high spatial resolution data have become crucial to characterize seascape morphology at local scales, for use in spatial ecology, maritime spatial planning, and navigation. In the last ten years, the advents in remote sensing technology have given birth to satellites with image acquisitions of higher frequency and lower pixel size, e.g., Landsat 8 (30-m and 16 days, Sentinel-2 (10-m and 5 days). The exemplar of the two latter satellite missions—owing to their open, free, and public data access policy—has allowed new scientific developments and operational applications in coastal SDB. In parallel, the Digital Globe's commercial constellation of WorldView and Quickbird satellites has been also offering sub-meter spatial resolution and revisit times of a single day, yet at a high and elusive cost for many institutions.

The majority of the available satellite platforms provide remotely sensed data at an either infrequent temporal resolution or expensive data provision. This gap starts to be filled in 2013 by a new company bridging the gap between the high spatial and high temporal resolution of satellite remote sensing data. Planet Labs, Inc. (http://planet.com) has successfully built and launched 281 CubeSats since 2013 at various phases. Now (2019), it has more than 148 satellites in sun-synchronous orbit which image nearly all off the global land surface and coastal marine surfaces at 3–5-m resolution daily. As such they provide near real-time imagery to the private industry, academic domain, and governmental organizations. The satellites are the so-called CubeSats 3U—about the size of a wine box (10 → 10 → 30 cm) carrying a four-band multispectral camera and power/downlinking equipment. Having small size and being built at lower costs, the CubeSats have the potential to overcome tradeoffs among high spatial and temporal resolution because of the multi-satellite constellation approach. This is linked to the mass production of the hardware and low launch costs using various platforms, driving to affordable solutions for commercial satellite companies as well as non-profit and research institutes. A drawback related to the image quality is that the multispectral imagery is acquired using inexpensive sensors at different batch productions with variable radiometric quality, consistency, and signal-to-noise ratio in comparison to the space agency-funded missions (Landsat and Sentinel series) and the commercial platforms (e.g., Marxan Technologies and WorldView satellites) [16]. So far, the CubeSat have limited use in the natural environment mainly due to image quality related to the user needs and the among satellites cross-sensor calibration approaches. In the seascape community, even if the constellation has great potential in transforming coastal remote sensing, few studies have come out so far [17,18].

The objective of the current study is the first utilization of CubeSat imagery to calculate Satellite-derived Bathymetry for a selected site in Crete, Greece, using a plethora of single images from the same month. Implementing the selected images, we apply the empirical method by Lyzenga [9], which requires only in-situ depth soundings. Based on the best fitted training model, we proceed by applying low pass filters for the enhancement of radiometric anomalies at neighboring pixels, and the calculation of bathymetry for the depth zones according to the International Hydrographic Organization (IHO). The latter approach provides insights into the suitability of the CubeSats for spaceborne bathymetry and of the results to the requirements of the IHO for the inclusion of such products in the production line of navigation charts.

2. Materials and Methods

2.1. Study Site and Insitu Data

Obros Gyalos is a small protected cove (Figure 1) located at the prefecture of Chania in the island of Crete; it lacks a proximity to rivers and anthropogenic activities. The cove has been selected for the creation of a SCUBA diving park of Crete due to its unique seascape morphology. There was a detailed bathymetric survey among the activities for its establishment during summer of 2017 [19]. In total, 9954 bathymetric points have been collected; these have been split into two parts for calibration and validation. Prior to the random spatial split using the "*Subset features*" tool in ArcGIS 10.5, which randomly splits the dataset into two parts based on percentages, an aggregation of the in-situ data has been performed to match the spatial resolution of the satellite data (3-m). Points have been converted to 3-m pixels using the MEAN function (the mean of the attributes of all the points within the cell) of the corresponding tool "*Point to Raster*" in ArcGIS 10.5. After that, the resulting raster dataset has been converted into points, producing the final dataset of 4756 points, split into 2854 points for training and 1902 points for validation.

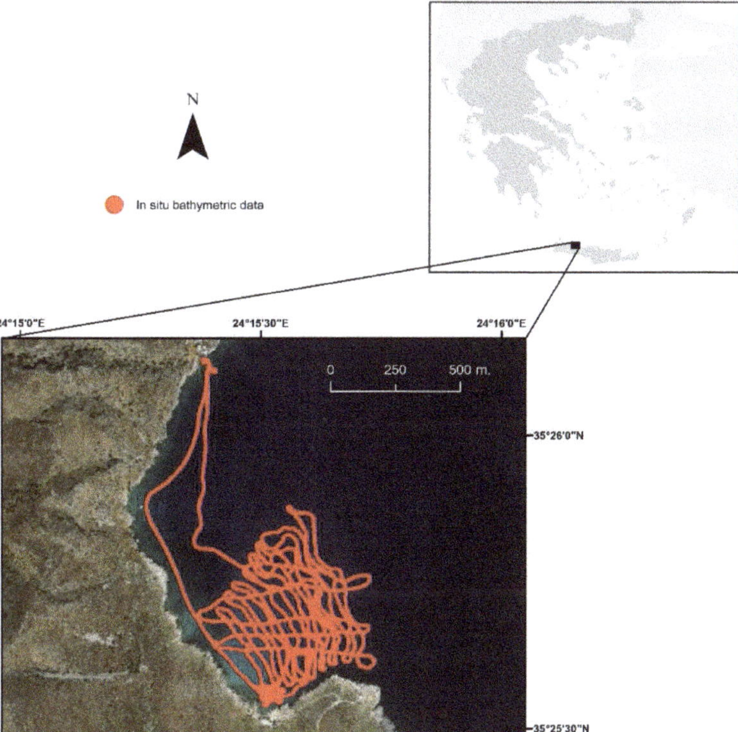

Figure 1. Obros Gyalos study site at Apokoronas area; with red dots the collected in-situ data using a single beam echosounder.

2.2. Satellite Remote Sensing Data

Five PlanetScope images at the level 3B were used in the current research (Figure 2, Table 1). Before the application of the empirical bathymetric models, all the images had been cropped to the defined study area. Images have been obtained for free under the Planet Education and Research program [20]. PlanetScope level 3B image is an orthorectified Ortho Scene Product, while the pixel value is scaled to Top-of-Atmosphere (TOA) radiance (at-sensor); a surface reflectance (SR) product is also available, reducing the need for atmospherically correcting the orthorectified PlanetScope images, projected to a Universal Transverse Mercator (UTM) cartographic projection [21]. The use of five PlanetScope images was necessary as a control for variations in the radiometric quality of PlanetScope images, and in the atmospheric and water surface conditions (sunglint, wave formation). While signal-to-noise ratio defines the quality of an image band, the current distribution of the PlanetScope CubeSats lacks this information in the metadata, thus we are not able to evaluate the quality of each selected image based on this metric. A multi-image assessment approach limits the chance factor in the conclusions regarding the performance of PlanetScope images—due to the radiometric quality and the atmospheric/water surface conditions of a single image. From the satellite images, the spectral bands blue (455–515 nm), green (500–590 nm), and red (590–670 nm) have been used for the regressions, while the near-infrared (780–860 nm) has been used for masking the land.

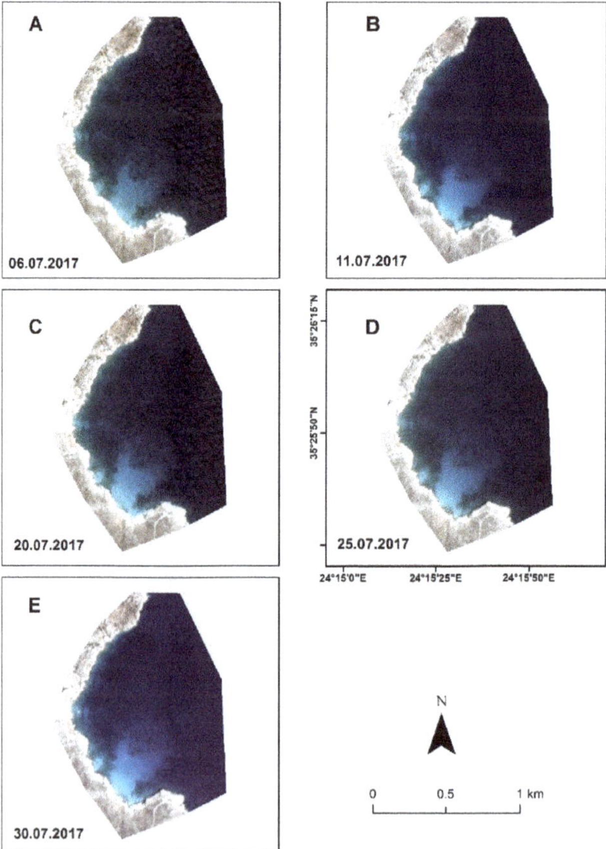

Figure 2. The selected daily images of the PlanetScope. All panels (**A–E**) share scale and north arrow. All maps share the same grid as of panel **D**.

Table 1. The PlanetScope CubeSat images used in the analysis.

Scene ID	Date	Time (UTC)	Sun Azimuth	Sun Elevation
20170706_082033_103e_3B_AnalyticMS_SR	6 July 2017	8:20	106.24	59.69
20170711_082041_1012_3B_AnalyticMS_SR	11 July 2017	8:20	107.14	59.27
20170720_082118_101d_3B_AnalyticMS_SR	20 July 2017	8:21	109.4	58.37
20170725_082251_1011_3B_AnalyticMS_SR	25 July 2017	8:22	111.02	57.78
20170730_082318_102e_3B_AnalyticMS_SR	30 July 2017	8:23	112.69	57.17

2.3. Empirical Satellite-Derived Bathymetry (SDB)

Satellite-derived Bathymetry started during the 1970s with empirical methods that used spectral bands at the visible wavelength; blue, green, and red have been widely used as the independent variables in a multiple regression approach, and the in-situ data, the depth soundings with known depths, as the dependent variables. Lyzenga [9] (hereafter Lyzenga85) was the first that developed the equation of the estimation of bathymetry through the aforementioned approach assuming that the relationship between the log-transformed bands and known depth via multiple regression is linear. The coefficients of the regression are applied to the satellite data for the calculation of the bathymetry [19]. For the five images, multiple linear regressions, using the package "*car*" of R [22], have been performed for the calculation of the coefficients. Based on the lower value of the corrected Akaike Information Criterion (AICc) [23,24] calculated by using the R package "*AICcmodavg*" [25], the selected image has been further analyzed by applying a low pass filter of 3 × 3 to reduce the potential radiometric anomalies between pixels. Two regressions have been applied using the training data; the first for the depth zone between 0–10 m and the second for the zone between 10–25 m; the respected coefficients have been applied to the selected image and for each depth zone the metrics "coefficient of determination (R^2)", "Standard Error (SEz)" and the "Root Mean Square Error (RMSEz)" have been estimated using the validation points. The comparison of the RMSEz value with the Category Zone of Confidence (CATZOC) values provides insights into the Zones of Confidence and the reliability of the bathymetry product for implementation in navigation charts produced by hydrographic offices [15].

3. Results

The availability of the full archive of the PlanetScope imagery allows us the selection of suitable images within the same month, setting two criteria: The absence of sunglint which poses an extra processing step and eventually could introduce additional noise to the resulting deglinted images; this could in turn reduce the suitability of the images due to the already known low signal-to-noise ratio [18]; and the cloud free scenes avoiding cloud masking. Thus, the only difference between the selected images is a slightly visible wavy water surface caused by local winds formed during the morning of the day of acquisition. No sedimentation in the water column was observed, while the bottom cover is mainly composed by two types, bright sandy bottom, and rocky formations (Figure 3). Images have been selected with approximately five days interval. By applying the method described in 2.3 using the subset of training in-situ data (n = 2854) and based on the AICc values (Table 2), the image of 11 July 2017 has been selected for further analysis.

The multiple regression results of the training, based on the coefficient of determination and the standard error on the RAW (Surface reflectance) and the transformed (3 × 3 median low pass filter) results are presented in Table 3, while the validation results in Figures 4 and 5, and the produced bathymetric map after the combination of the two different results from the two predefined bathymetric zones in Figure 6.

Table 2. Comparison of AICc values calculated by the multiple linear regressions for the SDB estimation of the in-situ data (training) and the log-transformed spectral bands (Blue-Green-Red) of the PlanetScope imagery at full depth range (0–25 m), (Modnames = Model names, AICc = Corrected AIC, Delta_AICc = delta AIC, AICcWt = weight of AICc, LL = Log-liklehood).

ID	Modnames	AICc	Delta_AICc	AICcWt	LL	R^2
1	sdb.glm0706	12497.5	1128.5	8.584×10^{-246}	−6243.7	0.84
2	sdb.glm0711	11368.9	0	1	−5679.4	0.89
4	sdb.glm0720	12143.5	774.6	6.203×10^{-169}	−6066.7	0.86
5	sdb.glm0725	12989.2	1620.3	0	−6489.6	0.81
6	sdb.glm0730	12583.9	1214.9	1.51×10^{-264}	−6286.9	0.84

Table 3. The results from the multiple regressions using the training data on the RGB spectral bands of the image of 11 July 2017 in the depth zones 0–10 m and 10–25 m.

Regression Statistics	RAW (0–10 m)	3 × 3 (0–10 m)	RAW (10–24 m)	3 × 3 (10–24 m)
Multiple R	0.93	0.94	0.9	0.92
R Square	0.86	0.88	0.81	0.84
Adjusted R Square	0.86	0.88	0.81	0.84
Standard Error	0.72	0.66	1.63	1.48
Observations	702	702	2152	2152

Figure 3. The two seabed cover types in the project area; above: sandy soft bottom, below: carbonated rocky surfaces partially covered by brown macroalgae.

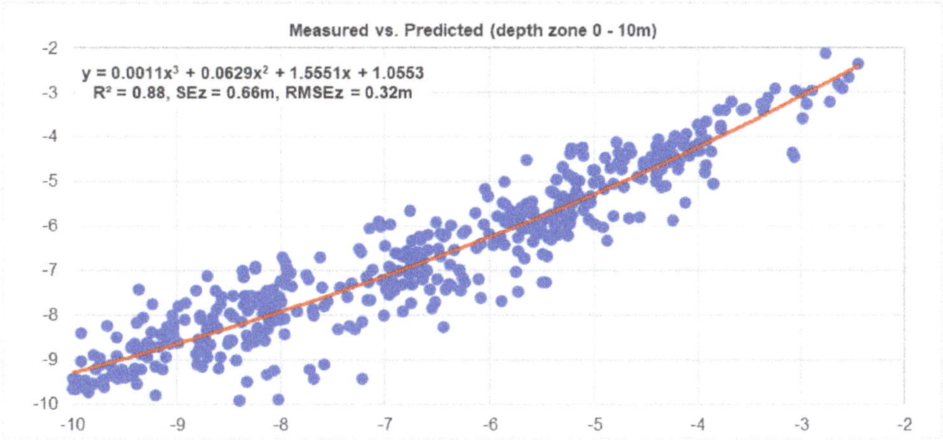

Figure 4. The validation plot for the depth zone of 0–10 m of in-situ depth points (x-axis) against image-derived bathymetries (y-axis) implementing the Lyzenga85 model on the image of 11 July 2017. 3rd order polynomial equation has been applied.

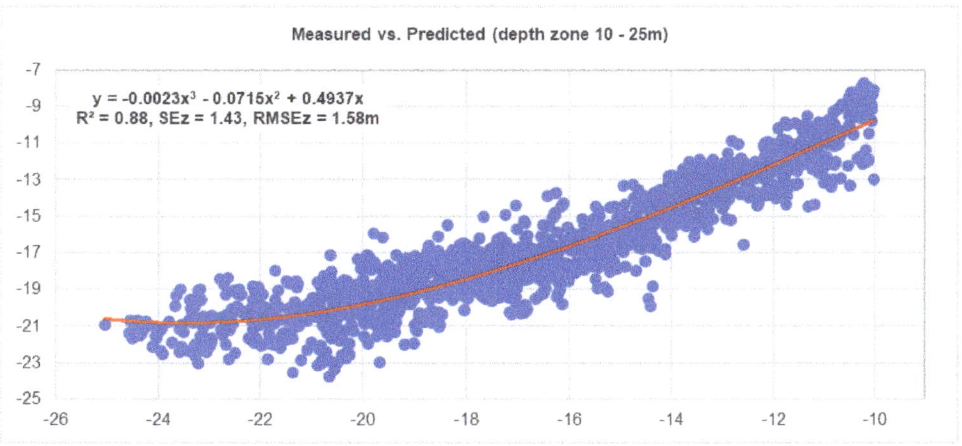

Figure 5. The validation plot for the depth zone of 10–25 m of in-situ depth points (x-axis) against image-derived bathymetries (y-axis) implementing the Lyzenga85 model on the image of 11 July 2017. 3rd order polynomial equation has been applied.

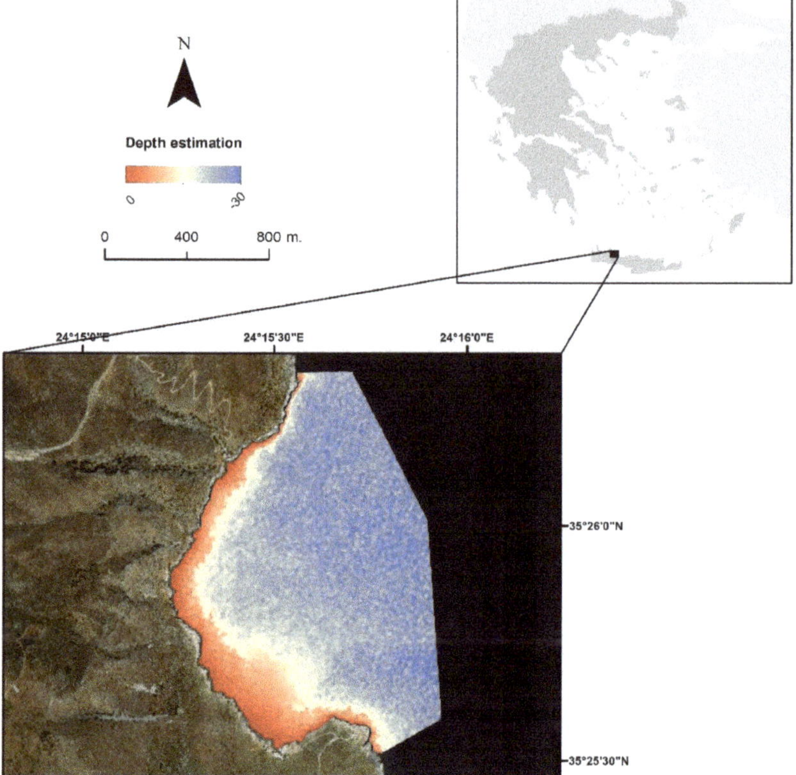

Figure 6. The Satellite-derived bathymetry estimated by the combination of the two-depth zones on the image of 11 July 2017.

4. Discussion

4.1. The CubeSats and the Performance of the Models

PlanetScope constellation is a unique Earth Observation fleet with daily revisit and almost 3-m pixel size, providing new insights into the observation and monitoring of nearly any place on Earth; this allows for new endeavors in mapping, change detection, and monitoring of both land and coastal zone areas. It is worth mentioning that this constellation has been based on CubeSats; they have been built using inexpensive electronics meaning that they are not direct inter-comparable as regards to the radiometric quality, consistency and signal-to-noise ratios of the commercial and agency-funded satellites [26]. Data from the fleet have been already implemented for tracking vegetation dynamics, hydrological applications, Digital Elevation Model creation, seabed cover and coral reef mapping [17,18,27–29]. The results from the recent studies are promising, while global projects like the Allen Coral Atlas—aiming at mapping the swallow (<15 m) global coral reefs using solely PlanetScope imagery [30]—show the importance of the availability of high spatial and temporal resolution satellite data for mapping and monitoring areas where persistent cloud cover is a barrier either for the open access data from USGS Landsat and Copernicus Sentinel-2 or for the data coming from the commercial satellite platforms; the latter have higher spatial resolution (<2 m) and better signal-to-noise ratio, but the data flow is in sparse intervals, making a cloud or almost cloud-free image a challenge.

The limitations of the present study are method and technology-wise. First, our implemented methodology incorporates an in-situ bathymetric dataset of 4726 points to both train and validate the satellite-derived bathymetry model. This violates Tobler's first law of geography—geographically neighboring observations have a tendency to be alike the ones further apart—introducing a statistical bias in our approach. Nevertheless, spatially independent field-based calibration and testing datasets are sparse due to reasons of practical nature, i.e., the high acquisition cost of such datasets in situ. Technology-wise, there are two main limitations that we could generalize for the majority of the coastal aquatic scientific and operational developments and applications with Planet's CubeSats: (a) Radiometric differences between the different PlanetScope CubeSats; (b) low signal-to-noise ratio for high-accuracy detection and mapping of the coastal benthos. We expect the present and near-future deployments of new Planet satellite sensors featuring higher radiometric quality to overcome the aforementioned technological limitations and improve the accuracy and efficiency of similar approaches. The present study is the first one that tries to understand the performance of PlanetScope CubeSats in the estimation of bathymetry in a mixed-bottom seascape using available in-situ data—collected for different purposes; this is carried out by applying empirical-based statistical relationships between the log-transformed, water-penetrating bands (most frequently at blue and green wavelengths but also including the red wavelength) and acoustic-derived in-situ depth data.

Having access to the full archive of the imagery, we selected five images over July 2017, to account for the effect of sea surface conditions that change due to waves and also for the radiometric consistency of the different flocks, as the CubeSats have been released in varying time frameworks since 2013. From the application of the Lyzenga85 empirical method in the five images and based on the AICc values from the training models (Table 2), we selected and further employed the image with the lower AICc value and the best R^2 (Figure 2B). The visual examination of the selected image (Figure 2B) lacks a wavy surface and visible noise in comparison to the other images that can possibly be caused by high altitude haze and the older electronics that the first flocks have been made of. The application of a low pass filter (3 × 3 median) to the surface reflectance imagery improves the results of the training of the method (Table 3) as it smooths possible pixel anomalies. The bathymetric zonation of the multiple regressions into two groups (0–10 m and 10–25 m) has been selected to understand the accuracies of the SDB in relation to the requirements of the IHO in the inclusion of bathymetry data in navigation maps and the corresponding ZOC that the depth accuracies fall into [3]. By splitting the analysis into these two zones, coefficients from the multiple regressions are applied into the respective bathymetric zones, while the final map is the unification of the results from the two zones.

4.2. *Are PlanetScope Suitable for Data Inclusion in Navigation Maps?*

The results from the application of the calculated coefficients are very promising for the value of the PlanetScope imagery in estimating bathymetry and its subsequent integration in navigation charts compliant with the IHO requirements and the fit of them into ZOCs. The results are also comparable with the Copernicus Sentinel-2 based SDB from the same site, as has been shown by [16] using the Google Earth Engine platform. For the Zone of 0–10 m (Figure 4), the estimated R^2, using a 3rd order polynomial equation, is 0.88, the SEz is 0.66 m, and the estimated RMSEz is 0.32 m while for the zone 10 m–25 m R^2 is 0.88 with SEz at 1.43, and RMSEz at 1.58 m. According to the IHO, the Zone of Confidence A1 requires data in the zone 0–10 m with a depth accuracy of ± 0.6 m and a position accuracy of ±5 m, while the zone A2 a depth accuracy of ±0.6 m and a position accuracy of ±20 m. For Zone B, it requires for the same depth as in A2, an accuracy of ±1.2 m and a position accuracy of ±50 m. For the depth zone 10–30 m, the ZOC A1 requires ±0.8 m, the ZOC A2 requires ±1.2 m and the ZOC B requires ±1.6 m. The results from the current study suggest that the PlanetScope data can fit into the ZOC A2 and possibly into the ZOC A1. This fit is also supported by the position accuracy of the satellite images, according to the recent study of [31], which shows that the geolocation performance of the PlanetScope's Level 3A product is good and the absolute geolocation performance is set by a max $RMSE_x$ = 5.18 m, $RMSE_y$ = 4.21 m and CE(90) = 9.93 m, respectively.

While the current study includes only one site and is targeted at understanding the behavior of the different selected PlanetScope CubeSat images from the same site covering one-month period, more studies are required to robustly conclude for the suitability of the employed Earth Observation data in the calculation of bathymetry. Studies that will include different satellites of very high-resolution data and high signal-to-noise ratio with images from the same period (e.g., within the same month) will allow a full benchmark on the radiometric quality of the PlanetScope data for the calculation of SDB. Other approaches like the cloud-based one presented in [19] will allow the full scientific and operational exploitation of the Planet archive for seabed mapping and monitoring without the laborious analysis per image and the visual inspection of the "best image" for further processing and analysis; this will be achieved through the utilization of multi-temporal (weekly, monthly, seasonal) PlanetScope image composites, which will reduce technological and environmental issues, both intra and inter-sensor, e.g., low signal-to-noise ratio, varying radiometric quality, atmospheric, water surface and column, and seabed conditions.

There is an increasing need in the charting of shallow coastal waters across the globe at fine scales that will allow the creation of new navigation maps, but also the update of old productions. IHO has started to adopt the SDB approach and the IHO S-44 standards are currently under revision. The revised one will be updated to include the ability to exploit new technologies for the update of nautical charts when that is possible under the specifications required, as the use of traditional means, like the hydroacoustics, in shallow uncharted waters is dangerous for the equipment and the per km^2 charted area is much more costly in comparison to the SDB approach.

5. Conclusions

The present study is the first one that attempts to examine the performance of the PlanetScope CubeSats in calculating Satellite-derived Bathymetry and explores whether the results fit into the requirements of IHO for nautical maps of navigation. The availability of almost daily satellite images in the archive allows the selection of the most suitable data based on cloud coverage, water surface conditions, and intense visible sunglint—an asset of the high spatial and temporal satellite constellation. The results from the applied empirical method with the intermediate preprocessing steps are promising and show that it has great potential for coastal bathymetry estimation, especially in the shallow waters. However, given that one site has been tested, more work is needed to understand the nature of PlanetScope CubeSats in estimating SDB by including several sites distributed in different water bodies that cover both case I and II waters and systematic collection of in-situ soundings that correspond to each month. Also, the analysis of monthly and annual image composites, as these are provided by Planet for commercial use, will support the elimination of issues related to the absence of cloud mask information, and the low signal-to-noise ratios of the PlanetScope imagery. All in all, given that the technology of CubeSats is improved (i.e., higher signal-to-noise ratios and more spectral bands in the visible wavelengths), we expect, in the next decade, a boom of fleets that can be eventually exploited to carry out scientific and operational mapping and monitoring of the coastal aquatic environment at a fraction of the cost of "traditional" satellite platforms.

Author Contributions: D.P. conceived the idea, performed the analysis, and prepared the manuscript; D.T. supported the analysis and contributed to the manuscript. N.C. and P.R. contributed to the manuscript.

Funding: D.P. and N.C. are supported by the European H2020 Project 641762 ECOPOTENTIAL: Improving future ecosystem benefits through Earth Observations. D.T. is supported by a DLR-DAAD Research Fellowship (No. 57186656).

Acknowledgments: This work contributed to and was partially supported by the European H2020 Project 641762 ECOPOTENTIAL: Improving future ecosystem benefits through Earth Observations. Dimosthenis Traganos is supported by a DLR-DAAD Research Fellowship (No. 57186656). We would like to thank the two anonymous reviewers for their constructive comments that improve our work. We also would like to thank Planet's Ambassadors program as Planet satellite images (Planet Team, 2017) are provided through this program.

Conflicts of Interest: The authors declare no conflict of interest.

References

1. Hamylton, S.M.; Hedley, J.D.; Beaman, R.J. Derivation of High-Resolution Bathymetry from Multispectral Satellite Imagery: A Comparison of Empirical and Optimisation Methods through Geographical Error Analysis. *Remote Sens.* **2015**, *7*, 16257–16273. [CrossRef]
2. Walbridge, S.; Slocum, N.; Pobuda, M.; Wright, D.J. Unified Geomorphological Analysis Workflows with Benthic Terrain Modeler. *Geosciences* **2018**, *8*, 94. [CrossRef]
3. International Hydrographic Organization (IHO). S-57 Supplement No. 3—Supplementary Information for the Encoding of S-57 Edition 3.1 ENC Data. International Hydrographic Organization: Monaco, 2014. Available online: https://www.iho.int/iho_pubs/standard/S-57Ed3.1/S-57_e3.1_Supp3_Jun14_EN.pdf (accessed on 5 January 2019).
4. Saylam, K.; Hupp, J.R.; Averett, A.R.; Gutelius, F.W.; Gelhar, B.W. Airborne lidar bathymetry: Assessing quality assurance and qualitycontrol methods with Leica Chiroptera examples. *Inter. J. Remote Sens.* **2018**, *39*, 2518–2542. [CrossRef]
5. Sánchez-Carnero, N.; Aceña, S.; Rodríguez-Pérez, D.; Couñago, E.; Fraile-Jurado, P.; Freire, J. Fast and low-cost method for VBES bathymetry generation in coastal areas. *Estuar. Coast. Shelf Sci.* **2012**, *114*, 175–182. [CrossRef]
6. Lyzenga, D. Passive remote sensing techniques for mapping water depth and bottom features. *Appl. Opt.* **1978**, *17*, 379–383. [CrossRef]
7. Lyzenga, D. Remote sensing of bottom reflectance and Water attenuation parameters in shallow water using aircraft and Landsat data. *Int. J. Remote Sens.* **1981**, *2*, 71–82. [CrossRef]
8. Dixon, T.H.; Naraghi, M.; McNutt, M.K.; Smith, S.M. Bathymetric prediction from SEASAT altimeter data. *J. Geophys. Res.* **1983**, *88*, 1563–1571. [CrossRef]
9. Sandwell, D.T.; Smith, W.H. Marine gravity anomaly from Geosat and ERS 1 satellite altimetry. *J. Geophys. Res. Solid Earth* **1997**, *102*, 10039–10054. [CrossRef]
10. Olson, C.J.; Becker, J.J.; Sandwell, D.T. A new global bathymetry map at 15 arcsecond resolution for resolving seafloor fabric: SRTM15_PLUS. In Proceedings of the AGU Fall Meeting Abstracts, San Francisco, CA, USA, 15–19 December 2014.
11. Conger, C.L.; Hochberg, E.J.; Fletcher, C.H.; Atkinson, M.J. Decorrelating remote sensing color bands from bathymetry in optically shallow waters. *IEEE Trans. Geosci. Remote Sens.* **2006**, *44*, 1655–1660. [CrossRef]
12. Lyzenga, D.R. Shallow-water bathymetry using combined lidar and passive multispectral scanner data. *Int. J. Remote Sens.* **1985**, *6*, 115–125. [CrossRef]
13. Lyzenga, D.R.; Malinas, N.P.; Tanis, F.J. Multispectral bathymetry using a simple physically based algorithm. *IEEE Trans. Geosci. Remote Sens.* **2006**, *44*, 2251–2259. [CrossRef]
14. Philpot, W.D. Bathymetric mapping with passive multispectral imagery. *Appl. Opt.* **1989**, *28*, 1569–1578. [CrossRef] [PubMed]
15. Kanno, A.; Tanaka, Y.; Kurosawa, A.; Sekine, M. Generalized Lyzenga's Predictor of Shallow Water Depth for Multispectral Satellite Imagery. *Mar. Geod.* **2013**, *36*, 365–376. [CrossRef]
16. Houborg, R.; McCabe, M.F. High-Resolution NDVI from Planet's constellation of earth observing nano-satellites: A new data source for precision agriculture. *Remote Sens.* **2016**, *8*, 768. [CrossRef]
17. Traganos, D.; Cerra, D.; Reinartz, P. Cubesat-Derived Detection of Seagrasses Using Planet Imagery Following Unmixing-Based Denoising: Is Small the Next Big? Available online: https://doi.org/10.5194/isprs-archives-XLII-1-W1-283-2017 (accessed on 20 April 2019).
18. Wicaksono, P.; Lazuardi, W. Assessment of PlanetScope images for benthic habitat and seagrass species mapping in a complex optically shallow water environment. *Inter. J. Remote Sens.* **2018**, *39*, 5739–5765. [CrossRef]
19. Traganos, D.; Poursanidis, D.; Aggarwal, B.; Chrysoulakis, N.; Reinartz, P. Estimating Satellite-Derived Bathymetry (SDB) with the Google Earth Engine and Sentinel-2. *Remote Sens.* **2018**, *10*, 859. [CrossRef]
20. Planet. Planet Education and Research Program. Available online: https://www.planet.com/markets/education-and-research/ (accessed on 10 March 2019).
21. Planet. Planet Imagery Product Specification. 2018. Available online: https://assets.planet.com/docs/Combined-Imagery-Product-Spec-Dec-2018.pdf (accessed on 10 March 2019).

22. Fox, J.; Weisberg, S. *An {R} Companion to Applied Regression, Second Edition*; Sage: Thousand Oaks, CA, USA, 2011; Available online: http://socserv.socsci.mcmaster.ca/jfox/Books/Companion (accessed on 10 March 2019).
23. Akaike, H. Information Theory and an Extension of the Maximum Likelihood Principle. In *Selected Papers of Hirotugu Akaike*; Parzen, E., Tanabe, K., Kitagawa, G., Eds.; Springer Series in Statistics (Perspectives in, Statistics); Springer: New York, NY, USA, 1998.
24. Burnham, K.; Anderson, D. *Model Selection and Multimodal Inference*; Springer: New York, NY, USA, 2002.
25. Mazerolle, M.J. AICcmodavg: Model Selection and Multimodel Inference Based on (Q)AIC(c). R Package Version 2.2-1. Available online: https://cran.r-project.org/package=AICcmodavg (accessed on 20 April 2019).
26. Chénier, R.; Faucher, M.-A.; Ahola, R. Satellite-Derived Bathymetry for Improving Canadian Hydrographic Service Charts. *ISPRS Int. J. Geo-Inf.* **2018**, *7*, 306. [CrossRef]
27. Cooley, S.W.; Smith, L.C.; Stepan, L.; Mascaro, J. Tracking Dynamic Northern Surface Water Changes with High-Frequency Planet CubeSat Imagery. *Remote Sens.* **2017**, *9*, 1306. [CrossRef]
28. Ghuffar, S. DEM Generation from Multi Satellite PlanetScope Imagery. *Remote Sens.* **2018**, *10*, 1462. [CrossRef]
29. Asner, G.P.; Martin, R.E.; Mascaro, J. Coral reef atoll assessment in the South China Sea using Planet Dove satellites. *Remote Sens. Ecol. Conserv.* **2017**, *3*, 57–65. [CrossRef]
30. Allen Coral Atlas. 2019. Available online: http://www.allencoralatlas.com/ (accessed on 5 January 2019).
31. Lemajic, S.; Vajsová, B.; Aastrand, P. New Sensors Benchmark Report on PlanetScope: Geometric Benchmarking Test for Common Agricultural Policy (CAP) Purposes. Available online: http://publications.jrc.ec.europa.eu/repository/bitstream/JRC111221/jrc_technical_report_planetscope-final_2.pdf (accessed on 29 May 2019).

Article

Comparison of Pixel- and Object-Based Classification Methods of Unmanned Aerial Vehicle Data Applied to Coastal Dune Vegetation Communities: Casal Borsetti Case Study

Michaela De Giglio [1,*], Nicolas Greggio [2], Floriano Goffo [1], Nicola Merloni [3], Marco Dubbini [4] and Maurizio Barbarella [1]

1 Civil, Chemical, Environmental and Materials Engineering Department-DICAM, University of Bologna, Viale Risorgimento 2, 40136 Bologna, Italy; floriano.goffo@studio.unibo.it (F.G.); maurizio.barbarella@unibo.it (M.B.)

2 Interdepartmental Centre for Environmental Science Research (CIRSA), Lab. IGRG, BiGeA Department, University of Bologna, Via S. Alberto 163, 48100 Ravenna, Italy; nicolas.greggio2@unibo.it

3 Scientific High School A. Oriani of Ravenna, Via C. Battisti, 2, 48121 Ravenna, Italy; nicolamerloni@Isoriani.istruzioneer.it

4 DiSCi, Geography Sec., University of Bologna, Piazza San Giovanni in Monte 2, I-40124 Bologna, Italy; marco.dubbini@unibo.it

* Correspondence: michaela.degiglio@unibo.it

Received: 6 May 2019; Accepted: 12 June 2019; Published: 14 June 2019

Abstract: Coastal dunes provide the hinterland with natural protection from marine dynamics. The specialized plant species that constitute dune vegetation communities are descriptive of the dune evolution status, which in turn reveals the ongoing coastal dynamics. The aims of this paper were to demonstrate the applicability of a low-cost unmanned aerial system for the classification of dune vegetation, in order to determine the level of detail achievable for the identification of vegetation communities and define the best-performing classification method for the dune environment according to pixel-based and object-based approaches. These goals were pursued by studying the north-Adriatic coastal dunes of Casal Borsetti (Ravenna, Italy). Four classification algorithms were applied to three-band orthoimages (red, green, and near-infrared). All classification maps were validated through ground truthing, and comparisons were performed for the three statistical methods, based on the k coefficient and on correctly and incorrectly classified pixel proportions of two maps. All classifications recognized the five vegetation classes considered, and high spatial resolution maps were produced (0.15 m). For both pixel-based and object-based methods, the support vector machine algorithm demonstrated a better accuracy for class recognition. The comparison revealed that an object approach is the better technique, although the required level of detail determines the final decision.

Keywords: vegetation mapping; dunes; unmanned aerial system; pixel-based classification; object-based classification

1. Introduction

Sand dunes are key environmental elements of coastal systems. They represent one of the few natural barriers that can defend the inland territories from extreme high tides, storms, and tsunamis, by absorbing the wave energy. Moreover, they have a significant role in the coastline dynamics, by balancing the erosion and/or accretion phenomena [1]. Dunes also constitute unique habitats and represent corridors that connect the diverse neighboring ecosystems [2]. However, over the last half century, dune systems have undergone habitat loss, and the coastal dunes of Mediterranean areas are among the most vulnerable ecosystems as they are seriously threatened by urbanization and mass tourism, particularly those of the north Adriatic region [3–6]. In a detailed study by Sytnik et al. [3], they reported on the large modifications to the coastal ecosystem of the Casal Borsetti dunes (Ravenna, Italy) study area over the last century. The majority of these modifications have directly affected the local dune systems, while being aimed at the mitigation of erosion phenomena and development of beach touristic infrastructures.

Human disturbance has had effects on the structure, composition, and function of plant communities, which are sensitive indicators of the state of these environments [7,8]. In effect, coastal sand dunes shift between states of activity and stability across the seasons and years [9,10]. Some dunes remain 'stable', perhaps for many decades or centuries, whereas others are 'dynamic', and can maintain an equilibrium between cyclical mobility and stability [11]. Dunes can become active over their entire surface, in particular in areas of present or past vegetation disturbance [12]. As reported by Fabbri et al. [13] (and references therein), many factors can affect dune dynamics, including the sand supply, dune/beach exchanges, beach topography and fetch effects, vegetation species and cover, climate, marine meteorological conditions, and human impact.

Vegetated dunes are fixed and stable, while bare sand dunes are more prone to sand mobilization and erosion. Indications of dune mobility are traditionally based on climatic variables, such as rainfall, temperature, and evapotranspiration, although, except for anthropogenic factors, wind energy is the only limiting factor for the vegetation cover [14].

Beach–dune systems include the dune, its vegetation cover and coastal geomorphology, and the local dynamics (e.g., wind regime, beach typology, erosion/accumulation rate), and these continuously interact with each other. Plant communities in the dune vegetation communities select and promote their own preferred environmental conditions, which stabilizes the dunes [15]. The vegetation communities and cover reflect the conservation state of the dunes and its analysis is used to study the phases of dune evolution and to gather information about entire coastal systems [16–18].

Nowadays, data on dune topography, reflectance, and vegetation cover are usually achieved through direct field sampling [19,20], photointerpretation of aerial and satellite ortho-imagery [21], light detection and ranging (LiDAR) point cloud analysis [22,23], terrestrial laser-scanning surveys [24], and ground differential global positioning system (GPS) measurements [25]. Furthermore, hyperspectral and multispectral aerial and space-based data have been applied to detect different degrees of activity and vegetation cover density [16]. The main limitations of these last systems are their high costs for a high resolution and their insufficient availability during the periods of interest [26]. Moreover, although these images provide useful information on global and regional scales, some processes need multi-temporal observations at a local scale [27].

In recent years, sensors installed on unmanned aerial vehicles (UAVs) have offered many technical and economic advantages for coastal sand dune monitoring [24,28]. To date, most studies have been aimed at the evaluation of coastal systems, and of dunes in particular, using digital RGB (red, green, blue) cameras to build accurate digital surface models with a high spatial resolution [29]. Instead, in the present study, we investigated the use of a UAV equipped with a camera that acquired images at red, green, and near-infrared wavelengths. Pixel-based and object-based approaches [11] were tested and compared for the recognition of the vegetation communities that were growing along the coastal dune system. The advantages and disadvantages of these methods were also investigated. Therefore, the aims of this study were to (i) verify the applicability of multispectral data collected using a UAV platform for the identification/discrimination of fragmented and interspersed coastal dune vegetation communities and (ii) determine the best performing classification method between pixel-based and object-based approaches for multispectral data in the case of disturbed dune vegetation communities.

2. Materials and Methods

2.1. Study Area

The investigated dune is located near the small touristic village of Casal Borsetti (Ravenna, Italy; Figure 1a). In light of the complexity, variability, and vulnerability of the small dune belt considered, a detailed description of both the regional settings and the study area is required.

The territory surrounding the city of Ravenna has been strongly anthropized and industrialized, and in the relevant natural areas, intensive agriculture and tourist facilities coexist [30]. Since the 1960s, the establishment of tourist infrastructures directly along the coastline of this area has resulted in widespread dune damage and destruction. Where the dunes have not been completely flattened to the ground, they have undergone huge fragmentation [4,31]. The natural coastal system has been substituted with the new structures that are directly exposed to climate-change effects, such as a rise in the level of the sea, sea storms, flooding events, and marine erosion [32]. As reported by Sytnik et al. [6], the sector where the study area lies has shown the highest rates of coastal erosion of the last six decades.

Nowadays, all of the remaining dunes are included within the Po Delta Regional Park, as a Special Protection Area (Directive 2009/147/EC "Birds") and a Site of Community Importance (Directive 92/43/EEC "Habitat"). The Regional Park is also included in the UNESCO World Heritage list.

The dune investigated is 350 m long and 60 m wide. The mean dune elevation is 2.5 m above sea level (a.s.l.), and the maximum elevation is 3.5 m a.s.l. (to the north of the site). It is one of the last stretches of dunes with psammophytic vegetation, which is very important for the conservation of coastal biodiversity [33,34].

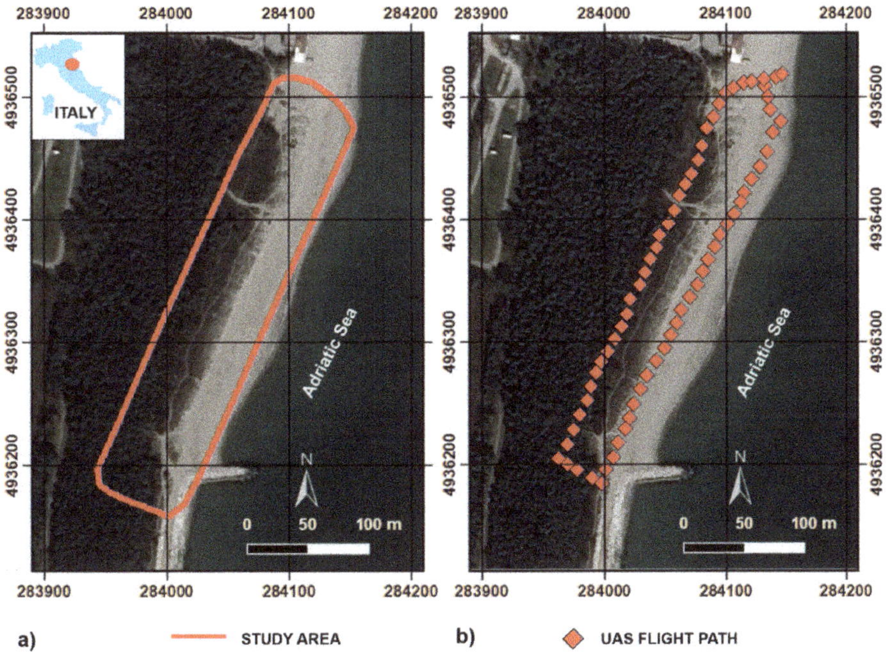

Figure 1. (**a**) Study area: Casal Borsetti dune, Ravenna, Italy. (**b**) Unmanned aerial vehicle flight path. The red squares correspond to the centers of acquisition (WGS84/ UTM zone 33N) (Image background: Google Earth).

While limiting themselves to the vegetation, many authors have described the typical zonation for the north Adriatic coastal dunes based on strips parallel to the coastline [35,36]. In agreement with the definitions of Directive 92/43/EEC "Habitat", and in light of the recent publication of Merloni et al. [37], from the sea going inland, the zonation consists of annual pioneer species, embryo dunes, white dunes, *Malcolmietalia* grassland, grey dunes, shrubby plant communities, and coastal pine woods (Figure 2).

The annual pioneer species described as *Salsolo kali–Cakiletum maritimae* (EUNIS Code B1.12; hereafter referred to as CA) grow close to the shoreline (within a few meters), and they aid in the formation of the dunes where the organic matter brought by the sea accumulates. The main ephemeral species are *Cakile maritima*, *Salsola tragus*, and *Chamaesyce peplis* [37,38]. Moving inland, the dunes start to become semi-stable, and the embryo dune vegetation described as *Agropyretum* (*Echinophoro spinosae–Elymetum farcti*; EUNIS Code B1.3; hereafter referred to as AG) includes psammophilous perennial plants, such as *Elytrigia juncea* (=*Agropyron junceum*), *Echinophora spinosa*, and *Calystegia soldanella*, and these capture the sand as it is moved by the winds, providing the vertical growth of the dune. The white dune community is present in the inner areas where the dunes become more stable, which is also described as *Echinophoro spinosae–Ammophiletum australis* (EUNIS Code B1.3; hereafter referred to as AM). This habitat usually covers 50% to 60% of the total area, and represents a semi-permanent stage, as the roots of *Ammophila arenaria*, *Echinophora spinosa*, and *Eryngium maritimum* form dense felts that promote dune consolidation. Behind the white dunes, where salt winds, coastal erosion, and burial by sand do not affect the vegetation, there are the grey dunes (*Tortulo–Scabiosetum*; EUNIS Code B1.4; hereafter referred to as GD), which are colonized by perennial species, such as *Lomelosia argentea*, *Fumana procumbens*, and *Teucrium polium*, and which have a significant carpet of mosses and lichens (e.g., *Tortula ruraliformis*, *Cladonia convoluta*).

A complication with respect to the theoretical distribution so far described is seen here by *Malcolmietalia* grassland (EUNIS Code B1.4; hereafter referred to as MG), which arises where trampling, salty winds, and disturbance occur. This plant community is mixed with AG, AM, and GD, and it can cover large surfaces [35]. The main diagnostic species are the annual plants *Silene canescens* and *Vulpia membranacea*, which are typically found alongside allochthonous species, such as *Ambrosia coronopifolia*.

A rapid change occurs when moving further inland, where shrubby plant communities settle into depressions where they replace the grey dune vegetation (EUNIS Code B1.63; hereafter referred to as the J habitat), e.g., *Juniperus communis* and *Phillyrea angustifolia*. This narrow belt of shrub is in continuity with the coastal pine woods (EUNIS Code B1.7; hereafter referred to as the P habitat), where *Pinus pinaster* and *Pinus pinea* are the dominant species.

Along the Casal Borsetti dunes in particular, this vegetation succession is often fragmented, with each becoming interspersed with the others; this has generated an atypical vegetation mosaic. Consequently, five vegetation classes were established for the technical classification requirements here, which represent the most significant evolutionary stages of these dunes. For graphical reasons, the previously reported EUNIS Codes are henceforth substituted by the following vegetation community abbreviations: "Bare sand and *Cakiletum*" (BSCA), "*Agropyretum* and *Ammophyletum*" (AGAM), GD, MG, and "Coastal shrub and arboreal formations" (CSAF) (Figure 2).

The BSCA class represents the merging of bare sand areas and *Cakiletum*, where *Cakiletum* species are always <5% of the entire coverage, even under conditions of naturalness and in the absence of disturbance. Moreover, *Cakiletum* habitats are systematically swept during summer beach cleaning operations, which destroys all of the growing plants.

Agropyretum has a relatively high presence, even if it also partly suffers from cleaning and trampling activities. Due to these anthropic disturbances, its main species (i.e., *Agropyron junceum*) grows into the next formation, the *Ammophiletum*. These two communities have thus been merged into the "Perennial herbaceous vegetation of the embryonic and white dunes" class (i.e., AGAM). *Ammophyletum* is not abundant, even if *Ammophila arenaria* grows luxuriantly in small areas. This union is also justified from an ecological point of view, because these two represent the perennial herbaceous vegetation that is typical of both embryonal and white dunes distributed along all Mediterranean littoral areas (*Ammophiletalia australis*) [33].

The GD is the third class considered. GD is found as a homogeneous strip that is almost totally covered in mosses (*Tortulo–Scabiosetum*).

The MG vegetation species are widespread within different habitats, both for AGAM and GD, and they tend to cover even large surfaces, because of both natural and anthropogenic disturbance. Indeed, the MG settlement is mainly linked to the frequent passage of people, which creates erosion of the perennial vegetation cover.

Finally, the J and P habitats have been merged into the CSAF, which mainly includes *Pinus pinaster*, *Juniperus communis*, *Eleagnus sp.*, *Pyracantha coccinea*, *Tamarix gallica*, and *Quercus ilex*. An example of a *Populus* × *canadensis* tree (located in the northern area of the dunes) is also included in this class. This tree is considered to be a naturalized neophytic species that mostly grows in sandy soils that are damp for most of the year, such as riverbeds or around sandy quarries. However, it can sometimes also be found in 'back-dune' environments, although it is not typical of this kind of environment.

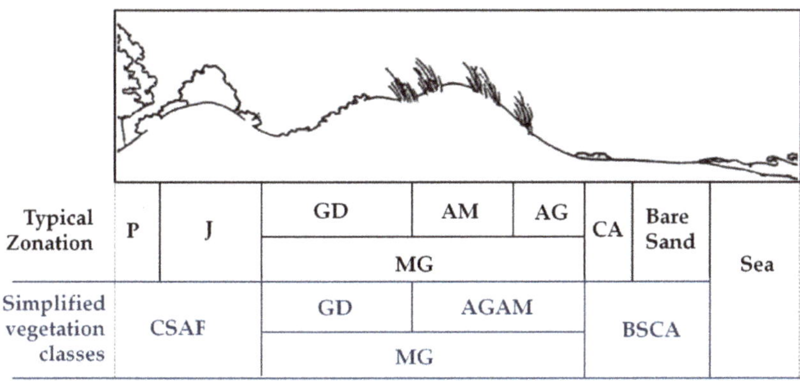

Figure 2. Schematic representation of the typical vegetation zonation [34] and the five simplified vegetation classes adopted in the present study after the in-situ botanical survey. For abbreviations, see main text.

2.2. Data Acquisition and Analysis

This analysis of the vegetation communities was based on a three-band orthoimage obtained through a photogrammetric pipeline from a dataset acquired using a UAV. The UAV platform used was an ESAFLY A2500 hexacopter (SAL Engineering, Modena, Italy). It was equipped with a commercial multispectral camera (Tetracam ADC Micro) that acquired images in the green, red, and near-infrared (NIR) bands, centered respectively at 550, 650, and 800 nm. The camera had only one sensor (Aptina CMOS; 6.55 mm × 4.92 mm; pixel size, 3.12 micron), which was screened with a filter array (Bayer RGB) in a 'checkerboard' pattern [39]. Moreover, the lens on the multispectral camera had an optical low-pass filter that stopped the blue band, but it did not have a filter to stop NIR. Through the combination of the filter in the lens with the filter array, each pixel can capture only one band between the green, red, and NIR bands, relating to its position in the checkerboard. For each pixel, it was possible to reconstruct the values of the two missing bands by interpolation of the corresponding measured values in the adjacent pixels [39]. The proprietary software PixelWrench2 (PW2) provided with the multispectral camera was used to manage this operation.

As the camera was based on the rolling-shutter acquisition system with a total frame creation time of a few milliseconds, the images were acquired with a drone translation speed of 4 m/s and with a constant flight altitude of 80 m above ground level, giving a ground sample distance of 0.03 m. These technical choices avoided blur-motion effects, thus avoiding subsequent problems in the image processing. Furthermore, the UAV had a stabilization system that consisted of a gimbal stabilized with two brushless motors on two axes (i.e., roll, pitch), with mechanical and magnetic cardanic damping and inertial reference with high speed, which allowed for stable framing [40].

The hexacopter was equipped with a single-frequency global navigation satellite system (GNSS) receiver with a code solution. It was used for both the control and definition of the programmed flight path (with longitudinal overlapping and side slap of 80%), and for the coordinate definition of the multispectral chamber centers of acquisition (Figure 1b). These approximate camera positions were used as estimated solutions for the reconstruction of the exterior image orientation in the photogrammetric pipeline, using the approach based on the structure from motion algorithms [41]. Five ground control points (GCPs) were positioned on the ground (i.e., four targets plus the master station), to optimize and increase the accuracy of the self-calibration process for an estimation of the camera interior orientation parameters (i.e., focal length, main point, lens distortion parameters), for a definition of the parameters of the external orientation, and also to obtain more accurate georeferencing. The five targets were sufficient for the optimization process through the bundle adjustment algorithms. Indeed, as recently published by Sanz-Ablanedo et al. [42], to reach a maximum level of relative accuracy, assessed in terms of the root mean squared error/ground sample distance, 2.5 GCPs are sufficient for every 100 images. In this case, the total number of images used was <100, and of these, <30 images were aligned longitudinally and consecutively. The target coordinates were measured using a geodetic dual-frequency GNSS receiver (Topcon GB500) in rapid-static mode. The International GPS Service for Geodynamics permanent station of Medicina (Bologna, Italy) and the European Reference for Quality Assured Breast Screening and Diagnostic Services permanent station of Porto Garibaldi (Ferrara, Italy) were used to define the positions of the GCPs in the WGS84 system. The chosen projection system is UTM 33N [43].

The data obtained from the survey were raster images (ground sample distance, 0.03 m) which were composed of a single matrix of digital numbers (DN) and stored in a raw format. These files were pre-processed in PW2 to reconstruct the information for the three bands, and exported as single tri-band TIFF (Tagged Image File Format) images. The PW2 was calibrated to account for the actual exposure conditions using a RAW image of the calibration tag, acquired under the same lighting conditions as the studied images. This procedure does not convert the sensor output to reflectance [40], and therefore, the subsequent analysis was based on the DN values. The tri-band TIFF images were then processed using the photogrammetric pipeline implemented in Agisoft Photoscan Professional (Agisoft LLC, St. Petersburg, Russia). In the first step, the approximate position and orientation from the GNSS and inertial measurement unit of the drone were associated with each image. In this way, a sparse point cloud model of the scene was created. Through this model, the external orientation parameters of each individual frame were recalculated. A preliminary dense cloud model was then created. The information relating to photogrammetric GCPs was then entered, with manual collimation of each GCP identified for each individual image. The dense cloud model was linearly transformed using seven similarity transformation parameters, which only compensated for linear model misalignment. The next optimization phase then removed non-linear deformations of the model and provided accurate geo-referencing based on the known GCP coordinates [24]. Through the constraints defined by the GCPs by means of the bundle adjustment algorithm, this step allowed a recalculation of the parameters of external and internal (self-calibration) orientation. The dense cloud model was then recreated. The optimization was used to ensure correct scaling and geo-location, to improve the camera interior and exterior parameters, and to correct for any systematic error and/or block deformation. Successively, a polygon mash was generated based on the previously built dense point cloud. Finally, the digital surface model and the orthoimage were generated with a resolution of 0.15 m. The orthoimage produced maintained the three channels (i.e., green, red, NIR) that were essential for the later vegetation analysis [43].

To perform the classification, the areas of interest of five vegetation classes were defined by both direct botanical field surveys and photo-interpretation. The areas of interest surveyed in the field were measured with dual-frequency GNSS instrumentation (Leica GPS1200) using the real-time kinematic

(RTK) technique. The differential corrections were received through the ItalPoS Network service [44]. The ItalPos network provides a network-based (N)RTK system, and specifically, the real-time service is based on the master-auxiliary approach, MAX [45]. The adopted cartographic reference system for all of the data produced is WGS84-UTM33N.

Finally, to collect the ground truth control points for the map classification validation, another NRTK survey (same equipment and procedure as described above) with botanical support was conducted in-situ, with the identification of 300 points of defined vegetation classes (Figure 3). The final accuracy of the NRTK coordinates was 6 cm to 8 cm in planimetry and 8 cm to 10 cm in altimetry. These values are very acceptable considering that the size of the image cell was 15 cm × 15 cm.

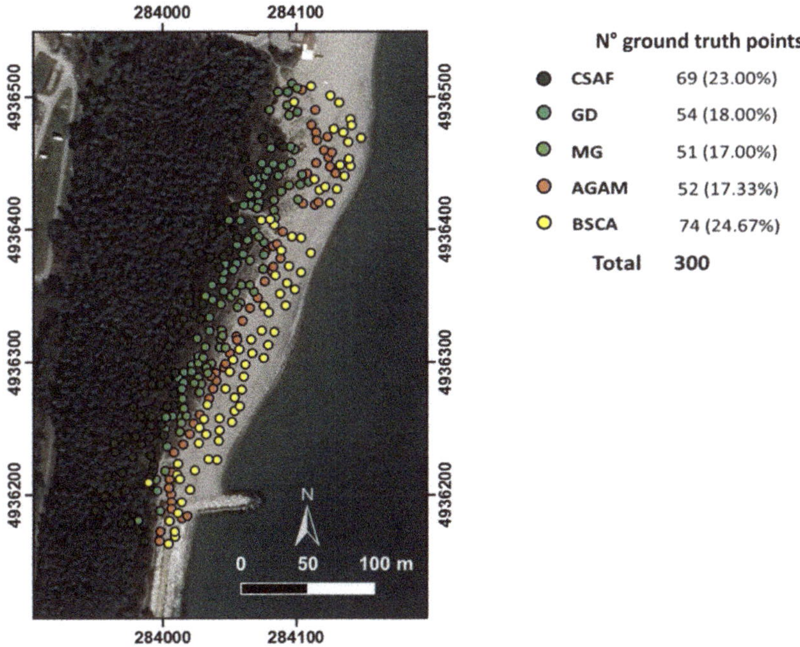

Figure 3. Map of the ground truth points used to classify the validation of results. For abbreviations, see main text.

2.3. Classification Methods

In this study, both pixel-based and object-based classification methods were applied to define the dune vegetation communities. The pixel-based classifier considers the information for the spectral signature of the individual pixels. In contrast, the object-based method classifies objects, i.e., groups of pixels with relatively homogeneous properties, that are created in a preliminary phase known as segmentation. These have intrinsic features, like information derived from the direct spectral observation and geometric properties, and contextual features that describe the relationships between multiple objects [46]. In both approaches, supervised classification algorithms were used. They both required previous knowledge of the vegetation in the study area. Beyond the complicated real field situation, the class selection was also driven by the technical feasibility of discrimination with automatic methods.

The normalized difference vegetation index was extracted from the multi-band orthoimages before the classification. The analysis started by applying the most used pixel and object classifiers, as the maximum-likelihood (ML) and nearest-neighbor (NN) algorithms [47,48], respectively, using the same areas of interest. The pixel-based classifications were conducted using the ENVI software, and the object classifications using the eCognition software.

The ML method assumes that pixels of each class belong to a multivariate normal distribution and defines the probability density function for each class. Each pixel is assigned to the class that it has the highest probability of being in [49]. Comparatively, the NN classifier uses spatial features to classify the object, based on the closest training examples and on the class in which its neighbors have been classified. The NN classification required a previous segmentation step. This phase was performed using a multilevel segmentation approach, and in addition to the tri-bands orthoimage, the normalized difference vegetation index and digital surface models were used as input layers. The multiresolution segmentation algorithm was used to generate the image objects. Four levels of segmentation were applied, which defined the color and shape parameters, with the scale parameter increased at each level (Figure 4). The scale parameter was set to define the level of heterogeneity. Each level was evaluated by photo-interpretation. The optimized level was generated by combining the object levels. The optimized level comprised large segments in homogeneous areas and distinctively smaller image objects that represented small-scale structures and heterogeneous regions [50,51]. The class sample shapes were imported in the eCognition software to identify the objects that corresponded to the areas of interest. Therefore, the sample level was created by chess-board segmentation. Finally, the classification procedure was applied to image objects at the optimized level.

To create a comparable classification with pixel-based products, only the mean green, mean red, and mean NIR features were used.

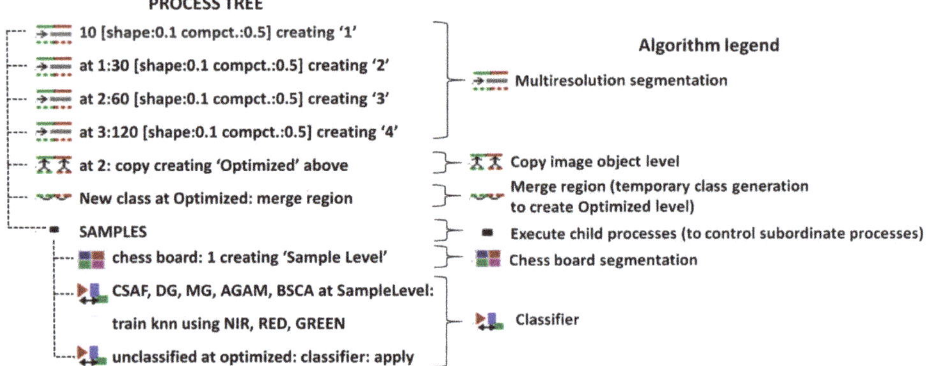

Figure 4. The eCognition process tree used for the segmentation and object classification steps.

To improve the comparison quality, the support vector machine (SVM) algorithm was applied for both pixel-based and object-based classifiers. The SVM method seeks to determine the optimal separating hyperplane between classes by focusing on the training cases (vector support) that are placed at the edge of the class descriptors [52]. Training cases other than support vectors are discarded. In this way, fewer training samples are effectively used. Therefore, a high classification accuracy is achieved with small training sets [53]. The SVM used for classification has the advantages of solving sparse sampling and nonlinear and global optimum problems, compared to other classifiers for satellite imagery classification [54].

The implementation of SVM in the ENVI software based on the pairwise classification strategy is a multiclass classification method that combines all of the comparisons for each pair of classes [55]. The SVM pixel (SVMPi) classification output represents the decision values of each pixel for each class, which are used for the probability estimations. Among the available mathematical kernel functions, the linear function was applied. For the penalty parameter field (Harris Geospatial solution, ENVI 5.2), the value of 100 was used, which represents a parameter that controls the trade-off between allowing training errors and forcing rigid margins. Based on the same theoretical approach, using the eCognition software, the equal function and parameters were applied at the optimized level to perform the corresponding object classification (SVMObj).

Once the classifications were performed, their accuracies were evaluated. The agreement between each classification and ground truth was assessed, as represented by 300 control points.

The pixel-by-pixel comparison provided the confusion matrix with dimensions equal to the number of classes, the overall accuracy, and the Kappa coefficient (*K*). For the object-based approach, the classification result was exported in a raster format to carry out the comparison. The *K* compares global accuracy with an expected global accuracy, taking into account random chance. $K > 0.80$ represents strong agreement and $K < 0.40$ represents poor agreement [56]. The user accuracy, producer accuracy, and *K-Conditional* were also calculated from a single confusion matrix. The first provides the probability that a random pixel extracted from among those belonging to class *i* in the reference belongs to class *i* in the classification. The producer accuracy, instead, defines whether a randomly chosen pixel among those belonging to class *j* in the classification also belongs to class *j* in the reference data [57]. As for *K*, *K-Conditional* represents the agreement between the reference pixels and those classified, as calculated for each class.

2.4. Statistical Classification Comparison

To determine the better performing classification method between pixel-based and object-based classification, three tests were carried out: TEST 1, TEST 2, and TEST 3.

TEST 1 established the significance of the difference in the accuracy between two maps with independent Kappa coefficients. Once establishing the null hypothesis, that the expected *K* values of the two statistics considered for each comparison (1 = first algorithm; 2 = second algorithm) were the same (i.e., no significant difference), the Student's *t* was applied [47,58], as in Equation (1):

$$z = \frac{K_1 - K_2}{\sqrt{\sigma_1^2 + \sigma_2^2}}, \qquad z \sim N(0,1) \tag{1}$$

where σ_1^2 and σ_2^2 represent the estimated variances of the derived *K* coefficients.

To calculate the required variance values associated with the *K* and *K-Conditional* coefficients, a Fortran program was implemented using the σ formula proposed by Rossiter (2004) [59]. Considering that *z* follows a normal normalized distribution and the significance level $\alpha = 0.10$, with a consequently confidence limit of 1.65, the hypothesis was accepted for the test statistic *z* of $|z| \leq 1.65$.

TEST 2 evaluated the significance of the difference between two independent proportions [60], using Equation (2), which takes into account the correction for continuity [58]:

$$z = \frac{\left|\frac{x_1}{n_1} - \frac{x_2}{n_2}\right| - \frac{1}{2}\left(\frac{1}{n_1} + \frac{1}{n_2}\right)}{\sqrt{p(1-p)\left(\frac{1}{n_1} + \frac{1}{n_2}\right)}}, \qquad z \sim N(0,1) \tag{2}$$

where x_1 and x_2 are the numbers of correctly allocated cases in two samples of size n_1 and n_2, respectively, and $p = (x_1 + x_2)/(n_1 + n_2)$. The statistical significance of the difference between two classification maps is verified through *z*, which follows a normal normalized distribution, in the same way as with the previous comparison of *K*.

TEST 3 was based on McNemar's test [60], which is suitable for comparisons of related samples. Equation (3) takes into account the correction for continuity [61]:

$$X_1^2 = \frac{\left(|f_{12} - f_{21}| - 1\right)^2}{f_{12} + f_{21}} \tag{3}$$

This non-parametric test uses a confusion matrix, of 2 × 2 dimensions, in which f_{ij} indicates the frequency of sites lying in confusion element *i*, *j*, as reported in the example in Table 1.

Table 1. The matrix elements used in Equation (3) [61].

		Classification 2	
		Correct	**Not Correct**
Classification 1	**Correct**	f_{11}	f_{12}
	Not correct	f_{21}	f_{22}

The McNemar statistic follows a chi-squared distribution X_1^2 with one degree of freedom, and its square root follows a normal normalized distribution. Therefore, the statistical significance of the difference between the two classification maps is evaluated as with the previous tests.

The continuity correction that is considered in TEST 2 and TEST 3 is particularly important when the sample size used is small [62].

In the first step, these tests were applied for the same classification technique at both the global and single class levels. Then, the tests were applied to compare the statistical significance of the accuracy between the two algorithms that demonstrated smaller errors in previous comparisons.

3. Results

3.1. Classification Results

Both of the classification algorithms recognized the five vegetation classes considered as irregular belts almost parallel to the coastline: BSCA, AGAM, MG, GD, and CSAF (see Figure 2).

For the pixel-based classification (Figure 5), the single vegetation strips showed nonhomogeneous coverage because many of the pixels were assigned to classes that were different from those of the membership (Figure 5, subplots a and b). The pixel distributions of the MG, GD, AGAM, and CSAF classes (i.e., all except for the BSCA class) were characterized by a 'salt and pepper' effect that makes the definition of the class contours difficult. This effect is particularly evident in the ML map over several zones, such as for the northern area, which was strongly affected by both natural and anthropic effects (Figure 5, subplots a1 and b1). Indeed, while the SVMPi showed greater sandy coverage, in the ML results, the MG and DG pixels prevailed in these areas. As can be seen from Figure 5, both of these pixel approaches recognized the shape of the tree (*Populus* × *canadensis*) that was included in the CSAF class (Figure 5, subplots a1 and b1, center left), although its composition was confused for both of these methods. The *Populus* × *canadensis* tree composition was more appropriate and homogeneous in the SVMPi result compared to the ML map, where several pixels were wrongly tagged as GD. In general, many pixels that belonged to the CSAF strip were erroneously classified as GD. The details shown in Figure 5, subplots a2 and b2, indicate other relevant errors, such as: (1) several BCSA pixels along the footpaths were incorrectly associated with MG or GD cover; (2) some of the AMAG pixels were erroneously recognized in proximity to the CSAF class, where the morphological and environmental conditions are not suitable for this vegetal community; and (3) bare sand footpaths were recognized as covered by GD pixels. In general, the most confused classes were AMAG, MG, and GD, while the BCSA class was more homogeneous and more correctly discriminated from the rest of the scene. The aforementioned examples of confusion were more evident for the ML class. The main difference between these two pixel-base results is the number of pixels classified as MG and GD. The area covered by the MG class in the SVMPi map was greater than the MG zone of the ML map, with the inverse situation seen for the GD class (Table 2).

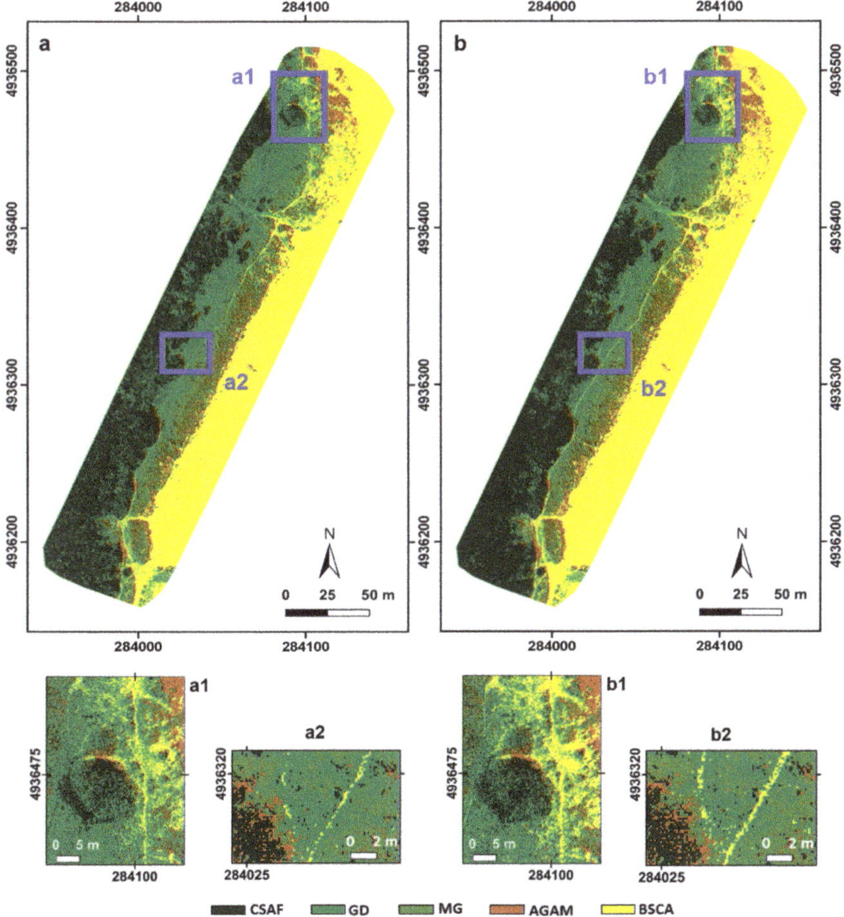

Figure 5. Pixel-based classification of the Casal Borsetti dune vegetation. (**a**) Maximum likelihood classification. (**b**) Support vector machine pixel classification. (**a1**,**b1**) 'Salt and pepper' effects and the *Populus* × *canadensis* tree. (**a2**,**b2**) Footpath classification.

For the application of the object-classification algorithms, the five vegetation classes were also identified (Figure 6, subplots a and b). The classes were more uniform than for the pixel-based results (Figure 5), and the confused zones were distributed along the perimeter of each class.

At the single-element level, the object maps showed less information. The tree located in the northern area (*Populus* × *canadensis*) was partially classified as CSAF, while several objects were associated with MG or GD (Figure 6, subplots a1 and b1). Moreover, it was only possible to identify this tree shape in the SVMObj map (Figure 6, subplot b1), because the NN classification (Figure 6, subplot a1) did not provide clear enough information. In addition, Figure 6, subplots a1 and b1, shows a greater presence of sand in the northern part of the dune compared to the pixel-based method (Figure 5, subplots a1 and b1).

Comparing Figure 5, subplots a2 and b2, with the details in Figure 6, subplots a2 and b2, another difference between the pixel and object methods can be seen for the sand footpaths. These were well-identified with the pixel-based method, while they were not recognized using the object-based approach.

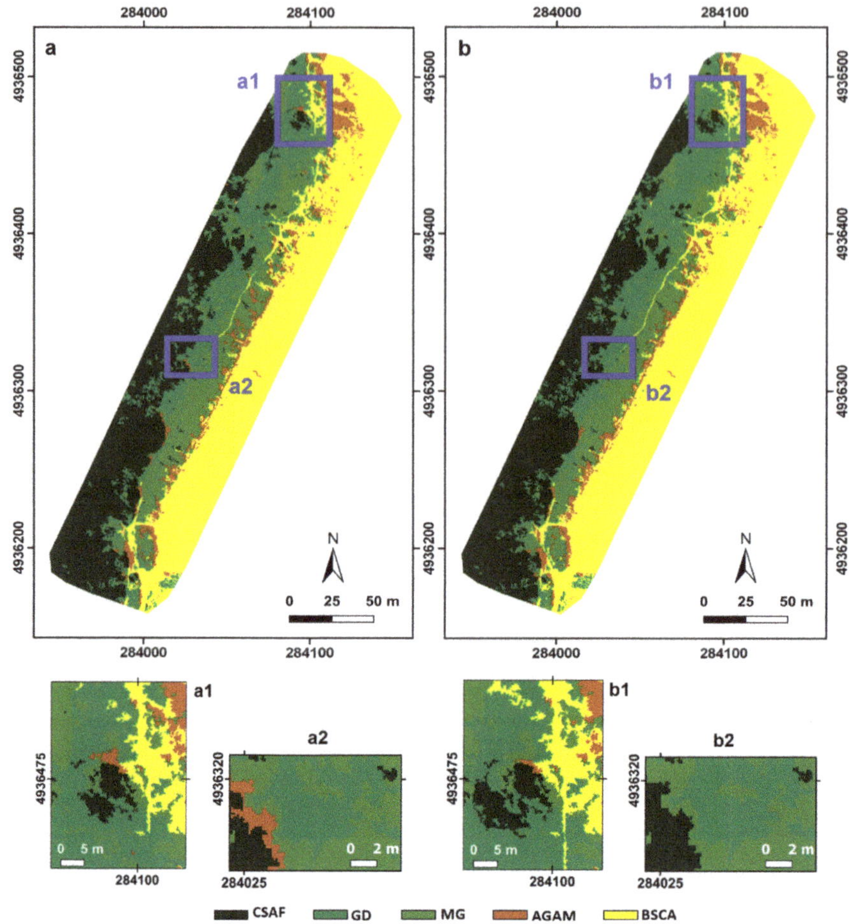

Figure 6. Object-based classification of the Casal Borsetti dune vegetation. (**a**) Nearest neighbor classification. (**b**) Support vector machine object classification. (**a1**,**b1**) *Populus* × *canadensis* tree classification. (**a2**,**b2**) Missed footpath classification.

Table 2. Vegetation community cover and extension for each classification methodology. For abbreviations, see main text.

Vegetation Community	Classification Methodology							
	ML		SVMPi		NN		SVMObj	
	Cover (%)	Extension (m^2)	Cover (%)	Extension (m^2)	Cover (%)	Extension (m^2)	Cover (%)	Extension (m^2)
BSCA	34.4	9763	36.1	10,251	36.6	10,412	36.8	10,469
CSAF	25.9	7349	28.5	8082	26.6	7558	26.8	7617
MG	12.8	3647	16.3	4628	19.1	5433	16.8	4782
GD	19.0	5398	10.8	3082	12.3	3511	14.3	4073
AGAM	7.9	2241	8.3	2359	5.4	1546	5.3	1520
Total	**100.0**	**28,398**	**100.0**	**28,402**	**100.0**	**28,460**	**100.0**	**28,460**

All previous considerations were confirmed by the confusion matrices given in Table 3, which provides a quantitative reading of the classification results. The ML and SVMPi results showed a lower global accuracy, with both <80%, while for the object classification, this was >80%. The global *K*s followed the same behavior as the global accuracies. In particular, considering the ML and SVMPi comparison and the NN and SVMObj comparison separately, the main diagonal values of the pixel classifications are almost always lower or equal to the corresponded values of the object classifications. For the ML map, the number of correct pixels is only higher than the NN classification for the GD class.

For the ML and SVMPi maps, the MG was the most confused class, with omission errors of 47.1% (24 missing pixels, of 51 true total pixels) and 43.1% (22/51 pixels missing), respectively (Tables 3 and 4). For the ML results, almost 42% of the MG omission error (10/24 pixels missing) is concentrated in the GD class (Table 3), while for the SVMPi map, the MG true pixels are distributed among the AGAM, GD, and CSAF covers (Table 3). Instead, the largest number of pixels associated with classes other than those of the reference is related to the GD class for ML, with a commission error of 46.1% (35 added pixels, of 76 total classified pixels; Table 4), and the MG class for SVMPi, with a commission error of 39.6% (19/48 pixels classified; Table 4). From Table 4, the producer accuracy was calculated. In the pixel-based algorithms, the class that was ranked most accurately was BSCA, with associated values of 86.5% for ML and 91.9% for SVMPi.

For the object-based classification results (Tables 3 and 4), the most confused class was GD in the NN map. For the 19 pixels not assigned (Table 4), 68.4% was classified as MG cover (Table 3), for the highest commission error (36.1%, i.e., 22/61 pixels classified; Table 4). Instead, for the SVMObj classification, the MG class showed the highest omission error (27.5%, i.e., 14/51 pixels missing), with almost 64.3% of these pixels classified as GD. The confusion between these classes was confirmed by the high commission error for GD (32.2%, i.e., 19/59 pixels classified; Table 4).

However, it should be emphasized that in the SVMObj results, the higher omission errors of the other classes did not differ much from the error related to MG. The BSCA producer accuracy achieved with the object-based techniques (93.2%), which was the same for NN and SVMObj, was the highest obtained among the several classes. However, some pieces of sand footpaths in the object-based classification maps were included in the GD or MG covers.

Table 3. Confusion matrices for each of the image classification methodologies. For abbreviations, see main text.

Classification Methodology	Vegetation Community	Ground Truth (pixels)						Overall Accuracy	*Kappa*
		BSCA	AGAM	GD	MG	CSAF	Total	(%)	
ML	BSCA	**64**	7	2	0	0	73	72.7	0.64
	AGAM	1	**34**	1	7	2	45		
	GD	7	3	**41**	10	15	76		
	MG	2	7	5	**27**	3	44		
	CSAF	0	1	5	7	**49**	62		
SVMPi	BSCA	**68**	4	3	1	0	76	76.3	0.70
	AGAM	1	**40**	1	6	2	50		
	GD	3	1	**39**	7	10	60		
	MG	2	6	7	**29**	4	48		
	CSAF	0	1	4	8	**53**	66		
NN	BSCA	**69**	6	3	1	0	79	80.3	0.75
	AGAM	1	**39**	2	3	1	46		
	GD	3	1	**35**	6	6	51		
	MG	1	5	13	**39**	3	61		
	CSAF	0	1	1	2	**59**	63		
SVMObj	BSCA	**69**	7	3	1	0	80	82.0	0.77
	AGAM	2	**41**	3	2	1	49		
	GD	3	1	**40**	9	6	59		
	MG	0	3	7	**37**	3	50		
	CSAF	0	0	1	2	**59**	62		
	Total	74	52	54	51	69	300		

Table 4. Commission and omission errors for each of the vegetation communities. For abbreviations, see main text.

Vegetation Community	Errors (pixels)							
	ML		SVMPi		NN		SVMObj	
	Commission	Omission	Commission	Omission	Commission	Omission	Commission	Omission
BSCA	9/73	10/74	8/76	6/74	10/79	5/74	11/80	5/74
AGAM	11/45	18/52	10/50	12/52	7/46	13/52	8/49	11/52
GD	35/76	13/54	21/60	15/54	16/51	19/54	19/59	14/54
MG	17/44	24/51	19/48	22/51	22/61	12/51	13/50	14/51
CSAF	13/62	20/69	13/66	16/69	4/63	10/69	3/62	10/69

3.2. Classification Results Comparison

The outcomes of TEST 1 applied to the pixel-based classification reveal that the differences in accuracy between the two pixel-based maps at both the global and single class level are not significant (Table 5). In addition, TEST 2 confirms that the ML and SVMPi classifications are statistically similar (Table 5).

For TEST 3, based on four combinations of correct and incorrect pixel frequencies in both of the maps considered, the classifications are globally different ($z = 2.65 > 1.65$). This is particularly due to the AGAM community, as the only class that does not pass the test. The analysis of the frequencies used for TEST 3 reveals that the ML method is the least accurate method for the AGAM class. Indeed, the calculations of TEST 3 show that almost 12% of the 52 ground truth pixels were classified incorrectly only for ML, while the same pixels were correct for SVMPi; vice versa, there are no pixels that were not correctly classified only for SVMPi. For the AGAM class, the lower accuracy for ML is also confirmed by the higher commission and omission errors compared with the corresponding SVMPi errors (Table 4).

As for the pixel-based classification, as for the object-based classification, the outcomes of TEST 1 show that the differences in the accuracy between the two object-based classification maps at both the global and single class level are not significant (Table 6). The TEST 2 results confirm that the NN and SVMObj classifications are similar (Table 6). Additionally, for TEST 3, these classifications are not different, except for the GD cover, at both the global and individual class levels. The frequency analysis reveals that the NN method is the least accurate method; indeed, almost 8% of the 54 GD ground truth pixels were incorrectly classified for NN, while the same pixels were correctly classified for SVMObj; vice versa, there were no pixels incorrectly classified only for SVMObj. For the GD class, the lower accuracy of NN is also confirmed by the higher omission errors compared with the corresponding SVMObj errors (Table 4). Indeed, in both object-based maps, the highest number of error classifications is between the GD and MG classes. However, the MG cover classifications are not statistically different.

From the previous comparisons, it emerges that the most accurate methods are SVMPi and SVMObj. Therefore, a new statistical comparison of these two methods was performed to define the most appropriate for this dune vegetation classification. The findings of TEST 1 show that the accuracy difference between these two maps is not significant at both the global and single class level (Table 7). The TEST 2 results confirm that the SVMPi and SVMObj classifications are similar (Table 7). Instead, for the findings of TEST 3, the two classifications are globally different ($z = 2.16 > 1.65$). At the single class level, all of the classes pass the test, except for the MG cover, with a z value the same as the confidence limit (1.65). The analysis of the frequencies reveals that almost 26% of the 51 MG ground truth pixels were classified incorrectly for SVMPi, while they were correct for SVMObj; vice versa, only 10% were correctly classified for SVMPi and not for SVMObj (Table 7).

Table 5. Test results for the pixel-based classification, as a comparison of ML and SVMPi. For abbreviations, see main text.

Vegetation Community	TEST1			TEST2	TEST3	Confidence
	K-ML	K-SVMPi	z (K)	z (Propor.)	z (McNemar)	Limit (α=0.10)
BSCA	0.82	0.89	−1.07	0.79	1.50	
AGAM	0.59	0.72	−1.32	1.08	2.04	
GD	0.68	0.65	0.24	0.22	0.50	**1.65**
MG	0.45	0.49	−0.35	0.20	0.50	
CSAF	0.63	0.70	−0.77	0.58	1.22	
Global	**0.64**	**0.70**	**−1.19**	**1.21**	**2.65**	

Table 6. Test results for the object-based classification, as a comparison of NN and SVMObj. For abbreviations, see main text.

Vegetation Community	TEST1			TEST2	TEST3	Confidence
	K-NN	K-SVMObj	z (K)	z (Propor.)	z (McNemar)	Limit (α=0.10)
BSCA	0.91	0.91	0.01	−0.33	0.00	
AGAM	0.70	0.75	−0.45	0.23	0.71	
GD	0.58	0.68	−0.10	0.84	1.79	**1.65**
MG	0.71	0.68	0.34	0.23	0.50	
CSAF	0.82	0.82	−0.01	−0.24	0.71	
Global	**0.75**	**0.77**	**−0.42**	**0.42**	**1.11**	

Table 7. Test results of the comparison of the SVMPi and SVMObj classifications. For abbreviations, see main text.

Vegetation Community	TEST1			TEST2	TEST3	Confidence
	K-SVMPi	K-SVMObj	z (K)	z (Propor.)	z (McNemar)	Limit (α=0.10)
BSCA	0.89	0.91	−0.29	0.00	0.00	
AGAM	0.72	0.75	−0.26	0.00	0.00	
GD	0.65	0.68	−0.24	1.39	0.00	**1.65**
MG	0.49	0.68	−1.77	0.00	1.65	
CSAF	0.70	0.82	−1.43	1.09	1.44	
Global	**0.70**	**0.77**	**−1.45**	**1.61**	**2.16**	

4. Discussion

Before discussing the results, some considerations regarding the particularity of the site and the instruments involved in the data acquisition need to be addressed. As described above for the study area description, the selected site is included in the Po Delta Regional Park, and there are restrictions regarding access, authorized activities, and management of the dunes. Despite this, during the summer, tourism results in the establishment of a lot of footpaths, which destroys the vegetation communities and fosters wind erosion of the dunes [3,31,37]. Therefore, the local vegetation communities are often fragmented and interspersed with each other, which provides an atypical vegetation composition in the area. Although each plant species provides its own spectral signature based on the growth period, geographic location, climatic conditions, and level of disturbance [63,64], working at the vegetation community level reduces the species diversification and provides more solid results [65,66]. In addition to this, the contemporary ground truth botanical survey allowed a reliable dataset of 'known' pixels to be established to validate the vegetation maps. In particular, the extreme dynamic environment represented by this dune system, its cycling, exposure to sea storms, erosion, salty winds, and floating groundwater from one side require an accurate selection of the aerial acquisition period, and from the other side, the achieved vegetation community distribution can be considered descriptive for a relatively medium-to-long period of time [17,67].

For the equipment used, the three standard bands available with the multispectral camera (i.e., green, red, NIR) are considered necessary and sufficient to recognize the vegetation [68], but not to discriminate between and examine in-depth the characteristics of the single species [69]. Furthermore, this camera has the following technical limitations: an overlay of portions of the bandwidth, a large bandwidth with a low-slope front of the filter, and the inability to discriminate between uninteresting portions of the bands [40].

For the classification results, all of the four algorithms recognized the five classes of vegetation considered. Despite this, the maps that were derived from the pixel-based methods showed that class boundaries are not well defined due to the 'salt and pepper' effect that was spread throughout the extension of the dunes. This effect was especially evident in the ML map, as well as in the portion of dunes where a combination of the wind and recurring anthropic passage has reduced the vegetation growth. Indeed, for the ML map, these areas were mostly covered by the MG and DG pixels, while the SVMPi classification revealed the wider sandy cover, thus better identifying the actual field conditions (Figure 5, subplots a1, b1). This result is explained by the relationship between the pixel size (15 cm) and the different plant components acquired [40]. Indeed, a pixel size of 15 cm can include more plant elements (e.g., leaves, flowers, small branches, shadows), or it can be homogeneously occupied by a single element. Therefore, its spectral signature might represent a medium signature or might not include the other components of the same class [11]. Furthermore, considering that the same type of object can be contained in many classes (e.g., sand, grass), some pixels might be classified differently compared to the surrounding pixels [70].

The object-based classification was less exposed to these problems compared to the pixel-based approach. The presence of some pixels with different vegetation covers has no influence on the correct class assignment [46,54]. As confirmation of this, the more extended BSCA class in the northern section of the study area is coherent with the actual situation, and reflects the higher anthropic disturbance due to the sand mobilization in the close-by bathing establishment.

If, on one hand, the object-based classification provides vegetation class uniformity, on the other hand, it does not allow the identification of some single elements that were identified with the pixel-based approach. For example, the *Populus* × *canadensis* tree shape was well-defined in the pixel-based approach, while it was not recognizable in the object-based approach. In the same way, some objects classified as CSAF appeared in their correct location, but the shape reconstruction was lost. Other missing information with the object-based approach was the identification of the sand footpaths, which were almost totally incorporated into the surrounding vegetated classes.

The confusion matrix analysis of the four classification methods confirms the greater accuracy of the object-based approaches (Table 3), due to the clearer class definition obtained. At the opposite end of the spectrum, the 'salt and pepper' effects reduce the accuracy of pixel-based methods [70].

Considering the pixel-based elaboration, the ML method shows a more coarse accuracy compared to SVMPi, as indirectly confirmed by the global TEST 3 result (Table 5). The AGAM class was differently recognized among the two pixel-based algorithms, but the most difficult class to identify was MG, due to its wide and complex presence in the dune vegetation structure. The MG and GD classes were often confused because of the de-structuring of the GD vegetation that was caused by the anthropogenic disturbance, which generates short biological growing cycles that are typical of the MG class. This problem is also seen for the object-based classification, but with an impact of only 14% on the mean class errors (Table 4). The errors of commissions and omissions with SVMPi were generally lower than those for ML (Table 4). In particular, a greater producer accuracy of the BCSA class was seen, especially along the footpaths, which appeared more defined and lengthened.

The differences in the classification between the NN and SVMObj methods are not significant, except for the GD class. However, the SVM algorithm was more accurate, and it should be the most reliable method to be applied to dune monitoring, especially in the case of characteristics similar to those of Casal Borsetti. The same conclusion was reported by Wang et al. [71] in a study where they compared pixel-based and object-based approaches in mangrove classification, as well as by Zhai et al. for a rubber plantation [72].

The choice between the SVM pixel-based and the SVM object-based techniques also depends on the level of detail required. According to TEST 3 (Table 7), at the global level, the results of these two techniques were different only because of the classification of the MG cover. However, this class is an opportunistic vegetation community that grows where disturbance is higher. Indeed, the z obtained for MG was the same as the threshold imposed for not passing the test (i.e., z = 1.65). Therefore, even if the SVMObj shows the best accuracy, as reported by Gao et al. [54], considering the accuracies of these two methods as similar does not induce significant errors, as already demonstrated through TEST 1 and TEST 2 (Table 7).

Based on the results of this study, a UAV equipped with a multispectral camera can be used for multi-year dune monitoring to provide a multitemporal classification of dune vegetation [73]. Moreover, based on the well-known relationships between dune vegetation communities and coastline status [74,75], both the equipment and methodology presented in this study can also be applied to future continuous monitoring of the coastline, in order to gather information on the coastal evolutionary status [76].

The authors should discuss the results and how they can be interpreted from the perspective of previous studies and the working hypotheses. The findings and their implications should be discussed in the broadest context possible. Future research directions may also be highlighted.

5. Conclusions

The aims of this study were to determine the applicability of multispectral data collected by a UAV platform for the identification/discrimination of fragmented and interspersed coastal dune vegetation communities, and to compare the pixel-based and object-based approaches to determine the better performing classification method.

The data acquired by the sensors installed on the UAV and elaborated with the SVM algorithm allowed the elaboration of reliable dune vegetation community maps with a high spatial resolution (0.15 m) and a global accuracy >80%. This system is cheaper and faster when compared to the traditional field surveys that are performed by botanical experts.

From the comparisons of the classification methods here, as NN, ML, SVMPi, and SVMObj, the SVM was the most accurate algorithm based on the statistical test results. From the numerical point of view, the SVMObj was the best performing approach. However, it has the disadvantage of including small elements (<1 m in size), such as single trees or footpaths, in the larger bordering classes, which impedes their suitable classification. In the case of limited extension, where the presence and variations of a single element can widely influence the final results, this aspect is a relevant limitation. For example, for the Casal Borsetti dune area, the protection of the relevant coastal habitat of the dune vegetation communities needs to be pursued by limiting the continuous human crossing. Therefore, the footpaths need to be clearly identified using the proposed methodology. Considering that the differences between SVMPi and SVMObj are not statistically significant, except for the MG class, the pixel technique is the most suitable for investigations that require greater levels of detail. The MG and GD classes are the most difficult to discriminate with a camera that only acquires the green, red, and infrared bands, as is the case for the camera used in this study. The entry onto the market of multi-spectral cameras with eight mono-band sensors (from visible to infrared) might also facilitate the more accurate identification of vegetation communities using a UAV.

Although the results obtained cannot be immediately generalized because they refer to a specific study site (i.e., Casal Borsetti dunes, Ravenna, Italy) that shows strong human disturbance and fragmented and interspersed dune vegetation communities, both the equipment used here and the classification approach appear portable, and can thus be applied to other dune sites.

Moreover, in light of the elevated spatial resolution of the vegetation maps produced, the authors believe that this system, the UAV data acquisition, and the SVM image classification approach, are sufficiently sensitive to allow multitemporal and continuous monitoring of dune evolution in the near future, in terms of erosion and progradation-dominant phenomena. Ongoing studies are collecting data for the monitoring of coastal erosive/progradation dynamics through dune vegetation communities.

Author Contributions: Conceptualization of the study, M.D.G., N.G., and N.M.; methodology, M.D.G., M.D., F.G., and M.B.; software, M.D.G., M.D., F.G., and M.B.; validation, M.B.; investigation, N.M., N.G., and M.D.G.; data curation, M.D.G., M.B., N.M., M.D., and F.G.; writing—original draft, M.D.G. and N.G.; writing—review and editing, M.D.G. and N.G.; supervision, M.B. and M.D.

Funding: This study received no external funding.

Acknowledgments: The authors would like to thank the Carabinieri for Biodiversity, Punta Marina Office, who allowed access to the protected natural dune areas, as the object of this study. Moreover, the authors would like to thank SAL ENGINEERING s.r.l. for the UAV survey, as well as the anonymous reviewer who helped to improve our manuscript quality.

Conflicts of Interest: The authors declare that they have no conflicts of interest.

References

1. Gómez-Pina, G.; Muñoz-Perez, J.J.; Ramirez, J.L.; Ley, C. Sand dune management problems and techniques, Spain. *J. Coast. Res.* **2002**, *36*, 325–332. [CrossRef]
2. Drius, M.; Malavasi, M.; Acosta, A.T.R.; Ricotta, C.; Carranza, M.L. Boundary-based analysis for the assessment of coastal dune landscape integrity over time. *Appl. Geogr.* **2013**, *45*, 41–48. [CrossRef]
3. Sytnik, O.; Stecchi, F. Disappearing coastal dunes: Tourism development and future challenges, a case-study from Ravenna, Italy. *J. Coast. Conserv.* **2015**, *19*, 715–727. [CrossRef]
4. Malavasi, M.; Santoro, R.; Cutini, M.; Acosta, A.T.R.; Carranza, M.L. The impact of human pressure on landscape patterns and plant species richness in Mediterranean coastal dunes. *Plant Biosyst.* **2016**, *150*, 73–82. [CrossRef]
5. Janssen, J.A.M.; Rodwell, J.S.; Criado, M.G.; Arts, G.H.P.; Bijlsma, R.J.; Schaminee, J.H.J. European Red List of Habitats: Part 2. Terrestrial and Freshwater Habitats. European Union. 2016. Available online: https://publications.europa.eu/en/publication-detail/-/publication/22542b64-c501-11e7-9b01-01aa75ed71a1/language-en (accessed on 10 April 2019).
6. Sytnik, O.; Del Río, L.; Greggio, N.; Bonetti, J. Historical shoreline trend analysis and drivers of coastal change along the Ravenna coast, NE Adriatic. *Environ. Earth Sci.* **2018**, *77*, 779. [CrossRef]
7. Provoost, S.; Jones, M.L.M.; Edmondson, S.E. Changes in landscape and vegetation of coastal dunes in northwest Europe: A review. *J. Coast. Conserv.* **2011**, *15*, 207–226. [CrossRef]
8. Sciandrello, S.; Tomaselli, G.; Minissale, P. The role of natural vegetation in the analysis of the spatio-temporal changes of coastal dune system: A case study in Sicily. *J. Coast. Conserv.* **2015**, *19*, 199–212. [CrossRef]
9. Masselink, G.; Pattiaratchi, C.B. Seasonal changes in beach morphology along the sheltered coastline of Perth, Western Australia. *Mar. Geol.* **2001**, *172*, 243–263. [CrossRef]
10. Anthony, E.J.; Vanhée, S.; Ruz, M.-H. Short-term beach–dune sand budgets on the north sea coast of France: Sand supply from shoreface to dunes, and the role of wind and fetch. *Geomorphology* **2006**, *81*, 316–329. [CrossRef]
11. Hantson, W.; Kooistra, L.; Slim, P.A. Mapping invasive woody species in coastal dunes in the Netherlands: A remote sensing approach using LIDAR and high-resolution aerial photographs. *Appl. Veg. Sci.* **2012**, *15*, 536–547. [CrossRef]
12. Levin, N.; Levental, S.; Morag, H. The effect of wildfires on vegetation cover and dune activity in Australia's desert dunes: A multisensor analysis. *Int. J. Wildl. Fire* **2012**, *21*, 459–475. [CrossRef]
13. Fabbri, S.; Giambastiani, B.M.; Sistilli, F.; Scarelli, F.; Gabbianelli, G. Geomorphological analysis and classification of foredune ridges based on Terrestrial Laser Scanning (TLS) technology. *Geomorphology* **2017**, *295*, 436–451. [CrossRef]
14. Tsoar, H. Sand dunes mobility and stability in relation to climate. *Phys. A Stat. Mech. Appl.* **2005**, *357*, 50–56. [CrossRef]
15. Miller, T.E.; Gornish, E.S.; Buckley, H.L. Climate and coastal dune vegetation: Disturbance, recovery, and succession. *Plant Ecol.* **2009**, *206*, 97–104. [CrossRef]

16. Duran, O.; Silva, M.; Bezerra, L.; Herrmann, H.; Maia, L. Measurements and numerical simulations of the degree of activity and vegetation cover on parabolic dunes in north-eastern Brazil. *Geomorphology* **2008**, *102*, 460–471. [CrossRef]
17. Durán, O.; Moore, L.J. Vegetation controls on the maximum size of coastal dunes. *Proc. Natl. Acad. Sci. USA* **2013**, *110*, 17217–17222. [CrossRef]
18. Mathew, S.; Davidson-Arnott, R.G.D.; Ollerhead, J. Evolution of a beach–dune system following a catastrophic storm overwash event: Greenwich Dunes, Prince Edward Island, 1936–2005. *Can. J. Earth Sci.* **2010**, *47*, 273–290. [CrossRef]
19. Schmidt, K.; Skidmore, A. Spectral discrimination of vegetation types in a coastal wetland. *Remote Sens. Environ.* **2003**, *85*, 92–108. [CrossRef]
20. Acosta, A.; Carranza, M.L.; Izzi, C.F. Combining land cover mapping of coastal dunes with vegetation analysis. *Appl. Veg. Sci.* **2009**, *8*, 133–138. [CrossRef]
21. Ajedegba, J.O.; Perotto-Baldivieso, H.L.; Jones, K.D. Coastal Dune Vegetation Resilience at South Padre Island, Texas: A Spatiotemporal Evaluation of the Landscape Structure. *J. Coast. Res.* **2019**, *35*, 534–544. [CrossRef]
22. Saye, S.; Van Der Wal, D.; Pye, K.; Blott, S.; Blott, S. Beach–dune morphological relationships and erosion/accretion: An investigation at five sites in England and Wales using LIDAR data. *Geomorphology* **2005**, *72*, 128–155. [CrossRef]
23. Brock, J.C.; Purkis, S.J. The Emerging Role of Lidar Remote Sensing in Coastal Research and Resource Management. *J. Coast. Res.* **2009**, *53*, 1–5. [CrossRef]
24. Mancini, F.; Dubbini, M.; Gattelli, M.; Stecchi, F.; Fabbri, S.; Gabbianelli, G. Using Unmanned Aerial Vehicles (UAV) for High-Resolution Reconstruction of Topography: The Structure from Motion Approach on Coastal Environments. *Remote Sens.* **2013**, *5*, 6880–6898. [CrossRef]
25. Suanez, S.; Cariolet, J.-M.; Fichaut, B. Monitoring of recent morphological changes of the dune of Vougot beach (Brittany, France) using differential GPS. *Shore Beach* **2010**, *78*, 37–47.
26. Casella, E.; Rovere, A.; Pedroncini, A.; Stark, C.P.; Casella, M.; Ferrari, M.; Firpo, M. Drones as tools for monitoring beach topography changes in the Ligurian Sea (NW Mediterranean). *Geo-Mar. Lett.* **2016**, *36*, 151–163. [CrossRef]
27. Anderson, K.; Gaston, K.J. Lightweight unmanned aerial vehicles will revolutionize spatial ecology. *Front. Ecol. Environ.* **2013**, *11*, 138–146. [CrossRef]
28. Guillot, B.; Pouget, F. UAV application in coastal environment, example of the Oleron island for dunes and dikes survey. *ISPRS Int. Arch. Photogramm. Remote Sens. Spat. Inf. Sci.* **2015**, *3*, 321–326. [CrossRef]
29. Gonçalves, J.; Henriques, R. UAV photogrammetry for topographic monitoring of coastal areas. *ISPRS J. Photogramm. Remote Sens.* **2015**, *104*, 101–111. [CrossRef]
30. Airoldi, L.; Ponti, M.; Abbiati, M. Conservation challenges in human dominated seascapes: The harbour and coast of Ravenna. *Reg. Stud. Mar. Sci.* **2016**, *8*, 308–318. [CrossRef]
31. Semeoshenkova, V.; Newton, A.; Contin, A.; Greggio, N. Development and application of an Integrated Beach Quality Index (BQI). *Ocean Coast. Manag.* **2017**, *143*, 74–86. [CrossRef]
32. Corbau, C.; Simeoni, U.; Melchiorre, M.; Rodella, I.; Utizi, K. Regional variability of coastal dunes observed along the Emilia-Romagna littoral, Italy. *Aeolian Res.* **2015**, *18*, 169–183. [CrossRef]
33. Piccoli, F.; Merloni, N.; Corticelli, S. *Carta della Vegetazione 1:25.000 della Stazione Pineta di San Vitale e Pialasse di Ravenna (Parco Regionale del Delta del Po)*; Regione Emilia-Romagna: Bologna, Italy, 1999.
34. Merloni, N.; Piccoli, F. Comunità vegetali rare e minacciate delle stazioni ravennati del Parco del Delta del Po (Regione Emilia-Romagna). *Fitosociologia* **2007**, *44*, 67–76.
35. Sburlino, G.; Buffa, G.; Filesi, L.; Gamper, U.; Ghirelli, L. Phytocoenotic diversity of the N-Adriatic coastal sand dunes—The herbaceous communities of the fixed dunes and the vegetation of the interdunal wetlands. *Plant Sociol.* **2013**, *50*, 57–77.
36. Valentini, E.; Taramelli, A.; Filipponi, F.; Giulio, S. An effective procedure for EUNIS and Natura 2000 habitat type mapping in estuarine ecosystems integrating ecological knowledge and remote sensing analysis. *Ocean Coast. Manag.* **2015**, *108*, 52–64. [CrossRef]
37. Merloni, N.; Rigoni, P.; Zanni, F. La vegetazione delle dune litoranee nella Riserva Naturale di Foce Bevano. In *Spiagge e dune dell'Alto Adriatico*; Corpo Forestale dello Stato: La Greca Arti Grafiche, Forlì, Italy, 2015.

38. Lazzari, G.; Merloni, N.; Saiani, D. Flora delle Riserve Naturali dello Stato nell'area costiera di Ravenna Parco Delta del Po-Emilia Romagna. *Quad. dell'Ibis* **2009**, *3*, 48.
39. TETRACAM Inc. *ADC Micro Camera User's Guide*; Tetracam Inc.: Chatsworth, CA, USA, 2013.
40. Candiago, S.; Remondino, F.; De Giglio, M.; Dubbini, M.; Gattelli, M. Evaluating Multispectral Images and Vegetation Indices for Precision Farming Applications from UAV Images. *Remote Sens.* **2015**, *7*, 4026–4047. [CrossRef]
41. Ullman, S. The interpretation of structure from motion. *Proc. R. Soc. Lond. Ser. B Biol. Sci.* **1979**, *203*, 405–426.
42. Sanz-Ablanedo, E.; Chandler, J.H.; Rodríguez-Pérez, J.R.; Ordóñez, C. Accuracy of Unmanned Aerial Vehicle (UAV) and SfM Photogrammetry Survey as a Function of the Number and Location of Ground Control Points Used. *Remote Sens.* **2018**, *10*, 1606. [CrossRef]
43. De Giglio, M.; Goffo, F.; Greggio, N.; Merloni, N.; Dubbini, M.; Barbarella, M. Satellite and unmanned aerial vehicle data for the classification of sand dune vegetation. *ISPRS Int. Arch. Photogramm. Remote Sens. Spat. Inf. Sci.* **2017**, *42*, 43–50. [CrossRef]
44. Castagneti, C.; Casula, G.; Dubbini, M.; Capra, A. Adjustment and transformation strategies of ItalPoSPermanent GNSS Network. *Ann. Geophys.* **2009**, *52*. Available online: http://hdl.handle.net/2122/5204 (accessed on 13 June 2019).
45. Leica Geosystem "White Paper: GPS Spider-NET—Take It to the MAX". Available online: https://www.smartnetna.com/documents/Leica_GPS_SpiderNET-Take_it_to_the_MAX_June2005_en.pdf (accessed on 14 April 2019).
46. Liu, D.; Xia, F. Assessing object-based classification: Advantages and limitations. *Remote Sens. Lett.* **2010**, *1*, 187–194. [CrossRef]
47. Gao, Y.; Mas, J.F. A Comparison of the Performance of Pixel-Based and Object-Based Classifications Over Images With Various Spatial Resolutions. *Online J. Earth Sci.* **2008**, *2*, 27–35.
48. Whiteside, T.G.; Boggs, G.S.; Maier, S.W. Comparing object-based and pixel-based classifications for mapping savannas. *Int. J. Appl. Earth Obs. Geoinf.* **2011**, *13*, 884–893. [CrossRef]
49. Lillesand, T.; Kiefer, R.W.; Chipman, J. *Remote Sensing and Image Interpretation*; John Wiley & Sons: New York, NY, USA, 2014.
50. Franci, F.; Spreafico, M. Processing of remote sensing data for the estimation of rock block size distribution in landslide deposits. In *Landslides and Engineered Slopes. Experience, Theory and Practice*; CRC Press: Boca Raton, FL, USA, 2016.
51. Taubenböck, H.; Esch, T.; Wurm, M.; Roth, A.; Dech, S. Object-based feature extraction using high spatial resolution satellite data of urban areas. *J. Spat. Sci.* **2010**, *55*, 117–132. [CrossRef]
52. Tzotsos, A.; Argialas, D. Support Vector Machine Classification for Object-Based Image Analysis. In *Object-Based Image Analysis*; Springer: New York, NY, USA, 2008; pp. 663–677.
53. Mercier, G.; Lennon, M. Support vector machines for hyperspectral image classification with spectral-based kernels. In Proceedings of the 2003 IEEE International Geoscience and Remote Sensing Symposium, IGARSS'03, Toulouse, France, 21–25 July 2003; Volume 1, pp. 288–290.
54. Gao, Z.; Liu, X. Support vector machine and object-oriented classification for urban impervious surface extraction from satellite imagery. In Proceedings of the 2014 Third International Conference on Agro-Geoinformatics, Beijing, China, 11–14 August 2014.
55. Wu, T.-F.; Lin, C.-J.; Weng, R.C. Probability Estimates for Multi-class Classification by Pairwise Coupling. *J. Mach. Learn. Res.* **2004**, *5*, 975–1005. [CrossRef]
56. Landis, J.R.; Koch, G.G. The measurement of observer agreement for categorical data. *Biometrics* **1977**, *33*, 159–174. [CrossRef]
57. Story, M.; Congalton, R.G. Accuracy assessment: A users perspective. *Photogramm. Eng. Remote Sens.* **1986**, *52*, 397–399.
58. Barbarella, M.; De Giglio, M.; Panciroli, L.; Greggio, N. Satellite data analysis for identification of groundwater salinization effects on coastal forest for monitoring purposes. *Proc. Int. Assoc. Hydrol. Sci.* **2015**, *368*, 325–330. [CrossRef]
59. Rossiter, D.G. *Statistical Methods for Accuracy Assessment of Classified Thematic Maps*; Technical Note; International Institute for Geo-Information Science and Earth: Enschede, The Netherlands, 2004.
60. Foody, G.M. Thematic map comparison. *Photogramm. Eng. Remote Sens.* **2004**, *70*, 627–633. [CrossRef]
61. Fleiss, J. *Statiscal Methods for Rates and Proportions*, 2nd ed.; Wiley: New York, NY, USA, 1981.

62. Neter, J.; Wasserman, W.; Whitmore, G.A. *Applied Statistics*; Allyn and Bacon: Boston, MA, USA, 1993.
63. Carter, G.A.; Knapp, A.K. Leaf optical properties in higher plants: Linking spectral characteristics to stress and chlorophyll concentration. *Am. J. Bot.* **2001**, *88*, 677–684. [CrossRef]
64. Walsh, S.J.; McCleary, A.L.; Mena, C.F.; Shao, Y.; Tuttle, J.P.; Gonzalez, A.; Atkinson, R. QuickBird and Hyperion data analysis of an invasive plant species in the Galapagos Islands of Ecuador: Implications for control and land use management. *Remote Sens. Environ.* **2008**, *112*, 1927–1941. [CrossRef]
65. Jiménez, M.; Díaz-Delgado, R. Towards a standard plant species spectral library protocol for vegetation mapping: A Case Study in the Shrubland of Doñana National Park. *ISPRS Int. J. Geo-Inf.* **2015**, *4*, 2472–2495. [CrossRef]
66. Hernandez-Santin, L.; Rudge, M.L.; Bartolo, R.E.; Erskine, P.D. Identifying Species and Monitoring Understorey from UAV-Derived Data: A Literature Review and Future Directions. *Drones* **2019**, *3*, 9. [CrossRef]
67. Jacobberger, P.A. Reflectance characteristics and surface processes in stabilized dune environments. *Remote Sens. Environ.* **1989**, *28*, 287–295. [CrossRef]
68. Peñuelas, J.; Filella, L. Technical focus: Visible and near-infrared reflectance techniques for diagnosing plant physiological status. *Trends Plant Sci.* **1998**, *3*, 151–156. [CrossRef]
69. Immitzer, M.; Vuolo, F.; Atzberger, C. First Experience with Sentinel-2 Data for Crop and Tree Species Classifications in Central Europe. *Remote Sens.* **2016**, *8*, 166. [CrossRef]
70. Yu, Q.; Gong, P.; Clinton, N.; Biging, G.; Kelly, M.; Schirokauer, D. Object-based Detailed Vegetation Classification with Airborne High Spatial Resolution Remote Sensing Imagery. *Photogramm. Eng. Remote Sens.* **2006**, *72*, 799–811. [CrossRef]
71. Wang, L.; Sousa, W.P.; Gong, P. Integration of object-based and pixel-based classification for mapping mangroves with IKONOS imagery. *Int. J. Remote Sens.* **2004**, *25*, 5655–5668. [CrossRef]
72. Zhai, D.; Dong, J.; Cadisch, G.; Wang, M.; Kou, W.; Xu, J.; Xiao, X.; Abbas, S. Comparison of pixel- and object-based approaches in phenology-based rubber plantation mapping in fragmented landscapes. *Remote Sens.* **2018**, *10*, 44. [CrossRef]
73. Van Iersel, W.; Straatsma, M.; Middelkoop, H.; Addink, E. Multitemporal Classification of River Floodplain Vegetation Using Time Series of UAV Images. *Remote Sens.* **2018**, *10*, 1144. [CrossRef]
74. Bitton, M.C.A.; Hesp, P.A. Vegetation dynamics on eroding to accreting beach-foredune systems, Florida panhandle. *Earth Surf. Process. Landf.* **2013**, *38*, 1472–1480. [CrossRef]
75. Feagin, R.A.; Figlus, J.; Zinnert, J.C.; Sigren, J.; Martínez, M.L.; Silva, R.; Smith, W.K.; Cox, D.; Young, D.R.; Carter, G. Going with the flow or against the grain? The promise of vegetation for protecting beaches, dunes, and barrier islands from erosion. *Front. Ecol. Environ.* **2015**, *13*, 203–210. [CrossRef]
76. Konlechner, T.M.; Kennedy, D.M.; Cousens, R.D.; Woods, J.L.D. Patterns of early-colonising species on eroding to prograding coasts; implications for foredune plant communities on retreating coastlines. *Geomorphology* **2019**, *327*, 404–416. [CrossRef]

Article

Capturing Coastal Dune Natural Vegetation Types Using a Phenology-Based Mapping Approach: The Potential of Sentinel-2

Flavio Marzialetti [1], Silvia Giulio [2,*], Marco Malavasi [3], Marta Gaia Sperandii [2], Alicia Teresa Rosario Acosta [2] and Maria Laura Carranza [1]

1 Envix-Lab, Departement of Biosciences and Territory, Molise University, Contrada Fonte Lappone, 86090 Pesche (Is), Italy; flavio.marzialetti@unimol.it (F.M.); carranza@unimol.it (M.L.C.)
2 Department of Sciences, Roma Tre University, Viale G. Marconi 446, 00146 Rome, Italy; martagaia.sperandii@uniroma3.it (M.G.S.); aliciateresarosario.acosta@uniroma3.it (A.T.R.A.)
3 Department of Applied Geoinformatics and Spatial Planning, Faculty of Environmental Sciences, Czech University of Life Sciences, Kamycka 129, 165 00 Prague 6, Czech Republic; malavasi@fzp.czu.cz
* Correspondence: silvia.giulio@uniroma3.it; Tel.: +39 328 5392876

Received: 20 May 2019; Accepted: 21 June 2019; Published: 25 June 2019

Abstract: Coastal areas harbor the most threatened ecosystems on Earth, and cost-effective ways to monitor and protect them are urgently needed, but they represent a challenge for habitat mapping and multi-temporal observations. The availability of open access, remotely sensed data with increasing spatial and spectral resolution is promising in this context. Thus, in a sector of the Mediterranean coast (Lazio region, Italy), we tested the strength of a phenology-based vegetation mapping approach and statistically compared results with previous studies, making use of open source products across all the processing chain. We identified five accurate land cover classes in three hierarchical levels, with good values of agreement with previous studies for the first and the second hierarchical level. The implemented procedure resulted as being effective for mapping a highly fragmented coastal dune system. This is encouraging to take advantage of the earth observation through remote sensing technology in an open source perspective, even at the fine scale of highly fragmented sand dunes landscapes.

Keywords: dune vegetation classification; coastal monitoring; multispectral satellite images; multi-temporal NDVI; pixels based supervised classification; Random Forest; harmonization

1. Introduction

Environmental monitoring is essential to identify and understand the structure, integrity and conservation status of different habitats forming landscape mosaics [1]. Next to traditional field-based techniques [2], Remote Sensing (RS) methods are useful tools for ecosystems monitoring, as they are able to capture a wide range of properties of vegetation in a standard and replicable way [3].

During the last few decades, satellite images have supported vegetation mapping and monitoring of wide landscapes [4], with a continuous improvement in spatial resolution and use of multi/hyperspectral sensors consistently boosting the performance of remotely sensed data for mapping highly fragmented areas [5–7]. In particular, for the interpretation of particularly complex or very fine-grained vegetation mosaics such as those commonly encountered on sand dunes, high-resolution data are an essential requirement [8,9]. Indeed, complex landscapes have long represented a challenge for vegetation mapping and multi-temporal monitoring applications, which still need further development [10,11]. In this context, several space agencies (e.g., ESA, NASA/USGS, CBERS, ISRO) deliver free remotely sensed products with several resolutions that represent a reliable support for different applications of ecosystem monitoring and management. Among these, the free access Sentinel-2 mission of ESA (European Space Agency), with a 10 m spatial resolution for bands in the spectral range of the blue (B2-490 nm), green (B3-560 nm), red (B4-665 nm) and infra-red (B8-842 nm) and a revisit rate of approximately five days, represents an important support for environmental mapping in the Mediterranean area [12,13]. The relatively recent release of Sentinel-2 mission (Sentinel-2A launched on 23 June 2016 and Sentinel-2B launched on 7 March 2017) still has several potentialities to be explored [14].

In order to achieve their full potential for accurate mapping of complex vegetation mosaics, RS techniques must be coupled with proper approaches able to capture spatial and temporal vegetation patterns [1]. For instance, the analysis of vegetation phenological properties, describing recurring biological events (e.g., seasonality) [15], is very effective for mapping intense seasonal biomass variations [16] as those characterizing Mediterranean landscapes. Among the remotely sensed data, vegetation indices, depicting ecosystem spectral properties, are very efficient tools to map vegetation and its temporal pattern [17,18]. Among these, Normalized Difference Vegetation Index (NDVI) [19,20] is a good proxy of canopy biomass [21], and its application for environmental monitoring is highly appreciated [22,23]. Furthermore, the monthly variation of NDVI values across an entire year proved to be a sound surrogate of ecosystem phenology [18,24,25], which allows for discriminating contiguous vegetation cover types featuring different seasonality [26–28].

Coastal dune landscapes are complex mosaics that develop in the transition zones between terrestrial and marine environments, occupying strips parallel to the seashore [29]. Along the sea-inland gradient, coastal dunes are ruled by a large variety of constraining environmental conditions, such as soil salinity, substrate instability, wind and marine aerosol [30–32]. By shaping the biomass levels, this gradient determines the occurrence of a mosaic of highly specialized and diverse plant communities coexisting in a relatively narrow area which represents a hotspot of exclusive biodiversity [32]. In spite of their high biodiversity value and complex ecosystem functioning, coastal dunes are among the most threatened ecosystems worldwide [33,34]. In the Mediterranean areas, the loss and degradation of coastal dune ecosystems have been particularly severe in the last few decades [34–36], with the main threats being urban expansion [10,37,38], coastal erosion [39] and invasion by alien species [40–43]. In order to prevent these and other endangered habitats from further degradation, all European Member States adopted the Council Directive 92/43/EEC (hereafter Habitats Directive, HD). By signing the HD, the States committed themselves to maintain, restore and monitor habitats and species of European conservation concern (listed in dedicated annexes) and to report their conservation status every six years. Each habitat type (mainly identified by plant communities) is characterized by specific biotic and abiotic factors [44]. In this light, innovative and scientifically sound instruments are needed for setting conservation priorities and providing management indications for the coastal dune habitat types [45–48]. Until now, coastal dunes mapping procedures have mostly been based on the integration of visual interpretation (e.g., photointerpretation of aerial imagery) and floristic data [10,49]. However, the use of these mapping procedures presents some shortcomings in linear and fragmented ecosystems characterized by low biomass such as coastal dunes. For instance, photo-interpretation is a time-consuming procedure and its results vary depending on the subjectivity of the interpreter, his experience and personal knowledge of mapped landscapes [50,51]. Moreover, as small patches (below the minimum mapping area) are neglected, the photo-interpreted maps can be limited for characterizing the coastal fine scale mosaics and related landscape processes [52,53].

In consideration of the above, the present work sets out to explore the potential of Sentinel-2 in capturing coastal dune natural vegetation types using a phenology-based mapping approach. In particular, by a multi-temporal analysis of NDVI images of coastal dunes in central Italy, we focused on two main questions: (i) does NDVI phenological profiles allow for identifying and correctly mapping different vegetation types distributed along a coastal zonation? (ii) does the product of a phenology-based classification agree with existing coastal dunes classification systems (i.e., photo interpreted land cover maps and 92/43/EEC habitat distribution)?

2. Materials and Methods

2.1. Study Area

The study was carried out on a representative tract of the Mediterranean coast, placed in the Tyrrhenian seashore of Central Italy (Lazio Region). The study area includes ca. 250 km of sandy coast, mainly formed by recent (Holocenic) dunes (Figure 1) [45]. These dunes are relatively simple in structure, with a single low dune ridge (lower than 10 m height) occupying a narrow strip (usually no more than 500 m with) along the seashore [32]. In this area, coastal plant communities mostly range from pioneer vegetation near the shoreline to Mediterranean shrubs on landward fixed dunes [33,45]. Previous studies discriminate eight different habitats occurring along this coastal zonation, of which two have been listed as priority (Table 1) [38,53]. In the analyzed littoral zone, human activities in the analyzed littoral zone have intensified in the course of the 20th century [10,54]. Specifically, during the last 60 years, the Tyrrhenian coast faced consistent processes of fragmentation [55], simplification [45] and biodiversity loss [56]. Nevertheless, the Tyrrhenian coast still hosts a good number of plant communities of conservation concern in Europe (92/43/EEC Habitats Directive; EEC, 1992) for which monitoring and conservation strategies must be improved.

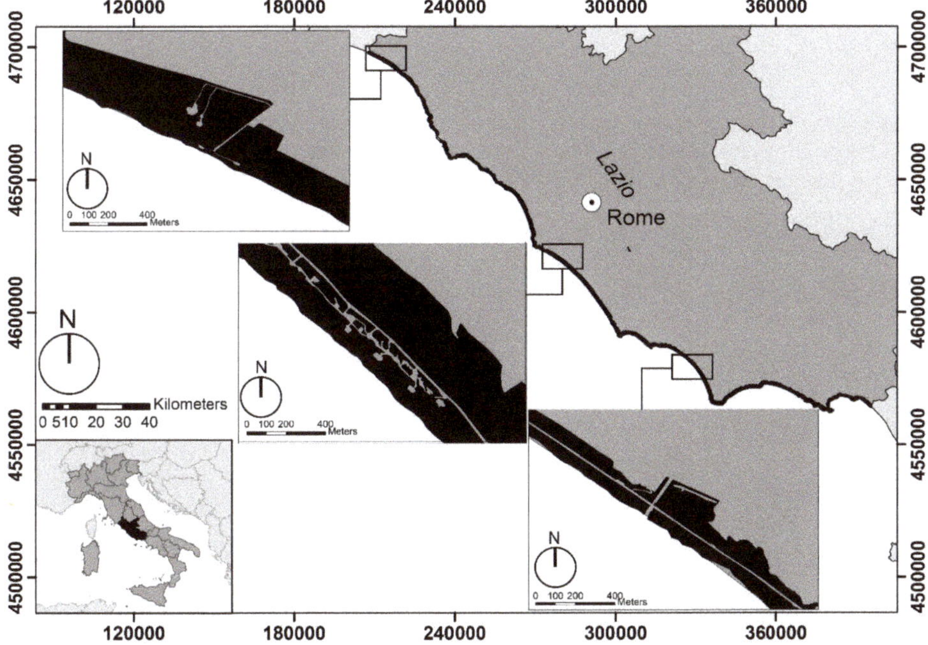

Figure 1. Study area. The coastline of the Lazio Region (Italy). Three zoomed examples at north, center and south are reported in black. Coordinates system: WGS84 UTM 33N (epsg: 32633).

Table 1. EC Habitat types (Habitat Directive 92/43/EEC) along with a brief description in terms of vegetation types [53]. Asterisk (*) indicates habitats with high priority for conservation.

EC Habitat	Name	Vegetation Types
1210	Annual vegetation of drift line (upper beach).	Pioneer annual vegetation characterizing the strandline zone of the beach.
2110	Embryonic shifting dunes (embryo dune).	Pioneer, perennial community of the low embryo-dunes dominated by *Elymus farctus*.
2120	Shifting dunes along the shoreline with *Ammophila arenaria* (mobile dune).	Seaward and semi-permanent cordons of dune systems dominated by *Ammophila arenaria* subsp. *australis*.
2210	*Crucianellion maritimae* fixed beach dunes.	Chamaephytic community of the inland side of fixed dunes dominated by *Crucianella maritima*.
2230	*Malcolmietalia* dune grasslands 2250	Annual, species-rich community colonized by small terophytes in dry, interdunal depressions of the coast.
2250*	Coastal dunes with *Juniperus* spp. (juniper scrub)	Shrub formations dominated by juniper on the fixed dunes.
2260	*Cisto- Lavanduletalia* dune sclerophyllous scrubs	Shrub formations dominated by sclerophyllous species
2270*	Wooded dunes with *Pinus pinea* and/or *Pinus pinaster*	Coastal dunes colonized by Mediterranean and Atlantic termophilous pines.

2.2. Methodology

We performed a phenology-based classification of coastal dune ecosystem following a sequence of steps (Figure 2): (1) multitemporal dataset collection and image preprocessing, (2) NDVI calculation and data processing, (3) data classification, (4) accuracy assessment, and (5) comparison of phenology-based classes vs. previous studies.

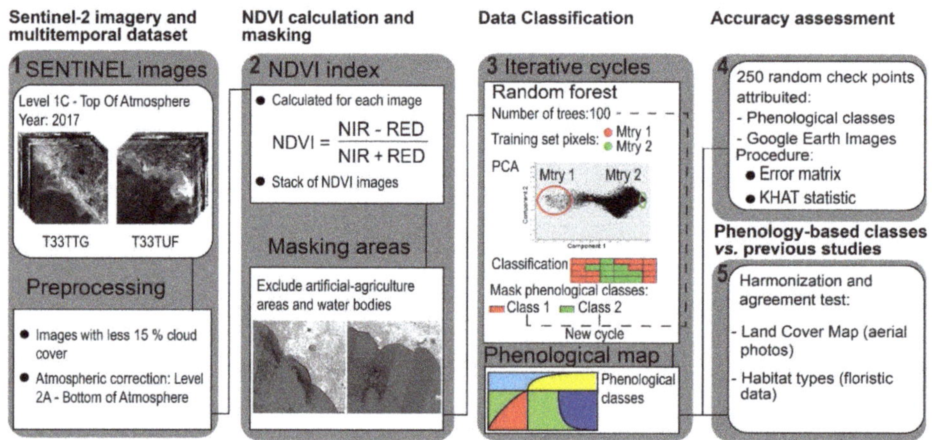

Figure 2. Workflow synthesizing the full mapping procedure of natural dune vegetation with Sentilnel-2 NDVI (Normalized Difference Vegetation Index) time series and Random Forest classification approach.

2.2.1. Sentinel-2 Imagery and Multitemporal Dataset

Sentinel-2 is a European wide-swath, multi-spectral imaging mission, constituted by a two-satellite platform: Sentinel-2A and Sentinel-2B [57]. The Multi Spectral Instrument (MSI) on-board Sentinel-2 can provide images with a temporal resolution of five days at the equator, and a 12-bit radiometric resolution from 492 nm to 1377 nm, which includes the Visible (VIS), Near Infra-Red (NIR), and Short Wave Infra-Red (SWIR) spectra (13 bands). Sentinel-2 images are freely downloadable from the Copernicus Open Access Hub [58]. In this study, we used the red band (R, around 665 nm in the

VIS spectrum) and the NIR band (around 833 nm), at 10 meters of resolution [12]. The study area is included in two Sentinel-2 dataset tiles (T33TTG, T33TUF).

We extracted monthly NDVI images recorded in the year 2017 for the two Sentinel-2 tiles describing the analyzed coast and we built a multi-temporal dataset (stack) (Table S1). For the analysis, we selected only those images with low cloud coverage (< 15%), excluding March, due to the excess in cloud coverage. The clean stack for the phenology-based classification was composed of 22 images (one per tile and month, excluding March). Only part of the downloaded data was already corrected from atmospheric noise, while the other part was rough. We corrected these others using Sen2Cor version 2.5.5 [59,60].

2.2.2. NDVI Calculation and Masking

For the entire stack, we calculated NDVI (Equation (1)) as follows:

$$\mathrm{NDVI} = \frac{\mathrm{NIR} - \mathrm{R}}{\mathrm{NIR} + \mathrm{R}}. \tag{1}$$

NDVI ranges from −1 to 1, with increasing values related with growing photosynthetic biomass [24]. The NDVI stack was masked using a fine scale land cover map (1:5000; SITR—Sistema Informativo Territoriale Regionale Lazio) in order to exclude from the classification, at least partially, urban areas, agriculture fields and water bodies.

2.2.3. Data Classification

We classified the resulting NDVI stack by implementing a Random Forest algorithm (RF) with a hierarchical logic, using ESA's Sentinel-2 toolbox—ESA Sentinel Application Platform 6.0 (SNAP). RF is a machine learning classification method that operates by constructing a multitude of decision trees [61,62]. RF algorithm is widely used for classification because of its speed, stability and ability to discriminate differences [63–65].

We cyclically performed different RF analyses, defining for each cycle two parameters: the training set of pixels (Mtry, see the next paragraph) and the number of trees (Ntree, in our case 100 runs per cycle). In order to maximize the efficacy of RF, in the training set, we used a number of pixels that was higher than the square root of the total of the pixels [61,62].

In each cycle, the identification of the training set was supported by a Principal Component Analysis (PCA) of the monthly NDVI values (pixels x monthly NDVI values matrix). We projected all the pixels in the PCA1 and PCA2 ordination space [66]. As objects that are close in the ordination space (similar component values) describe similar cover types [67], we therefore selected as the training set for classification the two furthest away clouds of pixels with maximum variance between them [68]. For each tree (run), the training set was split through a bootstrapping procedure in two groups: seed pixels (*inbag*), to build the classification tree, and validation pixels (*out-of-bag*), to estimate the classification performance. Then, all pixels were compared with the *inbag* set using the *Gini* inequality index [69] and assigned to a class based on the Lorenz Curve. The latter ranges from 0 (perfectly equality) to 1 (perfectly inequality) [69]. At the end of each run, RF conferred to each pixel an ordinal vote (the minimum value of Lorenz Curve). After the entire cycle of 100 runs, each pixel was definitively assigned to the more frequently attributed class [61]. In each cycle, the performance of classification was assessed by repeating the classification procedure on the basis of the *out-of-bag* data and comparing both classifications [61,70].

In each RF cycle, we classified the stack in two vegetation classes, and we carried on all the cycles where the *out-of-bag* error was <50%.

2.2.4. Accuracy Assessment

The accuracy of the phenology classification map was assessed through an error matrix (Table S2) calculating overall accuracy (Equation S1), producer's accuracy (Equation S2) and user's accuracy (Equation S3) and Kappa statistic ($0 \leq \mathrm{Kappa} \leq 100$; Equation S4) [71]. We based this assessment on

250 random checkpoints visually inspected on Google Earth images [72–74]. The positional errors of objects in Google Earth images are much lower compared to of the minimum spatial resolution of Sentinel-2 images [73] and their spatial resolution (~1m) is high enough to allow clear visual interpretation of the land cover [75]. In our case, the random checkpoints were inspected in Google Earth images for a 5 m radius buffer area (comparable scale of the 10 m pixel of Sentinel-2 images) to limit the scale mismatch errors. We built the error matrix, reporting the assigned phenology-based class in rows, and the Google Earth visual attribution in columns.

2.2.5. Phenology-Based Map vs. Previous Vegetation Studies

The error matrix was further used to compare the phenology-based map with two previous studies conducted in the same area. The first study reports an actual vegetation map produced at 1:5000 scale by visual interpretation of aerial orthophotos [76] and the second one refers to the natural dune vegetation types classified according to the Habitats directive (92/43/EEC; Table 1) and identified by floristic field data (Figure S1). To compare the phenology-based classes with the photointerpreted map, we used 250 random checkpoints, while the congruence with habitat types was tested using 135 floristic plots extracted from "RanVegDunes", a database of randomly distributed floristic surveys [77]. For testing the congruence of the phenology-based classification with the existing documents, we aggregated and homogenized the classes of the vegetation map (Table S4) and the habitat types (Table S5) on the basis of similarities in physiognomies and ecological conditions [78]. Then, we tested the correspondence among maps and defined their respective levels of agreement through the Kappa statistic (Table S3).

3. Results

3.1. Sentinel-2 NDVI Classification

The phenology-based classification allowed for identifying five vegetation classes organized in three hierarchical levels, each one characterized by a specific phenological pattern (Figure 3) referable to different mosaics of plant communities. The first level of classification distinguishes a class of Open Sand from the Vegetated class, the second level divides the Vegetation class in Herbaceous and Woody Vegetation, and the third level divides Herbaceous and Woody Vegetation classes into two further classes (the first in Sparse Herbaceous Vegetation (SHV) and Dense Herbaceous Vegetation and Ruderals (DHVR); the second in Sparse Woody Vegetation (SWV), and Dense Woody Vegetation (DWV)).

The Open Sand class (Figure 3a) is characterized by very low biomass. Monthly values are close to 0, except in summer, when NDVI is negative. The Open Sand class is extensively present along the whole coast as thin stripes even on quite urbanized coastal tracts.

The Sparse Herbaceous Vegetation class is characterized by low monthly NDVI values (Figure 3b1) that decrease towards 0 in summer. Sparse Herbaceous Vegetation is in contact with Open Sand and covers a narrow discontinuous strip of land between Open Sand and the inner vegetation classes. Sparse Herbaceous Vegetation preferentially occurs on well-preserved coastal tracts characterized by all phenology-based vegetation classes.

The Dense Herbaceous Vegetation and Ruderals class (Figure 3b2) is characterized by NDVI values slightly over 0.5 from November to April that decrease during summer. Dense Herbaceous Vegetation and Ruderals include a highly seasonal herbaceous vegetation. This class occurs close to the seashore in sectors exposed to environmental stress and on inner dune sectors characterized by high anthropic pressure.

The Sparse Woody class is composed by sparse evergreen vegetation (Figure 3c1) with high monthly NDVI values (> 0.5) that slightly decrease in summer. This class tends to occur at intermediate distances from the seashore and in the inner sectors of the dune in contact with densely wooded dunes.

Lastly, the Dense Woody Vegetation class presents high monthly NDVI values (> 0.7; Figure 3c2) throughout the year. It occurs in the back-dune zones corresponding to dense shrublands and forests.

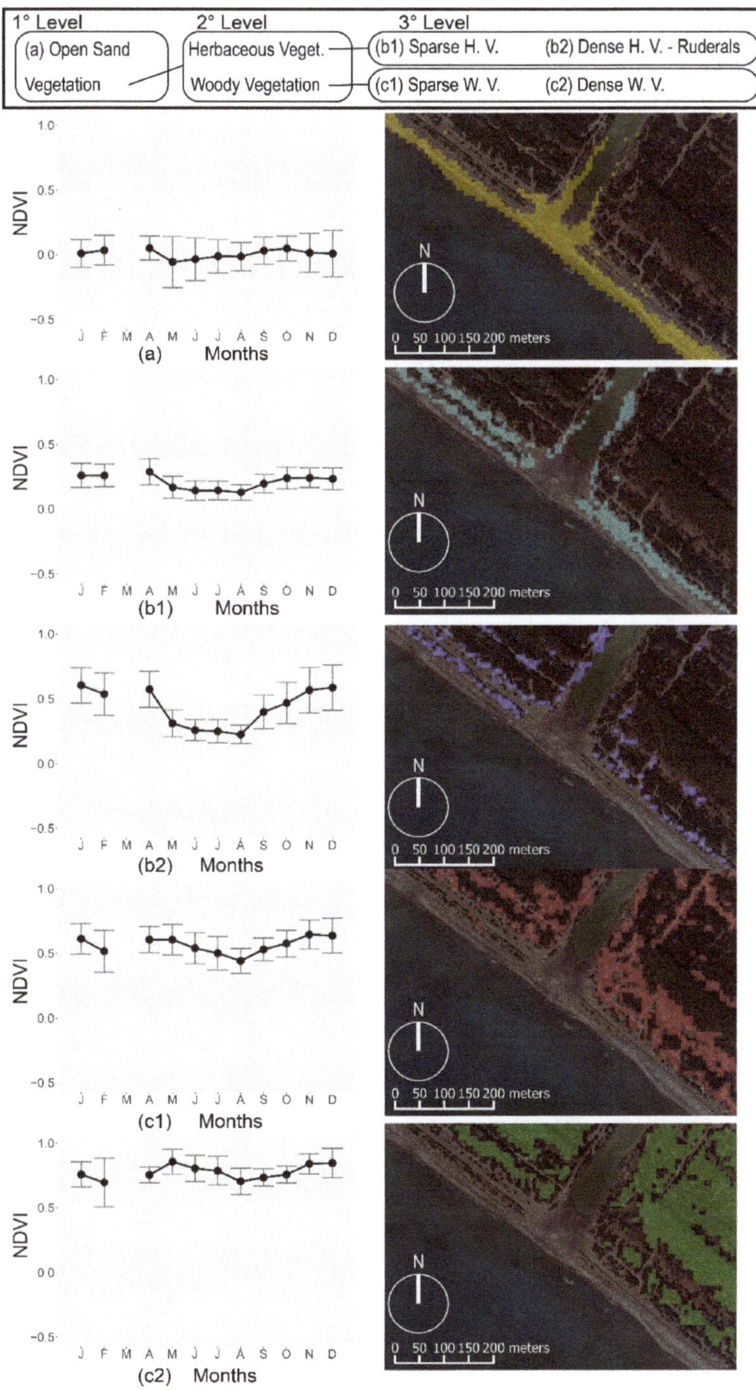

Figure 3. Cartesian diagrams (on the left) of NDVI values with average ± standard deviation (y-axis) for each month of the year except March (x-axis) of the five phenology-based classes identified by multitemporal classification of Sentinel-2 images, along with mapping examples (on the right) projected on Google Earth View.

3.2. Classification Accuracy Assessment

The classification of multi-temporal NDVI data (phenology-based classification) showed a good level of accuracy when compared with the visual interpretation of Google Earth images. The first level of classification (two classes: Vegetation and Open Sand) exhibited very high values of overall accuracy (96%) with both producer's and user's accuracy over 88% and Kappa statistic of 86% (Figure 4a).

Similarly, the overall accuracy of the second hierarchical level (Herbaceous and Woody Vegetation) was high (88%) with a moderate agreement among classes and reference data given by a Kappa statistic of 79% (Figure 4b).

All land cover classes identified at the third level of detail evidenced values of accuracy with producer's accuracy ranging between 69% (Herbaceous Vegetation) and 99% (Woody Vegetation). Similarly, the user's accuracy ranged between 86% (Woody Vegetation) and 90% (Open Sand and Herbaceous Vegetation) (Figure 4c).

The correspondence between the third hierarchical level of classification and the set of Google images resulted as moderate as underlined by both overall accuracy (79%) and Kappa statistic (71%) (Figure 4c). The producer's accuracy was adequate for all classes. In particular, Open Sand and Dense Woody Vegetation had the highest values of producer's accuracy (88% and 96%, respectively). This result defined an elevated precision of the map in these two classes. Moreover, both herbaceous classes showed moderate agreement (75% Sparse Herbaceous Vegetation, 65% Dense Herbaceous Vegetation). Finally, Sparse Woody Vegetation class featured the lowest value of producer's accuracy detected, even though the value showed moderate agreement (56%). The user's accuracy showed the higher values in Open Sand and Dense Herbaceous Vegetation classes (90% and 98%, respectively). Sparse Herbaceous Vegetation and Sparse Woody Vegetation had the lowest values (respectively 63% and 64%). Finally, Dense Woody Vegetation presented user's accuracy (75%).

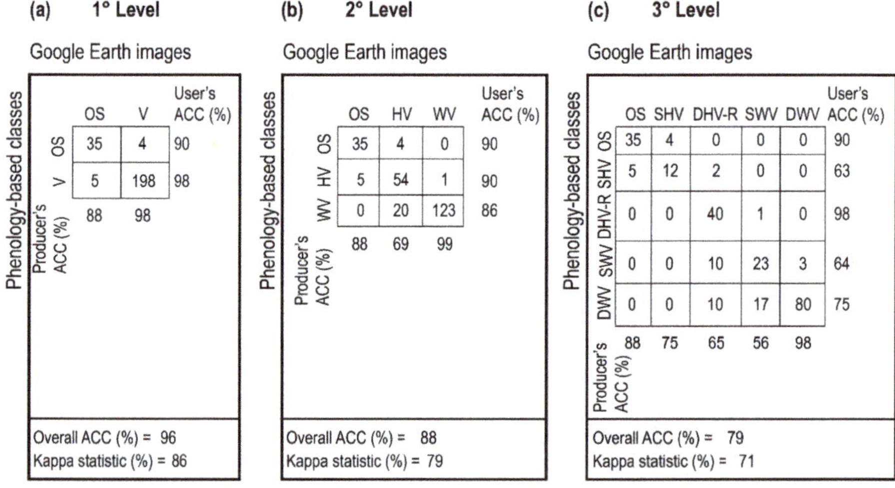

Figure 4. Error matrix, Accuracy (ACC) values, in particular Overall ACC, Producer's ACC, User's ACC, and Kappa statistic of all hierarchical levels of classification–phenology-based classes: Open Sand (OS), Vegetation (V), Herbaceous Vegetation (HV), Woody Vegetation (WV), Sparse Herbaceous Vegetation (SHV), Dense Herbaceous Vegetation (DHV-R), Sparse Woody Vegetation (SWV), and Dense Woody Vegetation (DWV).

3.3. Harmonization and Agreement Test with Existing Documents

The NDVI classification showed significant values of agreement with the photointerpreted map [76], and floristic field data with the habitat types [53] at both the first and the second hierarchical levels (Figure 5).

The first hierarchical level showed strong agreement values with the photointerpreted classification map, with 95% of overall accuracy and 83% of Kappa statistic (Table S6), and also the producer's and user's accuracy showed high agreement values (~100%). The agreement of the second classification level resulted in being quite significant for all of the classes, with high overall accuracy (80%) and Kappa statistic (66 %) depicting a moderate congruence between them. The user's accuracy for the three classes was over 77%. The producer's accuracy ranges between 58 % and 95% (Table S7).

Finally, at the third level of classification, the agreement test resulted in being moderate, with overall accuracy and Kappa statistic values being approximately 66% and 55%, respectively. However, values of user's accuracy were high only for Open Sand (94%) and Dense Woody Vegetation (85%), while the other classes showed values under 50%. The producer's accuracy ranged between 26% and 81% (Table S8).

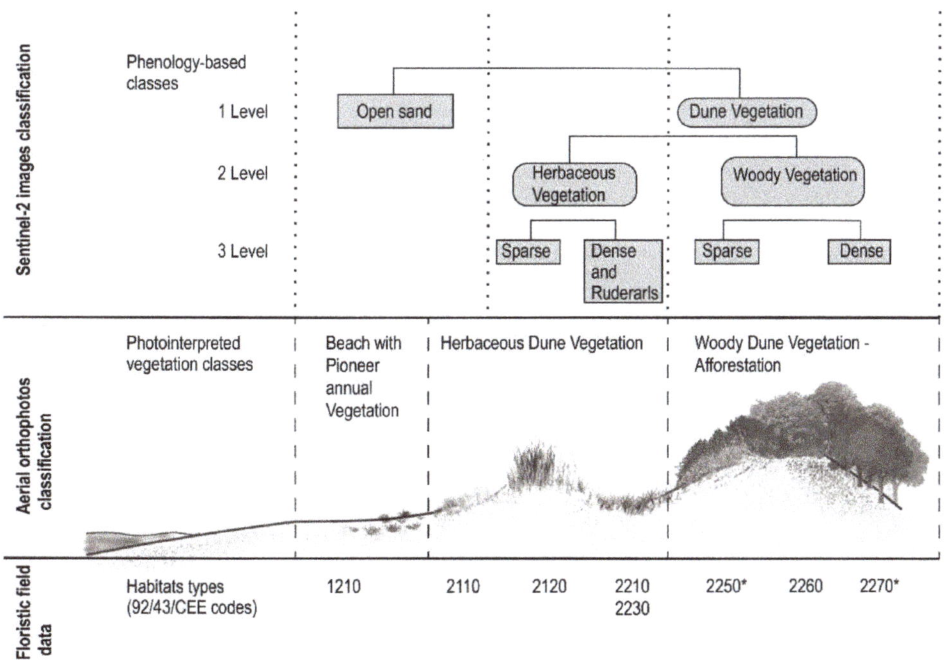

Figure 5. Cross sectional diagram indicating a typical coastal dune zonation along with their corresponding phenology-based classes, habitat types (92/43/EEC; for full habitat names see Table 1) and vegetation classes mapped in a previous study [76].

On the other hand, the agreement values of NDVI phenology-based classification and habitat types assigned by floristic data (Habitats directive 92/43/EEC) showed great differences among the hierarchical levels. At the first two levels, agreement values indicated moderate congruence between the classification systems. At the first hierarchical level, the overall accuracy was ~79% and the Kappa statistic denoted moderate agreement value. User's accuracy was relatively high for both classes, Open Sand (81%) and Vegetation (78%). Producer's accuracy ranged between 53% to 93% (Table S9).

The agreement test of the second hierarchical level indicated similar consistency, with overall accuracy ~71% and moderate value for the Kappa statistic (58%). The user's accuracy was relatively high for Open Sand and Woody Vegetation (81% and 90% respectively), and the Herbaceous Vegetation manifested moderate values (62%). Similarly, producer's accuracy ranged between 53% and 84% (Table S10). Finally, the third level of classification exhibited the lowest agreement values. The overall accuracy and Kappa statistic were approximately 53% and 43%, respectively. Moreover, user's accuracy was relatively high only for Open Sand (81%), and moderate agreement for Dense Herbaceous Vegetation and Ruderals (61%) and Dense Woody Vegetation (67%), with the other classes featuring values under 50%. Producer's accuracy ranged between 11% and 73% (Table S11).

4. Discussion

Our results suggest the analysis of remote sensed data (Sentinel-2 images) by a phenology-based classification as an effective approach for monitoring natural landscapes. Sentinel-2 images confirmed their high potential for vegetation mapping [12], while the multitemporal analysis of NDVI provided complementary and useful information, proving its convenience even in complex vegetation mosaics, that is to say, beyond their traditional field of application [18,26,28].

The phenology-based classification using a Random Forest algorithm on a Mediterranean wide coastal area allowed for identifying and mapping five vegetation classes organized in three hierarchical levels. Such classes, each one characterized by specific phenological and ecological features, exhibited high levels of accuracy and clearly depicted the coastal ecosystem zonation ranging from Open Sand, occurring near the seashore line, to Dense Woody Vegetation on the inner dunal sectors [45,79,80]. Sparse Herbaceous Vegetation and Sparse Woody Vegetation occurred discontinuously, while Open Sand and Dense Herbaceous Vegetation and Ruderals formed a continuous strip close to the seashore running along all the analyzed coast. Finally, Dense Woody Vegetation formed regular shaped patches, and, as previously observed [32,38], occurred in the back-dune zone.

The classification of multi temporal NDVI images, which was successfully used for land cover mapping [26,80], is extended here for vegetation mapping on Mediterranean coasts. The phenological analysis that allows for depicting vegetation seasonality [26,81,82] enabled to discriminate woody evergreen from herbaceous annual vegetation [26]. Furthermore, by exploring phenological spatial variations occurring in correspondence with biomass transitions, it was possible to distinguish between densely and sparsely vegetation formations and to identify edges [83,84] between vegetation classes occurring in the analyzed complex mosaic.

Our results respond to the scope of Sentinel mission [12,57] and give new evidence of its potential for monitoring and mapping coastal dunes with reduced costs and time efforts. The high temporal resolution that assures a continuous release of new clean and fine resolution images (~each 10 days) postulates Sentinel-2 as one of the most effective supports for phenology-based coastal dunes monitoring [18].

The good agreement between phenology-based classification and the photo-interpreted vegetation map [76] suggests the adequacy of Sentinel 2 multi-temporal NDVI classification for producing new vegetation maps of the coastal dune landscapes. Indeed, the phenology based map is quite consistent with the photo-interpreted vegetation map (scale 1:5000) produced using panchromatic digital aerial ortho-photographs with about one meter of resolution of the year 2008. Furthermore, the hierarchical nature of RF classification offers a good basis for comparing the new remotely sensed classification with existing documents produced with different methodological procedures, data sources and spatial resolution. Linking the new remotely sensed classes with previous maps and mapping supports (as aerial photos) is essential for building a long-term ecological series and for monitoring coastal dune landscapes across time [78,85].

The agreement test of NDVI classes and floristic data referable to EC habitats (92/43/EEC) was significant at the first and second hierarchical level for all classes. At the third level, only Open Sand, Dense Herbaceous Vegetation and Ruderals, and Dense Woody Vegetation showed a relative high congruence with vegetation plots, consequently showing the possibility to discriminate the habitats included in these classes. The lower agreement of NDVI classification and vegetation plots classified in habitat types is probably related with differences in the spatial resolution between the remotely sensed instrument (10x10 m) and the field floristic plots (2x2 m). Furthermore, the Mediterranean coastal dunes, naturally conformed by fine scale ecosystem mosaics [33,45], are often disturbed by fragmentation processes that alter their biodiversity [56]. This probably reduces the possibility of a match between conventional small floristic plots and the 10 m resolution of Sentinel-2 images. This scale mismatch is the principal restriction in this study, and the impossibility to discriminate the single habitat type through this phenology-based approach using 10 m resolution images limits its use to coarse vegetation classes (general physiognomies) [86,87]. In coastal dunes, both subtle variations of environmental factors and human pressure promote the formation of fine scale mosaics of habitats, some of them featuring similar physiognomies but differing in their floristic composition [88–93].

To deal with such shortcomings, the integration of sentinel data with finer resolution data and tools are advisable. For instance, the use of multispectral satellite images with higher spatial resolution, or the implementation of other classification methodologies as spectral unmixing algorithms able to quantify the percentage of different cover classes inside the single pixels [94,95] should improve the performance of the proposed classification procedure. In any case, the classification performance of each cycle, elaborated by out-of-bag pixels, estimated the uncertainty of the Random Forest result giving an idea of the presence of mixed pixels.

Overall, the potential of Sentinel-2 data in a phenology-based mapping was accomplished with a relative high degree of accuracy assessment and significant congruence with existing previous classifications. It is very promising in the discrimination of annual, deciduous and evergreen vegetation. Moreover, the integration of remotely sensed maps with field data could contribute to continuous update of coastal dune habitats maps, reducing costs and risks of delaying the periodical reporting requested by the Habitat Directive (92/43/EEC).

5. Conclusions

The performed phenology-based classification emphasizes the potential of Sentinel-2 images for mapping natural vegetation and extends its field of application to low biomass and highly fragmented systems as coastal dunes. The combined use of NDVI multi-temporal data, machine learning (Random Forest) algorithms, and a pixel-oriented approach allowed for adequately describing with high values of accuracy the complex mosaic of coastal dune vegetation.

The phenology-based classification approach with Sentinel-2 data proposed here is a time saving and more objective approach, complemented with open source earth observation data and implemented through free ESA software, effective and inexpensive instruments for coastal monitoring. Furthermore, the good levels of agreement of phenology-based with previous vegetation maps should allow for building long-term ecological series necessary for exploring and monitoring coastal ecosystems dynamics over time.

Nevertheless, there are several possibilities to improve this phenology-based classification and enhance its potential. For instance, the integration of phenological classification with LiDAR and other remotely sensed data or the implementation of spectral unmixing algorithms could improve the agreement with floristic filed data and should represent new research frontiers to explore.

From an applied perspective, the phenology-based vegetation classification provides relevant knowledge for coastal monitoring and management; therefore, we hope new studies exploring increasingly larger areas will be analyzed to further test the proposed classification and, at the same time, to provide homogeneous information for coasts in the Mediterranean.

Supplementary Materials: The following are available online at http://www.mdpi.com/2072-4292/11/12/1506/s1, Table S1: The total Sentinel-2 imagery dataset. It shows for each Sentinel-2 image the month, day, platform (Sentinel-2A or Sentinel-2B), the hour of acquisition, the cloud percentage, the processing level (top of the atmosphere—1C or bottom of the atmosphere—2A), and the tiles (T33TTG for Lazio north, T33TUF for Lazio south). Table S2: Example of error matrix. It is a contingency table (k x k array, where k is the number of classes in the classification). Equation S1: Overall accuracy, defined as the total of the correctly classified checkpoints on the total number of the checkpoints where nii indicates the number of checkpoints classified in the same category both in the satellite mapped classes and the Google Earth reference data; in other words, the elements of the major diagonal. Equation S2: Producer's accuracy: fraction of correctly classified checkpoints in all checkpoints of the produced classification. Equation S3: User's accuracy: fraction of correctly classified checkpoints in all checkpoints of the reference data. Equation S4: Kappa statistic ($\hat{K}$) computed as follows where n_ij is the number of observation in row i and column j, n_(i+) and n_(+j) are respectively the total number of observation of rows, and the second total number of observation of columns. Table S3: Classes of Kappa statistic interpretation. The Kappa statistic is a measure of the difference between the actual agreement of real objects observable on Google Maps with resulted classes, and an agreement due to chance (where real objects are compared with a random classification). Kappa varies between 0 and 100, where values close to 0 represent a poor agreement, and values close to 100 are indicated as excellent level of agreement. Figure S1: A subset of the photointerpreted vegetation map produced at 1:5000 scale by visual interpretation of aerial ortophotos, and a subset of the floristic field data classified according with the Habitats directive (92/43/EEC). Table S4: Nomenclature homogenization between the produced phenology-based map and the vegetation map. Table S5: Nomenclature homogenization of the EC habitats (92/43/EEC) types: 1210 (Annual vegetation of drift lines), 2110 (Embryonic shifting dunes), 2120 (Shifting dunes along the shoreline with *Ammophila arenaria*), 2210 (*Crucianellion maritimae* fixed beach dunes), 2230 (Malcolmietalia dune grasslands), 2250 (Coastal dunes with *Juniperus* spp.), 2260 (Cisto-Lavanduletalia dune sclerophyllous scrubs), 2270 (Wooded dunes with *Pinus pinea* and/or *P. pinaster*). Table S6: Results of the harmonization test (error matrix and Kappa statistic) between phenology-based classes in the first hierarchical level of classification and the photo–interpreted classification map. Table S7: Results of the harmonization test (error matrix and Kappa statistic) between phenology-based classes and the photo–interpreted classification map in the second hierarchical level of classification. Table S8: Results of the harmonization test (error matrix and Kappa statistic) between phenology-based classes in the third hierarchical level of classification and the photo–interpreted classification map. Table S9: Results of the harmonization test (error matrix and Kappa statistic) between phenology-based classes in the first hierarchical level of classification and habitats of conservation concern (Habitats Directive; 92/43/EEC; Table 1) assigned on 2 m floristic plots collected in the field. Table S10: Results of the harmonization test (error matrix and Kappa statistic) of between phenology-based classes in the second hierarchical level of classification and habitats of conservation concern (Habitats Directive; 92/43/EEC; Table 1) assigned on 2 m floristic plots collected in the field. Table S11: Results of the harmonization test (error matrix and Kappa statistic) between phenology-based classes in the second hierarchical level of classification and habitats of conservation concern (Habitats Directive; 92/43/EEC; Table 1) assigned on 2 m floristic plots collected in the field.

Author Contributions: All authors contributed substantially to the work: F.M. and M.L.C. conceived and designed the study; all authors collected the data; F.M., S.G., M.L.C. analyzed the data; F.M. led the writing of the manuscript. All authors contributed critically to the drafts and gave final approval for publication.

Funding: This research received no external funding.

Acknowledgments: The authors acknowledge the Principal Investigator(s) of the Sentinel-2 mission for providing datasets in the archive and the developers of SNAP software used for data analysis. Sentinel-2 images are freely downloadable from the Copernicus Open Access Hub (https://scihub.copernicus.eu/).

Conflicts of Interest: The authors declare no conflict of interest.

References

1. Díaz-Delgado, R.; Lucas, R.; Hurford, C. *The Roles of Remote Sensing in Nature Conservation. A Pratical Guide and Case Studies*, 1st ed.; Springer: Cham, Switzerland, 2017; pp. 1–318.
2. Adam, E.; Mutanga, O.; Rugege, D. Multispectral and hyperspectral remote sensing for identification and mapping of wetland vegetation: A review. *Wetl. Ecol. Manag.* **2010**, *18*, 281–296. [CrossRef]
3. Lawley, V.; Lewis, M.; Clarke, K.; Ostendorf, B. Site-based and remote sensing methods for monitoring indicators of vegetation condition: An Australian review. *Ecol. Indic.* **2016**, *60*, 1273–1283. [CrossRef]
4. Xie, Y.; Sha, Z.; Yu, M. Remote sensing imagery in vegetation mapping: A review. *J. Plant Ecol.* **2008**, *1*, 9–23. [CrossRef]
5. Adamo, M.; Tarantino, C.; Tomaselli, V.; Veronico, G.; Nagendra, H.; Blonda, P. Habitat mapping of coastal wetlands using expert knowledge and Earth observation data. *J. Appl. Ecol.* **2016**, *53*, 1521–1532. [CrossRef]
6. Wu, W.; Zhou, Y.; Tian, B. Coastal wetlands facing climate change and anthropogenic activities: A remote sensing analysis and modelling application. *Ocean Coast. Manage.* **2017**, *138*, 1–10. [CrossRef]
7. Betbeder, J.; Rapinel, S.; Corgne, S.; Pottier, E.; Huber-Moy, L. TerraSar-X dual-pol time-series for mapping of wetland vegetation. *ISPRS J. Photogramm. Remote Sens.* **2015**, *107*, 90–98. [CrossRef]
8. Rapinel, S.; Clément, B.; Magnanon, S.; Sellin, V.; Hubert-Moy, L. Identification and mapping of natural vegetation on a coastal site using a Worldview-2 satellite image. *J. Environ. Manage.* **2014**, *144*, 236–246. [CrossRef] [PubMed]

9. Zhang, L.; Baas, A.C.W. Mapping functional vegetation abundance in a coastal dune environment using a combination of LSM and MLC: A case study at Kenfig NNR, Wales. *Int. J. Remote Sens.* **2012**, *33*, 5043–5071. [CrossRef]
10. Malavasi, M.; Santoro, R.; Cutini, M.; Acosta, A.T.R.; Carranza, M.L. What has happened to coastal dunes in the last half century? A multitemporal coastal landscape analysis in Central Italy. *Landscape Urban Plan.* **2013**, *119*, 54–63. [CrossRef]
11. Valentini, E.; Taramelli, A.; Filipponi, F.; Giulio, S. An effective procedure for EUNIS and Natura 2000 habitat type mapping in estuarine ecosystems integrating ecological knowledge and remote sensing analysis. *Ocean Coast. Manage.* **2015**, *108*, 52–64. [CrossRef]
12. Drusch, M.; Del Bello, U.; Carlier, S.; Colin, O.; Fernandez, V.; Gascon, F.; Hoersch, B.; Isola, C.; Labertini, P.; Martimort, P.; et al. Sentinel-2: ESA's optical high-resolution mission for GMES operational services. *Remote Sens. Environ.* **2012**, *120*, 25–36. [CrossRef]
13. Chatziantoniou, A.; Petropoulos, G.P.; Psomiadis, E. Co-Orbital Sentinel 1 and 2 for LULC mapping with emphasis on wetlands in a mediterranean setting based on machine learning. *Remote Sens.* **2017**, *9*, 1259. [CrossRef]
14. Recanatesi, F.; Giuliani, C.; Ripa, M.N. Monitoring Mediterranean Oak decline in a peri-urban protected area using the NDVI and Sentinel-2 images: The case study of Castelporziano State Natural Reserve. *Sustainability* **2018**, *10*, 3308. [CrossRef]
15. Lloyd, D. A phenological classification of terrestrial vegetation cover using shortwave vegetation index imagery. *Int. J. Remote Sens.* **1990**, *11*, 2269–2279. [CrossRef]
16. Dudley, K.L.; Dennison, P.E.; Roth, K.L.; Roberts, D.A.; Coates, A.R. A multi-temporal spectral library approach for mapping vegetation species across spatial and temporal phenological gradients. *Remote Sens. Environ.* **2015**, *167*, 121–134. [CrossRef]
17. Huete, A.R. Vegetation indices, remote sensing and forest monitoring. *Geogr. Compass* **2012**, *6*, 513–532. [CrossRef]
18. Vrieling, A.; Meroni, M.; Darvishzadeh, R.; Skidmore, A.K.; Wang, T.; Zurita-Milla, R.; Oosterbeek, K.; O'Connor, K.; Paganini, M. Vegetation phenology from Sentinel-2 and field cameras for a Dutch barrier island. *Remote Sens. Environ.* **2018**, *215*, 517–529. [CrossRef]
19. Rouse, J.W.; Haas, R.H.; Schell, J.A.; Deering, D.W. Monitoring vegetation systems in the Great Plains with ERTS. In Proceedings of the Third ERTS Symposium, NASA, Washington, DC, USA, 10–14 December 1973; NASA SP-351. pp. 309–317.
20. Tucker, C.J. Red and photographic infrared linear combinations of monitoring vegetation. *Remote Sens. Environ.* **1979**, *8*, 127–150. [CrossRef]
21. Campbell, J.B.; Wynne, R.H. *Introduction to Remote Sensing*, 5th ed.; the Guilford Press: New York, NY, USA, 2011; pp. 1–622.
22. Stoms, D.M.; Hargrove, W.W. Potential NDVI as a baseline for monitoring ecosystem functioning. *Int. J. Remote Sens.* **2000**, *21*, 401–407. [CrossRef]
23. Honeck, E.; Castello, R.; Chatenoux, B.; Richard, J.P.; Lehmann, A.; Giuliani, G. From a vegetation index to a sustainable development goal indicator: Forest trend monitoring using three decades of earth observations across Switzerland. *Int. J. Geo-Inf.* **2018**, *7*, 455. [CrossRef]
24. Pettorelli, N.; Vik, J.O.; Mysterud, A.; Gaillard, J.M.; Tucker, C.J.; Stenseth, N.C. Using the satellite-derived NDVI to assess ecological responses to environmental change. *Trends Ecol. Evol.* **2005**, *20*, 503–510. [CrossRef] [PubMed]
25. Tomaselli, V.; Adamo, M.; Veronico, G.; Sciandrello, S.; Tarantino, C.; Dimopoulos, P.; Medagli, P.; Nagendra, H.; Blonda, P. Definition and application of expert knowledge on vegetation pattern, phenology, and seasonality for habitat mapping, as exemplified in a Mediterranean coastal site. *Plant Biosyst.* **2017**, *151*, 887–899. [CrossRef]
26. Helman, D.; Lensky, I.M.; Tessler, N.; Osem, Y. A phenology-based method for monitoring woody and herbaceous vegetation in Mediterranean forests from NDVI time series. *Remote Sens.* **2015**, *7*, 12314–12335. [CrossRef]
27. Sun, C.; Fagherazzi, S.; Liu, Y. Classification mapping of salt marsh vegetation by flexible monthly NDVI time-series using Landsat imagery. *Estuar. Coast. Shelf. Sci.* **2018**, *213*, 61–80. [CrossRef]

28. Bajocco, S.; Ferrara, C.; Alivernini, A.; Bascietto, M.; Ricotta, C. Remotely-sensed phenology of Italian forest: Going beyond the species. *Int. J. Appl. Earth Obs.* **2019**, *74*, 314–321. [CrossRef]
29. van der Maarel, E. Some remarks on the functions of European coastal ecosystems. *Phytocoenologia* **2003**, *33*, 187–202. [CrossRef]
30. Jones, L.; Sowerby, A.; Williams, D.L.; Jones, R.E. Factors controlling soil development in sand dunes: Evidence from a coastal dune soil chronosequence. *Plant Soil* **2008**, *307*, 219–234. [CrossRef]
31. Santoro, R.; Jucker, T.; Carranza, M.L.; Acosta, A.T.R. Assessing the effects of *Carpobrotus* invasion on coastal dune soils. Does the nature of the invaded habitat matter? *Community Ecol.* **2011**, *12*, 234–240. [CrossRef]
32. Bazzichetto, M.; Malavasi, M.; Acosta, A.T.R.; Carranza, M.L. How does dune morphology shape coastal EC habitats occurrence? A remote sensing approach using airborne LiDAR on the Mediterranean coast. *Ecol. Indic.* **2016**, *71*, 618–626. [CrossRef]
33. Acosta, A.T.R.; Carranza, M.L.; Izzi, C.F. Are there habitats that contribute best to plant species diversity in coastal dunes? *Biodivers. Conserv.* **2009**, *18*, 1087–1098. [CrossRef]
34. Schlacher, T.; Dugan, J.; Schoeman, D.S.; Lastra, M.; Jones, A.; Scapini, F.; McLachlan, A.; Defeo, O. Sandy beaches at the brink. *Divers. Distrib.* **2007**, *13*, 556–560. [CrossRef]
35. Cori, B. Spatial dynamics of Mediterranean coastal regions. *J. Coast. Conserv.* **1999**, *5*, 105–112. [CrossRef]
36. Sperandii, M.G.; Bazzichetto, M.; Gatti, F.; Acosta, A.T.R. Back into the past: Resurveying random plots to track community changes in Italian coastal dunes. *Ecol. Indic.* **2019**, *96*, 572–578. [CrossRef]
37. Couch, C.; Leontidou, L.; Petschel-Held, G. *Urban Sprawl in Europe: Landscapes, Land-use Change and Policy*, 1st ed.; Wiley-Blackwell: Oxford, UK, 2007; pp. 1–294.
38. Carranza, M.L.; Acosta, A.T.R.; Stanisci, A.; Pirone, G.; Ciaschetti, G. Ecosystem classification for EU habitat distribution assessment in sandy coastal environments: An application in central Italy. *Environ. Monit. Assess.* **2008**, *140*, 99–107. [CrossRef] [PubMed]
39. Jolicoeur, S.; O'Carroll, S. Sandy barriers, climate change and long-term planning of strategic coastal infrastructures, Îles-de-la-Madeleine, Gulf of St. Lawrence (Québec, Canada). *Landsc. Urban Plan.* **2007**, *81*, 287–298. [CrossRef]
40. Malavasi, M.; Acosta, A.T.R.; Carranza, M.L.; Bartolozzi, L.; Basset, A.; Bassignana, M.; Campanaro, A.; Canullo, R.; Carruggio, F.; Cavallaro, V.; et al. Plant invasions in Italy. An integrative approach using LifeWatch infrastructure database. *Ecol. Indic.* **2018**, *91*, 182–188. [CrossRef]
41. Sun, C.; Liu, Y.; Zhao, S.; Li, H.; Sun, J. Saltmarshes response to human activities on prograding coast revealed by a dual-scale time-series strategy. *Estuar. Coast.* **2017**, *40*, 522–539. [CrossRef]
42. Marzialetti, F.; Bazzichetto, M.; Giulio, S.; Acosta, A.T.R.; Stanisci, A.; Malavasi, M.; Carranza, M.L. Modelling *Acacia saligna* invasion on the Adriatic coastal landscape: An integrative approach using LTER data. *Nat. Conserv.* **2019**, *34*, 127–144. [CrossRef]
43. Janssen, J.A.M.; Rodwell, J.S.; García Criado, M.; Gubbay, S.; Haynes, T.; Nieto, A.; Sanders, N.; Landucci, F.; Loidi, J.; Ssymank, A.; et al. *European red list of habitats. Part 2. Terrestrial and freshwater habitats*, 1st ed.; European Commision: Luxembourg, 2016; pp. 1–40.
44. Evans, D. The habitats of the European Union Habitats directive. *Biol. Environ. Proc. R. Ir. Acad.* **2006**, *106B*, 167–173. [CrossRef]
45. Acosta, A.T.R.; Blasi, C.; Carranza, M.L.; Ricotta, C.; Stanisci, A. Quantifying ecological mosaic connectivity and hemeroby with a new topoecological index. *Phytocoenologia* **2003**, *33*, 623–631. [CrossRef]
46. Carboni, M.; Carranza, M.L.; Acosta, A.T.R. Assessing conservation status on coastal dunes: A multiscale approach. *Landsc. Urban Plan.* **2009**, *91*, 17–25. [CrossRef]
47. Doody, J.P. *Sand dune conservation, management and restoration*, 1st ed.; Springer: Dordrecht, The Netherlands, 2013; pp. 201–235.
48. Drius, M.; Jones, L.; Marzialetti, F.; De Francesco, M.C.; Stanisci, A.; Carranza, M.L. Not just a sandy beach. The multi-service value of Mediterranean coastal dunes. *Sci. Total Environ.* **2019**, *668*, 1139–1155. [CrossRef] [PubMed]
49. Acosta, A.T.R.; Carranza, M.L.; Izzi, C.F. Combining land cover mapping of coastal dunes with vegetation analysis. *Appl. Veg. Sci.* **2005**, *8*, 133–138. [CrossRef]
50. Drake, S. Visual interpretation of vegetation classes from airborne videography: An evaluation of observer proficiency with minimal training. *Photogramm. Eng. Remote Sens.* **1996**, *62*, 969–978.

51. Green, E.P.; Clark, C.D.; Edwards, A.J. Image classification and habitat mapping. In *Remote sensing handbook for tropical coastal management*, 1st ed.; Green, E.P., Mumby, P.J., Edwards, A.J., Clark, C.D., Eds.; Coastal Management Sourcebooks: Paris, France, 2000; pp. 141–154.
52. Drius, M.; Malavasi, M.; Acosta, A.T.R.; Ricotta, C.; Carranza, M.L. Boundary-based analysis for the assessment of coastal dune landscape integrity over time. *Appl. Geogr.* **2013**, *45*, 41–48. [CrossRef]
53. Sperandii, M.G.; Prisco, I.; Acosta, A.T.R. Hard times for Italian coastal dunes: Insights from a diachronic analysis base on random plots. *Biodivers. Conserv.* **2018**, *27*, 633–646. [CrossRef]
54. Carranza, M.L.; Drius, M.; Malavasi, M.; Frate, L.; Stanisci, A.; Acosta, A.T.R. Assessing land take and its effects on dune carbon pools. An insight into the Mediterranean coastline. *Ecol. Indic.* **2018**, *85*, 951–955. [CrossRef]
55. Malavasi, M.; Carboni, M.; Cutini, M.; Carranza, M.L.; Acosta, A.T.R. Land use legacy, landscape fragmentation and propagule pressure promote plant invasion on coastal dunes. A patch based approach. *Landsc. Ecol.* **2014**, *29*, 1541–1550. [CrossRef]
56. Malavasi, M.; Bartak, V.; Carranza, M.L.; Simova, P.; Acosta, A.T.R. Landscape pattern and plant biodiversity in Mediterranean coastal dune ecosystems: Do habitat loss and fragmentation really matter? *J. Biogeogr.* **2018**, *45*, 1367–1377. [CrossRef]
57. Bertini, F.; Brand, O.; Carlier, S.; Del Bello, U.; Drusch, M.; Duca, R.; Fernandez, V.; Ferrario, C.; Ferreira, M.H.; Isola, C.; et al. *Sentinel-2 ESA's Optical High-Resolution Mission for GMES Operational Services*; ESA Communications: Noordwjik, The Netherlands, 2012; pp. 1–70.
58. Copernicus Open Access Hub. Available online: https://scihub.copernicus.eu/ (accessed on 24 April 2018).
59. Luis, J.; Debaecker, V.; Pflug, B.; Main-Knorn, M.; Bieniarz, J.; Mueller-Wilm, U.; Cadau, E.; Gascon, F. Sentinel-2 Sen2cor: L2A processor for users. In Proceedings of the Living Planet Symposium 2016, Prague, Czech Republic, 9–13 May 2016.
60. European Space Agency. Sen2cor–version 2.5.5. 2018. Available online: http://step.esa.int/main/third-party-plugins-2/sen2cor/ (accessed on 24 April 2018).
61. Breiman, L. Random Forests. *Mach. Learn.* **2001**, *45*, 5–32. [CrossRef]
62. Gislason, P.O.; Benediktsson, J.A.; Sveinsson, J.R. Random forests for land cover classification. *Pattern Recognit. Lett.* **2006**, *27*, 294–300. [CrossRef]
63. Rodriguez-Galiano, V.F.; Ghimire, B.; Rogan, J.; Chica-Olmo, M.; Rigol-Sanchez, J.P. An assessment of the effectiveness of a random forest classifier for land-cover classification. *ISPRS J. Photogramm. Remote Sens.* **2012**, *67*, 93–104. [CrossRef]
64. Pal, M. Random forest classifier for remote sensing classification. *Int. J. Remote Sens.* **2005**, *26*, 217–222. [CrossRef]
65. Hakkenberg, C.R.; Peet, R.K.; Urban, D.L.; Song, C. Modeling plant composition as community continua in a forest landscape with LiDAR and hyperspectral remote sensing. *Ecol. Appl.* **2018**, *28*, 177–190. [CrossRef] [PubMed]
66. European Space Agency. SNAP–version 6.0. 2018. Available online: https://step.esa.int/main/ (accessed on 24 April 2018).
67. Lasaponara, R. On the use of principal component analysis (PCA) for evaluating interannaual vegetation anomalies from SPOT/VEGETATION NDVI temporal series. *Ecol. Model.* **2006**, *194*, 429–434. [CrossRef]
68. Brown, S.C.; Lester, R.E.; Versace, V.L.; Fawcett, J.; Laurenson, L. Hydrologic landscape regionalization using deductive classification and random forests. *PLoS ONE* **2014**, *9*, 1–20. [CrossRef] [PubMed]
69. Breiman, L.; Friedman, J.H.; Olshen, R.A.; Stone, C.J. *Classification and regression trees*, 3rd ed.; Chapman & Hall: Boca Ration, FL, USA, 1984; pp. 1–358.
70. Cutler, D.R.; Edwards, T.C., Jr.; Beard, K.H.; Cutler, A.; Hess, K.T.; Gibson, J.; Lawler, J.J. Random forests for classification in ecology. *Ecology* **2007**, *88*, 2783–2792. [CrossRef]
71. Congalton, R.G.; Green, K. *Assessing the accuracy of remotely sensed data. Principles and practices, 2nd ed*; CRC Press Taylor & Francis Group: Boca Raton, FL, USA, 2009; pp. 1–183.
72. Tilahun, A.; Teferie, B. Accuracy assessment of land use land cover classification using Google Earth. *AJEP* **2015**, *4*, 193–198. [CrossRef]
73. Mohammed, N.Z.; Landry, R., Jr. Assessing horizontal positional accuracy of Google Earth imagery in the city of Montreal, Canada. *Geod. Cartogr.* **2016**, *43*, 55–65.

74. Qin, Y.; Xiao, X.; Dong, J.; Zhang, G.; Roy, P.S.; Joshi, P.K.; Gilani, H.; Murthy, M.S.R.; Jin, C.; Wang, J.; et al. Mapping forests in monsoon Asia with ALOS PALSAR 50-m mosaic images and MODIS imagery in 2010. *Sci. Rep.* **2016**, *6*, 1–10. [CrossRef]
75. Dorais, A.; Cardille, J. Strategies for incorporating high-resolution Google Earth databases to guide and validate classifications: Understanding deforestation in Borneo. *Remote Sens.* **2011**, *3*, 1157–1176. [CrossRef]
76. Malavasi, M.; Santoro, R.; Cutini, M.; Acosta, A.T.R.; Carranza, M.L. The impact of human pressure on landscape patterns and plant species richness in Mediterranaean coastal dunes. *Plant Biosyst.* **2016**, *150*, 73–82. [CrossRef]
77. Sperandii, M.G.; Prisco, I.; Stanisci, I.; Acosta, A.T.R. RanVegDunes – A random plot database of Italian coastal dunes. *Phytocoenologia* **2017**, *47*, 231–232. [CrossRef]
78. Kosmidou, V.; Petrou, Z.; Bunce, R.G.H.; Mücher, C.A.; Jongman, R.H.G.; Bogers, M.M.B.; Lucas, R.M.; Tomaselli, V.; Blonda, P.; Padoa-Schioppa, E.; et al. Harmonization of the Land Cover Classification System (LCCS) with the General Habitat Categories (GHC) classification systems. *Ecol. Indic.* **2014**, *36*, 290–300. [CrossRef]
79. Forey, E.; Lortie, C.J.; Michalet, R. Spatial patterns of association at local and regional scales in coastal sand dune communities. *J. Veg. Sci.* **2009**, *20*, 916–925. [CrossRef]
80. Belgiu, M.; Csillik, O. Sentinel-2 cropland mapping using pixel-based and object-based time-weighted dynamic time warping analysis. *Remote Sens. Environ.* **2018**, *204*, 509–523. [CrossRef]
81. Mannel, S.; Price, M. Comparing classification results of multi-seasonal TM against AVIRIS imagery – seasonality more important than number of bands. *PFG* **2012**, *5*, 603–612. [CrossRef]
82. Immitzer, M.; Vuolo, F.; Atzberger, C. First experience with Sentinel-2 data for crop and tree species classififcation in central europe. *Remote Sens.* **2016**, *8*, 166. [CrossRef]
83. Jones, G.; Bunting, P.; Hurford, C. Mapping Coastal Habitats in Wales. In *The Roles of Remote Sensing in Nature. A practical guide and case studies*, 1st ed.; Díaz-Delgado, R., Lucas, R., Hurford, C., Eds.; Springer: Gewerbestrasse, Switerland, 2017; pp. 91–120.
84. Nagendra, H.; Lucas, R.; Honrado, J.P.; Jongman, R.H.G.; Tarantino, C.; Adamo, M.; Mairota, P. Remote sensing for conservation monitoring: Assessing protected areas, habitat extent, habitat condition, and threats. *Ecol. Indic.* **2013**, *33*, 45–49. [CrossRef]
85. Herold, M.; Woodcock, C.E.; di Gregorio, A.; Mayaux, P.; Belward, A.S.; Latham, J.; Schmullius, C.C. A joint initiative for harmonization and validation of land cover datasets. *IEEE Trans. Geosci. Remote Sens.* **2006**, *44*, 1719–1727. [CrossRef]
86. Ghimire, B.; Rogan, J.; Miller, J. Contextual land-cover classification: Incorporating spatial dependence in land-cover classification models using random forests and the Getis statistic. *Remote Sens. Lett.* **2010**, *1*, 45–54. [CrossRef]
87. Xu, Y.; Dickson, B.G.; Hampton, H.M.; Sisk, T.D.; Palumbo, J.A.; Prather, J.W. Effects of mismatches of scale and location between predictor and response variables on forest structure mapping. *Photgramm. Eng. Remote Sens.* **2009**, *75*, 313–322. [CrossRef]
88. Thompson, S.D.; Gergerl, S. Conservation implications of mapping rare ecosystems using high spatial resolution imagery: Recommendations for heterogeneous and fragmented landscapes. *Landsc. Ecol.* **2008**, *23*, 1023–1037. [CrossRef]
89. Santoro, R.; Jucker, T.; Prisco, I.; Carboni, M.; Battisti, C.; Acosta, A.T.R. Effects of trampling limitation on coastal dune plant communities. *Environ. Manag.* **2012**, *49*, 534–542. [CrossRef] [PubMed]
90. Feagin, R.A.; Sherman, D.J.; Grant, W.E. Coastal erosion, global sea-level rise, and the loss of sand dune plant habitats. *Front. Ecol. Environ.* **2005**, *3*, 359–364. [CrossRef]
91. Battisti, C.; Poeta, G.; Pietrelli, L.; Acosta, A.T.R. An unexpected consequence of plastic litter clean-up on beaches: Too much sand might be removed. *Environ. Pract.* **2016**, *18*, 242–246. [CrossRef]
92. Adam, E.; Mutanga, O.; Odindi, J.; Abdel-Rahman, E.M. Land-use/cover classification in a heterogeneous coastal landscape using RapidEye imagery: Evaluating the performance of random forest and support vector machines classifiers. *Int. J. Remote Sens.* **2014**, *35*, 3440–3458. [CrossRef]
93. Wu, J. Effects of changing scale on landscape pattern analysis: Scaling relations. *Landsc. Ecol.* **2004**, *19*, 125–138. [CrossRef]

94. Machín, A.M.; Marcello, J.; Hernández-Cordero, A.I.; Abasolo, J.M.; Eugenio, F. Vegetation species mapping in a coastal-dune ecosystem using high resolution satellite imagery. *GISci. Remote Sens.* **2018**, *56*, 210–232. [CrossRef]
95. Lucas, N.S.; Shanmugam, S.; Barnsley, M. Sub-pixel habitat mapping of a costal dune ecosystem. *Appl. Geogr.* **2002**, *22*, 253–270. [CrossRef]

Article

Sub-Pixel Waterline Extraction: Characterising Accuracy and Sensitivity to Indices and Spectra

Robbi Bishop-Taylor *, Stephen Sagar, Leo Lymburner, Imam Alam and Joshua Sixsmith

Geoscience Australia, Cnr Jerrabomberra Ave and Hindmarsh Drive, Symonston ACT 2609, Australia; Stephen.Sagar@ga.gov.au (S.S.); Leo.Lymburner@ga.gov.au (L.L.); Imam.Alam@ga.gov.au (I.A.); Joshua.Sixsmith@ga.gov.au (J.S.)

* Correspondence: Robbi.Bishop-Taylor@ga.gov.au; Tel.: +61-2-6249-9839

Received: 28 October 2019; Accepted: 6 December 2019; Published: 12 December 2019

Abstract: Accurately mapping the boundary between land and water (the 'waterline') is critical for tracking change in vulnerable coastal zones, and managing increasingly threatened water resources. Previous studies have largely relied on mapping waterlines at the pixel scale, or employed computationally intensive sub-pixel waterline extraction methods that are impractical to implement at scale. There is a pressing need for operational methods for extracting information from freely available medium resolution satellite imagery at spatial scales relevant to coastal and environmental management. In this study, we present a comprehensive evaluation of a promising method for mapping waterlines at sub-pixel accuracy from satellite remote sensing data. By combining a synthetic landscape approach with high resolution WorldView-2 satellite imagery, it was possible to rapidly assess the performance of the method across multiple coastal environments with contrasting spectral characteristics (sandy beaches, artificial shorelines, rocky shorelines, wetland vegetation and tidal mudflats), and under a range of water indices (Normalised Difference Water Index, Modified Normalised Difference Water Index, and the Automated Water Extraction Index) and thresholding approaches (optimal, zero and automated Otsu's method). The sub-pixel extraction method shows a strong ability to reproduce both absolute waterline positions and relative shape at a resolution that far exceeds that of traditional whole-pixel methods, particularly in environments without extreme contrast between the water and land (e.g., accuracies of up to 1.50–3.28 m at 30 m Landsat resolution using optimal water index thresholds). We discuss key challenges and limitations associated with selecting appropriate water indices and thresholds for sub-pixel waterline extraction, and suggest future directions for improving the accuracy and reliability of extracted waterlines. The sub-pixel waterline extraction method has a low computational overhead and is made available as an open-source tool, making it suitable for operational continental-scale or full time-depth analyses aimed at accurately mapping and monitoring dynamic waterlines through time and space.

Keywords: waterline extraction; sub-pixel; surface water mapping; coastal monitoring; data cube; contour extraction; water extraction; water indices; thresholding; remote sensing

---

1. Introduction

Accurately mapping the boundary between land and water (the 'waterline') is critical for tracking coastal change and managing water resources in an era characterised by the increasing environmental impacts of development and anthropogenic climate change. By providing repeated observations of dynamic coastal zones and inland waters through time, satellite remote sensing provides a powerful and cost-effective alternative to traditional land-based surveys or waterline extraction methods based on sporadically acquired aerial data [1]. Historically, the use of satellite data for operational environmental monitoring has been limited by data access restrictions, inconsistent data structures and formats, a requirement for time and resource-intensive pre-processing before analysis could be conducted, and rapidly increasing data volumes which made desktop-based analyses of large extents or long time-series impractical [2]. In recent years, the development of high-performance earth observation 'data cubes' such as the Open Data Cube [3,4] and Google Earth Engine [5] have revolutionised the analysis of extremely large and complex remote sensing datasets [2]. By organising freely available medium resolution satellite imagery such as Landsat or Sentinel-2 into spectrally and geometrically calibrated stacks of analysis-ready data, these platforms support the development of new automated and cost effective workflows for consistent, rapid and repeatable operational monitoring of environmental change across time and space [4]. Data cube approaches leveraging multi-decadal remote sensing time series have so far been used successfully in a wide range of coastal and inland water applications, including monitoring 33 years of coastal change across global sandy beach shorelines [6], mapping the topography and extent through time of threatened intertidal ecosystems [7–10], and tracking changes in the distributions of inland waterbodies at both continental and global scale [11,12].

Although data cube platforms have driven a paradigm shift in the maximum spatio-temporal scale of analyses that are possible using remote sensing, the ability to accurately monitor environmental change at fine spatial scales remains critical for many potential coastal and inland water applications based on accurately modelling the position of the waterline. These include monitoring coastal erosion along narrow, steep coastlines [13], modelling dynamic intertidal topography and geomorphic change [14,15], and assessing small changes in water levels within narrow or steep sided inland rivers and waterbodies [16,17]. Most remote sensing studies to date have mapped the boundary between land and water in the landscape using binary classifiers based on either machine learning [11,12,18], or thresholding either a single satellite band [19–21] or remote sensing water index such as the normalised difference water index or NDWI [7,22–27]. These methods are inherently limited to the resolution of the satellite sensor, and are unable to resolve changes in waterline positions occurring at a scale of less than a whole pixel (e.g., 10 m for Sentinel-2 or 30 m for Landsat; [28]). Although high resolution satellite data from commercial providers such as Planet Labs and DigitalGlobe are increasingly available for these applications [29,30], these sources of data are typically prohibitively expensive to implement across regional to global extents, and lack either the temporal depth or systematic revisit frequency that are available for medium resolution satellite programs with coarser pixel resolutions (e.g., Landsat imagery available since 1972, [31]). Accordingly, there is a pressing need for the development of operational methods for extracting waterline information from freely available medium resolution satellite imagery at spatial scales relevant to coastal and environmental management.

By collapsing a complex heterogeneous landscape into a binary land-water classification, whole-pixel approaches to waterline mapping potentially discard valuable information about subtle differences between the spectral characteristics of neighbouring pixels. This information can be used to obtain more nuanced insights into the structure and location of dynamic waterlines through time. Recently, new techniques have been developed to extract the location of the waterline at sub-pixel resolution from medium resolution satellite data by utilising a pixel's neighbourhood to optimise the position of the waterline. Pardo-Pascual [32,33] and Almonacid-Caballer [34] developed a two-step process where waterlines were first identified at the whole-pixel scale based on infrared Landsat bands, then adjusted to sub-pixel level by fitting a fifth-degree polynomial function to a 7 x 7 pixel neighbourhood around each whole-pixel point. The Laplacian of this function was then calculated to identify the maximum gradient of change. The location of this inflection point between land and water could vary within the extents of a pixel itself, allowing waterlines to be extracted at a higher accuracy relative to the underlying 30 m imagery resolution (e.g., 4.69 to 5.47 m root mean square error or RMSE, [32]). While these techniques clearly demonstrate the potential sub-pixel waterline mapping accuracy that can be achieved using only medium resolution satellite data, they have thus far been implemented at relatively small scales (e.g., 800 m to 20 km of coastline, [32–34]) due to their computational intensity (e.g., processing times of up to 5 seconds per kilometer of coastline, [32]). This currently presents a challenge for operationally extracting large numbers of waterlines at continental scale across a full archive of satellite imagery spanning 30+ years.

Waterline mapping methods based on contour extraction provide a computationally efficient alternative to more intensive modelling approaches. Foody et al. [35] demonstrated that by fitting contours to a continuous 2D surface, the waterline could be positioned through the pixel rather than being constrained to pixel boundaries. This produces visually intuitive smooth waterlines that closely reproduce the true waterline position and shape without the distracting blocky pixel artefacts that affect whole-pixel approaches. While early applications of the technique applied waterline extraction to soft-classified layers where the exact proportion of water and land within each pixel was known (e.g., [35–38]), recent applications have instead used remote sensing water indices such as NDWI which can be calculated directly from open source remote sensing imagery [28,38,39]. These approaches have proven able to extract waterline positions with high levels of accuracy in sandy beach environments (e.g., up to 5.7 m against a 30 year validation dataset at Narrabeen Beach in eastern Australia, [38]). However, applying these extraction methods to remote sensing water indices implicitly assumes that index values respond consistently and linearly to underlying proportions of land and water within a pixel, an assumption that has been poorly tested to date [40,41]. In addition, previous studies have almost exclusively assessed the performance of these approaches within sandy beach environments [28,38,39,42] at the expense of other complex coastal or inland environments (e.g., tidal flats, wetland vegetation, artificial shorelines or rocky shorelines). A better understanding of how sub-pixel waterline extraction responds to variation in spectral properties of different environments and the selection and thresholding of water indices is critical for allowing these techniques to be scaled up from small scale applications to operational analyses at the regional, continental or global scales.

In this study, we provide a comprehensive evaluation of a contour-based method for extracting sub-pixel accuracy waterlines that can be rapidly applied to standard water indices with minimal additional effort, enabling integration with large-scale automated remote sensing workflows (Figure 1). We use a synthetic landscape approach to test the accuracy and precision of extracted sub-pixel waterlines across contrasting environments (sandy beaches, artificial shorelines, wetland vegetation, tidal mudflats and rocky shorelines), and assess how the performance of the method is affected by water index selection and commonly used thresholding approaches. We test the method using high resolution Worldview-2 imagery to verify our experimental findings in a complex, real-world coastal case study, and use our results to make best-practice recommendations for the future application of sub-pixel approaches for mapping waterlines consistently across time and space.

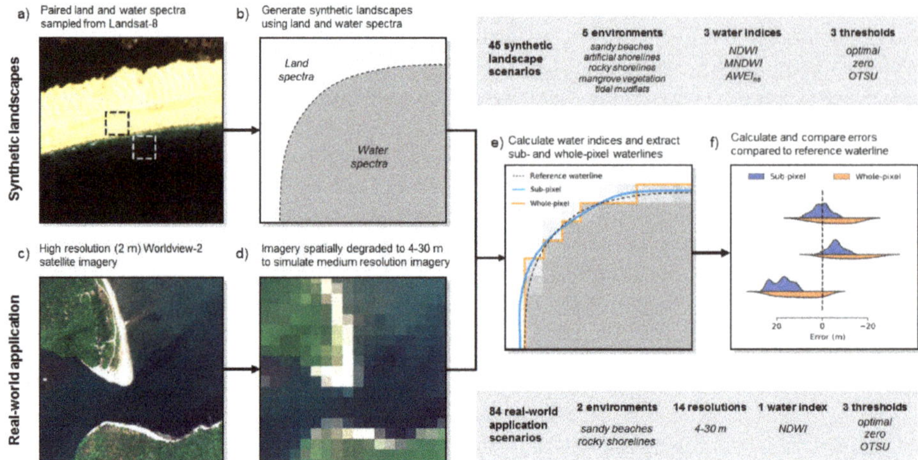

Figure 1. Overview of the experimental design followed in this study, including the synthetic landscape component based on spectra extracted from Landsat-8 (**a–b**), and the real-world application component based on high resolution WorldView-2 satellite imagery (**c–d**). Sub-pixel waterlines for both components of the study were extracted for multiple environment, water index, threshold and resolution scenarios (**e**), and statistically compared against traditional whole-pixel waterlines (**f**).

2. Materials and Methods

We conducted a synthetic landscape experiment to evaluate how accurately and precisely extracted waterlines could reproduce true waterline positions under controlled conditions, without the influence of environmental or sensor-related factors (e.g., tides, sensor noise, white water, cloud cover). We generate a simplified but environmentally plausible synthetic landscape using a hyperbolic tangent shape equation [43]. This equation was originally developed to empirically describe the shape of headland bay beach shorelines, and produces a range of steep to shallow curvatures which are suitable for evaluating sub-pixel waterline extraction method performance. The equation is defined as:

$$y = \pm a \ tanh^{\ m} \ (bx) \tag{1}$$

where y = distance across shore in metres; x = distance alongshore in metres; and a (units of length), b (units of 1/length), and m (dimensionless) are empirically-determined coefficients. We used values derived by Moreno and Kraus [43] from a study of 46 beaches in Spain and North America to generate the curve, and converted this function into a two-dimensional 1200 by 600 array where each cell represented a 1.0 x 1.0 m resolution 'land' or 'water' pixel.

2.1. *Influence of Environment on Waterline Extraction Performance*

2.1.1. Sample Spectra and Index Calculation

To assess subpixel waterline extraction performance across a range of common land-water boundaries, we extracted a set of typical spectra from freely available Landsat 8 OLI imagery available within the 30 year Digital Earth Australia archive produced by Geoscience Australia [3,4]. This data is available as atmospherically and terrain corrected 'analysis-ready' data processed to surface reflectance, allowing reliable spectra to be extracted with no additional processing or calibration required [44]. We focused on five commonly studied environments to explore the influence of contrasting spectral properties on waterline extraction performance: a) sandy beaches [1,6,28,33,34,38,42], b) artificial shorelines [32,45], c) rocky shorelines [10,29,45], d) wetland vegetation [46–48], and

e) tidal mudflats [7,10,19,49]. Paired samples of land and neighbouring water spectra were extracted from cloud free imagery along the Australian coastline by taking the mean value for each satellite band within each sample region (Figure 2). Imagery acquired at low tide was selected in the case of the tidal mudflat environment. As it would be impractical to analyse the full range of natural spectral variability for any of these individual environments, the sample spectra were intended to serve as a set of spectrally unique examples rather than an exhaustive and representative sample of natural conditions. For each environment, spectral values were assigned to the 'land' and 'water' pixels of the synthetic landscape to create a multispectral array with six spectral bands (red, green, blue, near-infrared, shortwave infrared 1, shortwave infrared 2) broadly shared by frequently used sources of open-source satellite imagery (e.g., Landsat TM, ETM+, OLI and Sentinel 2 MSI).

Using this array as a high-resolution (i.e., 1.0 m) reference dataset, we then simulated a typical medium resolution satellite dataset (e.g., Landsat) by spatially aggregating the higher-resolution data to a spatial resolution of 30 m using an 'average' aggregation rule. This reduced resolution 30 m dataset was then used as the basis for waterline extraction. Initially, we computed a series of common water extraction indices. The normalised difference water index (NDWI, [50]) is one of the most commonly used remote sensing water indices, having been applied to facilitate waterline delineation in a wide range of papers across inland [23,24,26] and coastal environments [7,10,25,51]. This index ranges from -1.0 (land) to 1.0 (water), and uses the ratio of visible green and near-infrared (NIR) reflectance to separate water pixels from land based on water's high reflectance of visible green light and low reflectance of NIR, and the high reflectance of NIR by dry soil and terrestrial vegetation:

$$NDWI = \frac{(Green - NIR)}{(Green + NIR)} \tag{2}$$

Recently, a modification of the NDWI which substitutes short-wave infrared (SWIR) in place of NIR has seen increasing popularity for waterline extraction [22]. This modified normalised difference water index (MNDWI) has been suggested as a more accurate alternative to NDWI, particularly in environments affected by white water from surf or high levels of turbidity [1,33]. However, this index can produce poor results in intertidal environments where wet substrate remaining after high tide is often mapped as open water [19,20].

$$MNDWI = \frac{(Green - SWIR\ 1)}{(Green + SWIR\ 1)} \tag{3}$$

The automated water extraction index (AWEI) has been proposed to address these limitations of both NDWI and MNDWI. Unlike NDWI and MNDWI, which are based on the relative ratio of green and near-infrared or shortwave radiation, AWEI is based on the sum of multiple spectral bands, and has been shown to produce higher water identification accuracy across a broad range of coastal and inland water environments by maximising spectral contrast between water and land classes [52]. While two formulations of the AWEI have been proposed to deal with extreme shadows in mountainous environments, we focused on the 'non-shadow' variant ($AWEI_{ns}$) in this study due to the typically low relief of the five environments being assessed:

$$AWEI_{ns} = 4 \times (Green - SWIR\ 1) - (0.25 \times NIR + 2.75 \times SWIR\ 2). \tag{4}$$

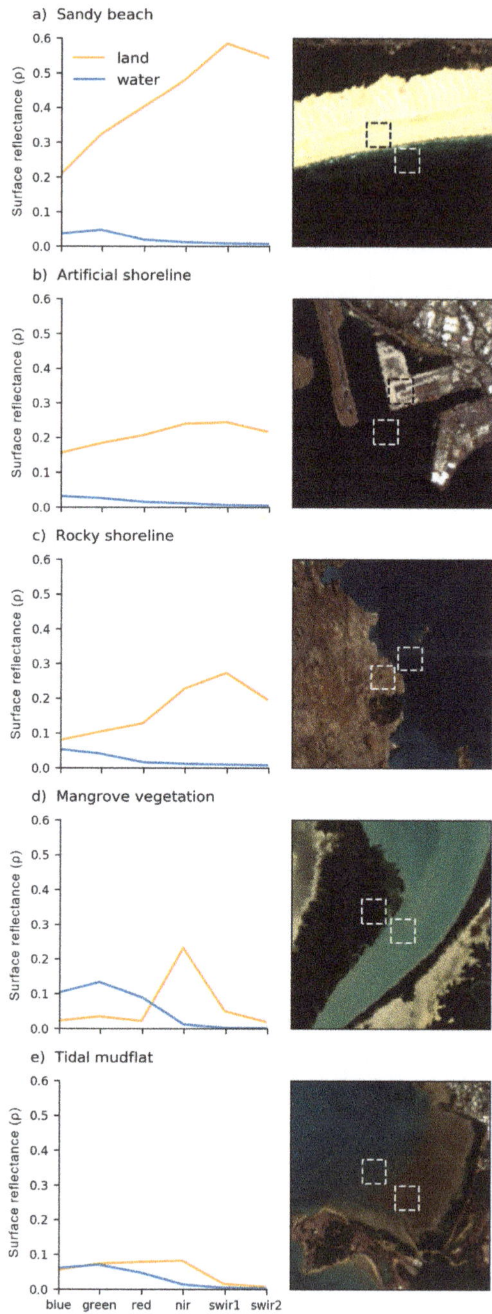

Figure 2. Paired samples of land and neighbouring water surface reflectance spectra from five contrasting environments along the Australian coastline (**a–e**). Spectra were extracted from cloud-free Landsat 8 OLI imagery from the Digital Earth Australia archive [3,4].

2.1.2. Waterline Extraction

We extracted sub-pixel resolution waterlines for each environment and water index using the marching squares with linear interpolation contour extraction algorithm implemented in the Python skimage.measure.find_contour function [53,54]. Contour-based methods aim to detect the boundary between two classes based on a two-dimensional input surface (for example, precisely mapping the position of a given elevation contour from a digital elevation model). When used to extract sub-pixel waterlines from a continuous water index surface, we make the assumption that low or medium resolution pixels along the land-water boundary represent mixed pixels, and that water index values (e.g., NDWI) for these pixels directly reflects the relative proportion of water and land within those pixels. It should therefore be possible to compare the relative water index values of two neighbouring pixels, and use this comparison to more precisely locate the boundary between land and water than simply drawing a line directly along their pixel boundaries. The marching squares with linear interpolation algorithm linearly interpolates between the water index values of neighbouring pixels to map out the precise location of the waterline according to a specified threshold value [53]. For example, where a land pixel has a water index value that is similar to its neighbouring water pixel (indicating the land pixel may contain a significant proportion of water), this will result in the output waterline position being shifted in an inland direction. Theoretically, this allows the location of the derived waterline to be identified at a precision that exceeds the resolution of the input satellite imagery [53].

As waterlines derived from remote sensing water indices can be highly sensitive to the threshold value selected to separate land from water, we extracted subpixel waterlines for three unique thresholds: an 'optimal' threshold, a 'zero' water index threshold, and an automatically derived threshold. The 'optimal' threshold represented the threshold that most closely replicated the high resolution reference shoreline, and was identified for each environment and water index scenario by iterating through a wide range of threshold scenarios (from–1.0 to 1.0 in 0.01 increments for NDWI and MNDWI, and from –2.0 to 2.0 in 0.01 increments for $AWEI_{ns}$ given its larger range of possible values), and selecting the water index that produced the lowest RMSE compared to the high resolution reference waterline. This threshold scenario was intended to compare the best-possible waterline extraction performance in an application where the ideal water index threshold was known. As knowing the ideal threshold ahead of an analysis is often not possible, the 'zero' water index threshold scenario used a constant zero threshold to extract waterlines, a procedure which is commonly used in operational waterline extraction applications conducted across large spatial extents [7,10]. Finally, procedures for automatically deriving threshold values based on the histogram distribution of water index values have been increasingly used to provide improved waterline extraction performance in complex heterogeneous environments where a single threshold (i.e., 0) would produce poor results [6,39,42]. We used Otsu thresholding implemented by the Python function skimage.filters.threshold_otsu to calculate an automatic threshold for each environment and water index [54,55]. This image processing approach identifies a threshold value that best separates a histogram of water index values into two distinct classes (e.g., water and land, [55]).

Subpixel shorelines extracted for each environment, water index and threshold scenario were converted to vector line features to facilitate comparisons against the 1.0 m high-resolution reference waterline. To provide a comparison dataset against which to compare the accuracy and precision of sub-pixel resolution waterlines, we additionally extracted matching waterlines for each scenario using a traditional whole-pixel thresholding approach (henceforth, 'whole-pixel' waterlines). This involved identifying all pixels with a water index equal to or above the given threshold as 'water' pixels and all values less than the threshold as 'land' pixels. This thresholded dataset was then polygonised, and a vector line feature extracted along the boundary of the two classes.

2.1.3. Statistical Comparison

We evaluated the ability of our two extracted waterlines to reproduce the true waterline position by computing distances (errors) between the 1.0 m high-resolution reference waterline and each set of sub-pixel and whole-pixel waterlines. At 1.0 m intervals along the reference waterline, we calculated Euclidean distance to the nearest point on both the sub-pixel and whole-pixel waterlines using the distance method from the shapely Python package [56]. Distances were assigned a direction (water- or land-ward offset from the reference waterline) based on whether the comparison point fell in a 'land' or 'water' pixel in the synthetic landscape: if the comparison point fell in a 'land' pixel, this distance was assigned a negative value to infer a land-ward offset. These distance errors were calculated for each environment, water index and threshold scenario, and compared using split violin plots to visualize error distributions between the sub-pixel and whole-pixel water extraction approaches. We additionally calculated two summary statistics to evaluate the extracted waterlines: root mean square error (RMSE) assessed the absolute accuracy of the derived waterlines compared to the reference waterline, while standard deviation allowed us to compare overall precision, or how closely the extracted waterlines replicated the overall shape of the reference shoreline even if the lines were consistently offset in a water or land-ward direction (e.g., a shoreline that closely followed the relative shape of the reference shoreline would have a low variability in error values).

2.2. *Real-World Application Using WorldView-2*

Although the synthetic data-based modelling framework above allowed us to rapidly test and isolate the performance of sub-pixel waterline extraction across a wide range of environmental and image processing scenarios, the theoretical approach represented a significant simplification of the complex heterogeneous environments typically found in satellite imagery. To complement the theoretical analysis, we assessed sub-pixel waterline extraction using high resolution satellite imagery. We obtained a largely cloud-free 2.0 m resolution WorldView-2 (WV-2) image (acquired 13th November 2010, 00:40 UTC) for a complex coastal region in north-western Queensland that contained extensive sandy beaches and rocky shoreline environments (Figure 3). This six band multispectral image (red, green, blue, yellow, near infrared 1, near infrared 2) was processed to surface reflectance using a bidirectional reflectance distribution function (BRDF) and topographically corrected MODTRAN-based atmospheric correction as detailed in Li et al. [44]. As WV-2 lacks the shortwave infrared bands required to compute MNDWI and $AWEI_{ns}$, we focused on NDWI (using the NIR 1 band) as the water index used to extract waterlines.

To provide a high-resolution reference shoreline, we extracted a vector shoreline at the imagery's native 2.0 m resolution based on a consistent zero NDWI threshold. While this consistent threshold did not account for the different coastal environments present in the image, using a consistent threshold allowed us to provide a single point-of-truth within the modelling framework that could be compared consistently against lower resolution shorelines, and eliminated any subjectivity associated with manually digitising shorelines. This reference waterline was stratified into sandy beaches and rocky shorelines by visually inspecting the underlying true colour WV-2 imagery, allowing us to compare results across different coastal environments.

To eliminate the confounding influence of tidal processes and coastal change between image acquisitions, we simulated lower resolution satellite imagery using spatially degraded versions of the WV-2 image itself rather than obtain co-incident imagery from other satellite platforms (e.g., Landsat or Sentinel 2). We progressively spatially degraded the 2.0 m resolution WV-2 image from 4 m to 30 m (in 2 m increments) using the average aggregation rule applied in the theoretical approach. Each aggregated image was used to compute NDWI, and extract sub-pixel and whole-pixel waterlines for the optimal, zero and automated Otsu index threshold scenarios above. Otsu thresholding was conducted separately on the two coastal environments (sandy beaches and rocky shorelines) by first buffering the reference shoreline by 50 m, and using this polygon as a mask to ensure that water and land classes were equally represented in the NDWI layer (an assumption of the Otsu process). Waterlines for each scenario were then statistically compared against the reference waterline by computing Euclidean distance errors at 1 m intervals along the extracted waterlines, and assessing RMSE and standard deviation error across the range of spatial resolutions for each of the two coastal environments.

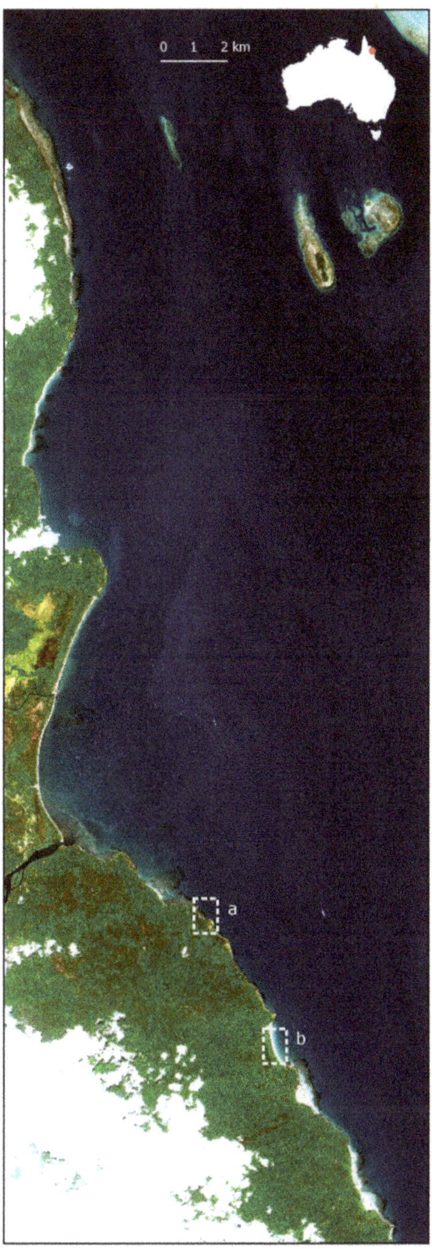

Figure 3. The 2 m resolution WorldView-2 image for the Cape Tribulation region in north-western Queensland used for the real-world application component of this study indicate regions featured in Figure 8.

All analyses in this study were conducted on Australia's National Computing Infrastructure's Virtual Desktop infrastructure (8 vCPUs, 32GB RAM). Code was written in the Python 3.6 programming language based on open-source functions from the OpenDataCube (https://www.opendatacube.org/), xarray [57], NumPy [58], SciPy [59], scikit-image [54] and pandas [60] libraries. All figures were generated using xarray, Matplotlib [61] and seaborn [62] data visualization libraries.

To allow the analyses presented here to be applied to future studies, functions for rapidly extracting waterlines from large multi-dimensional satellite data arrays are provided as an open-source toolset (Supplementary Materials).

3. Results and Discussion

3.1. Sub-Pixel Waterline Extraction Performance

We evaluated waterline extraction performance based on our synthetic landscape approach by comparing the distribution of distances in metres from the modelled to the reference shoreline (errors), using root mean squared error (RMSE) to evaluate whether errors were tightly distributed around 0 (Figure 4). Across 38 of our 45 synthetic landscape scenarios, RMSE for sub-pixel waterline extraction was lower compared to the whole-pixel approach, a clear result that indicated the method was able to more accurately represent the reference waterline position. This was particularly true for 'optimal' threshold scenarios, where sub-pixel waterlines were between 2.1 to 5.8 times more accurate than equivalent whole-pixel waterlines across all landscapes and water indexes (Table 1, Figure 4). Sub-pixel waterline accuracy was the highest using the $AWEI_{ns}$ index, where all waterlines extracted from the 30 m resolution data using 'optimal' thresholds had an RMSE of 1.50–1.51 m (Table 1, Figure 4). All seven scenarios where sub-pixel waterlines had higher RMSE than whole-pixel waterlines occurred using 'zero' thresholding (see Section 3.3, Effect of Water Index Threshold below).

Calculating RMSE allowed us to compare waterline extraction accuracy in absolute terms. However, many applications of waterline extraction such as coastal erosion and inland water level monitoring require an ability to monitor the relative location or shape of the shoreline consistently across time. To quantify how precisely waterline extraction approaches could reproduce the shape of the waterline, we calculated the standard deviation of errors between the modelled and reference waterline, assuming that a low variance in errors indicated that modelled shorelines were always positioned a consistent distance away from the true shoreline position. In all 45 scenarios, sub-pixel waterline errors exhibited significantly lower variance compared to whole-pixel waterlines, indicating a far better reproduction of waterline shape at a scale of less than one pixel (Table 1, Figure 4). Although this result was most pronounced across 'optimal' threshold scenarios where standard deviations were up to 5.5 times lower than whole-pixel waterlines, lower variances (i.e., 1.9 to 4.6 times lower) were also found across all 'zero' threshold scenarios (Table 1, Figure 4). This was the case even when sub-pixel waterlines had higher RMSE due to being clearly offset from the reference waterline position. Importantly, this result suggests that the resulting sub-pixel waterlines are likely to better reflect the relative waterline shape, even when offset land or sea-ward from the true waterline position due to an inappropriate choice of water index threshold. This result indicates that even a zero threshold approach based on a non-optimal water index threshold combined with sub-pixel waterline extraction can be suitable for applications where the priority is consistently monitoring trends in relative water line positions through time. As we illustrate in the following sections, selecting the appropriate thresholds and indexes both spatially and temporally to enable like-for-like comparison is a non-trivial exercise.

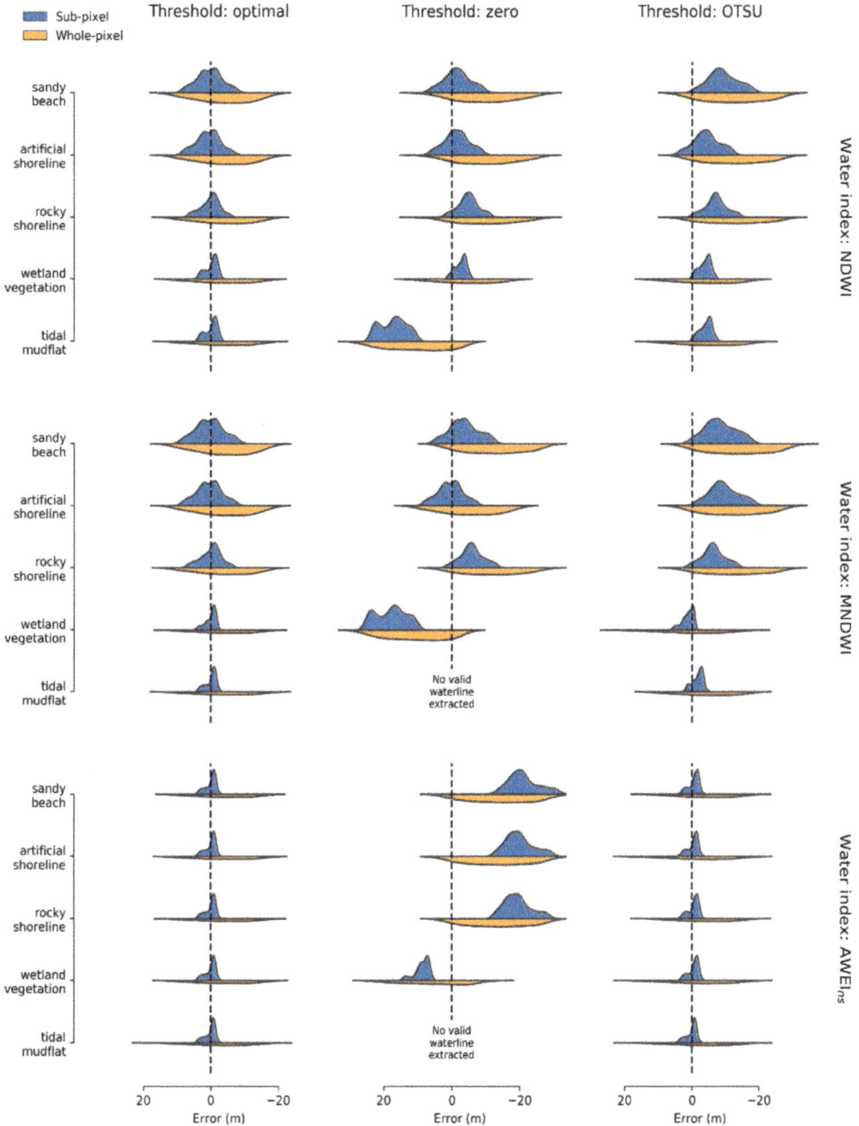

Figure 4. Error distributions compared to the reference waterline for sub-pixel (blue or dark) and whole-pixel (orange or light) waterline extraction approaches across five sample environments with contrasting spectral characteristics (y-axis). Errors were assessed for three water indices (vertical panels) and three thresholding approaches (horizontal panels). Errors distributed around 0 indicate a good reproduction of the absolute waterline position, while tightly distributed errors represent a good reproduction of the relative shape of the waterline (even if waterlines were offset by a constant distance from the reference waterline). Two tidal mudflat scenarios failed to extract waterlines due to the inability of a 0 MNDWI or $AWEI_{ns}$ water index to separate turbid water from wet intertidal mud.

Table 1. Root mean squared error (RMSE in metre units) and standard deviation (in metre units) of errors compared to the reference waterline for sub-pixel and whole-pixel waterline extraction approaches across five sample environments with contrasting spectral characteristics. Waterline extraction methods were compared using three water indices (normalised difference water index (NDWI), modified normalised difference water index (MNDWI), and automated water extraction index 'non-shadow' variant ($AWEI_{ns}$)) and thresholding strategies (optimal thresholding, zero thresholding and automated Otsu thresholding). Bold cells in the RMSE and standard deviation columns indicate the best modelling performance for each scenario (i.e., sub-pixel or whole-pixel waterlines).

			RMSE (m)		Standard Deviation (m)	
Spectra	**Water Index**	**Threshold**	**Sub-pixel**	**Whole-pixel**	**Sub-pixel**	**Whole-pixel**
Sandy beach	NDWI	Optimal	**3.74**	8.63	**3.73**	7.60
		Zero	**4.17**	12.85	**3.87**	8.35
		Otsu	**10.07**	14.93	**4.31**	8.39
	MDWI	Optimal	**3.99**	8.63	**3.97**	7.60
		Zero	**5.53**	14.93	**4.21**	8.39
		Otsu	**9.40**	16.24	**4.48**	8.80
	$AWEI_{ns}$	Optimal	**1.50**	8.63	**1.50**	7.59
		Zero	21.60	**16.94**	**4.37**	8.94
		Otsu	**1.58**	8.63	**1.49**	7.60
Artificial shoreline	NDWI	Optimal	**3.48**	8.63	**3.39**	7.60
		Zero	**3.89**	12.85	**3.61**	8.35
		Otsu	**5.63**	14.93	**3.77**	8.39
	MDWI	Optimal	**3.68**	8.63	**3.65**	7.60
		Zero	**3.68**	9.58	**3.65**	8.03
		Otsu	**10.32**	14.93	**4.28**	8.39
	$AWEI_{ns}$	Optimal	**1.51**	8.63	**1.51**	7.59
		Zero	20.39	**16.24**	4.06	8.80
		Otsu	**1.53**	8.70	**1.50**	8.27
Rocky shoreline	NDWI	Optimal	**2.73**	8.63	**2.72**	7.60
		Zero	**5.64**	12.85	**2.98**	8.35
		Otsu	**7.67**	14.93	**3.14**	8.39
	MDWI	Optimal	**2.93**	8.63	**2.93**	7.60
		Zero	**6.72**	14.43	**3.29**	8.22
		Otsu	**7.01**	14.93	**3.32**	8.39
	$AWEI_{ns}$	Optimal	**1.50**	8.63	**1.51**	7.59
		Zero	19.79	**16.24**	**3.90**	8.80
		Otsu	**1.57**	8.63	**1.49**	7.60
Wetland vegetation	NDWI	Optimal	**1.75**	8.63	**1.74**	7.59
		Zero	**3.06**	8.98	**1.67**	7.70
		Otsu	**4.04**	8.98	**1.71**	7.70
	MDWI	Optimal	**1.52**	8.63	**1.52**	7.59
		Zero	18.45	**13.41**	**4.32**	8.42
		Otsu	**2.08**	9.04	**1.62**	9.04
	$AWEI_{ns}$	Optimal	**1.51**	8.63	**1.51**	7.59
		Zero	**9.02**	10.05	**2.04**	8.85
		Otsu	**1.56**	8.70	**1.49**	8.27
Tidal mudflat	NDWI	Optimal	**1.75**	8.63	**1.74**	7.59
		Zero	17.68	**13.23**	**3.76**	8.60
		Otsu	**4.32**	9.58	**1.73**	8.03
	MDWI	Optimal	**1.55**	8.63	**1.55**	7.60
		Otsu	**2.22**	8.98	**1.53**	7.70
	$AWEI_{ns}$	Optimal	**1.51**	8.70	**1.51**	8.27
		Otsu	**1.51**	8.70	**1.51**	8.27

3.2. *Effect of Spectra and Water Index*

While sub-pixel waterlines more accurately and precisely reproduced reference waterline position and shape across the majority of our scenarios, we observed key differences by environment type which have important implications for remote sensing-based waterline extraction. 'Optimal' threshold results for our two normalised difference water indices (NDWI and MNDWI) showed a consistent trend of decreasing RMSE and standard deviation which coincided with a decrease in spectral contrast between water and land features (e.g., lower accuracy and precision along high contrast artificial shorelines and sandy beaches, and better accuracy and precision in low contrast tidal mudflats and wetland vegetation; Figure 4). For example, sub-pixel RMSE and standard deviation for NDWI waterlines improved from 3.74 and 3.73 m in sandy beach environments to 1.75 and 1.74 in tidal mudflats (Table 1). This effect could be qualitatively observed by comparing the shape of the shorelines generated for sandy beach and tidal mudflat environments (Figure 5). In tidal mudflats environments with low spectral contrast between wet mud and turbid water, sub-pixel waterlines closely follow the true shape of the reference shoreline along the entire length of the synthetic coastline. In comparison, sub-pixel waterlines in sandy beach environments displayed repeated undulating artefacts along the shallower curved region of the landscape (Figure 5). In these environments, high contrast between bright sand and dark water reduced the ability of the sub-pixel waterline extraction to resolve fine-scale curves or subtle undulations in waterline shape.

Although spectral contrast appeared to negatively affect both NDWI and MNDWI, the $AWEI_{ns}$ index did not reveal any relationship with spectral contrast. Accuracy and precision for $AWEI_{ns}$ remained relatively constant across all environments (Table 1, Figure 4), with no obvious step-like artefacts visible in the resulting waterlines (Figure 5). An explanation for this effect can be found in the non-linear response of certain water index values to the sub-pixel fractional coverage of water within each pixel. In environments with bright land features and dark water, a small increase in the proportion of land within a pixel can produce a large change in remote sensing indices like NDWI that are based on the ratio of visible and infrared radiation. To visualise this, we can calculate an NDWI value based on the weighted average spectra for each satellite band as the percentage of land within the pixel increases from 0 to 100% (Figure 6). In a low contrast environment such as a tidal flat, NDWI values decrease approximately linearly with increasing land. However, in a high contrast sandy beach environment, even a small increase in land coverage (e.g., from 0% to 10%) can result in a rapid decrease in NDWI. When used as the input for sub-pixel waterline extraction, this response drives the repeated undulating response of the waterline, with the resulting waterline position highly affected by a small change in the fractional coverage of bright sand. In contrast, the $AWEI_{ns}$ index is based on the sum of six spectral bands and linear coefficients, and responds linearly to increasing proportion of land or water within a pixel (Figure 6).

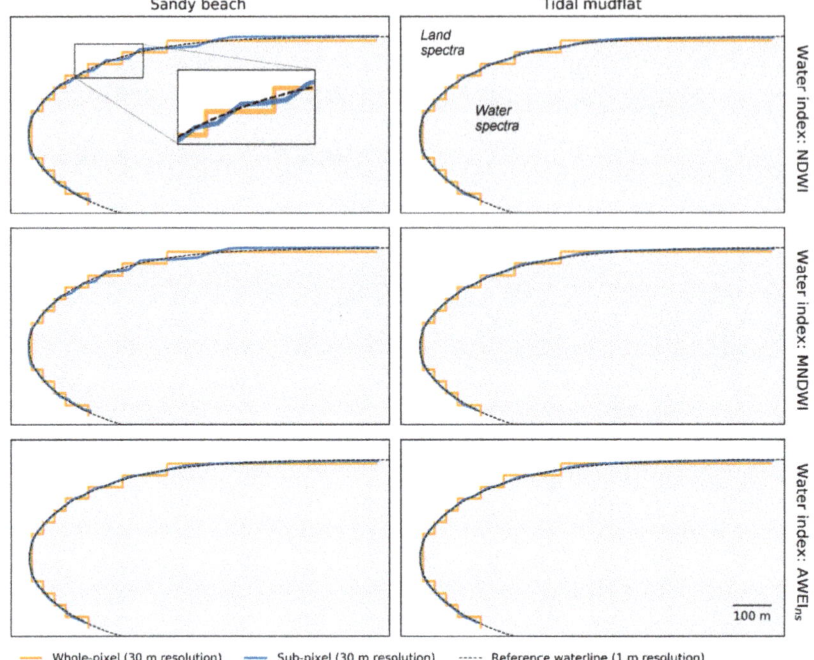

Figure 5. An example of sub-pixel (blue or dark) and whole-pixel (orange or light) waterlines extracted for three water indices (NDWI, MNDWI and $AWEI_{ns}$; vertical panels) across two contrasting environments: sandy beaches (left panels) and tidal mudflats (right panels). The reference waterline is shown by a dotted line. In high spectral contrast sandy beach environments, sub-pixel NDWI and MNDWI waterlines display repeated undulating artefacts, particularly along the shallower eastern curve of the synthetic beach landscape (see inset). Sub-pixel waterlines for the $AWEI_{ns}$ index and all water indices within low spectral contrast tidal flat environments more closely followed the reference shoreline shape with greatly reduced undulating artefacts.

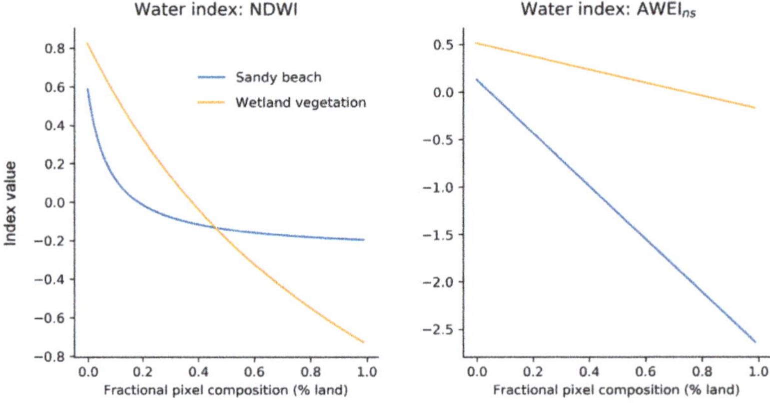

Figure 6. A comparison of the response of the NDWI (left) and $AWEI_{ns}$ (right) water indices to the fractional composition of land within a pixel. Normalised difference indices such as NDWI respond non-linearly to increasing proportion of land within a pixel, particularly within high spectral contrast sandy beach environments. In contrast, the $AWEI_{ns}$ index responds linearly regardless of spectral contrast between land and water.

3.3. *Effect of Water Index Threshold*

The 'optimal' water index threshold scenarios allowed us to assess the theoretical limits of sub-pixel waterline extraction performance under ideal conditions where the value of the best threshold value was known. However, threshold selection remains one of the key limitations of satellite-derived waterline extraction, particularly where reference data is not available to tailor the threshold to contrasting heterogeneous coastal or inland water environments [1,63]. This is frequently the case for operation analyses aimed at modelling waterlines across large spatial extents (e.g., at regional, continental or global scale). To evaluate how sub-pixel waterline extraction performs when the optimal value of the threshold is not known in advance, we also compared performance using two more realistic threshold selection strategies. The simplest 'zero' threshold scenario applied a consistent threshold of zero to all environments and metrics, in an approach which has been widely used in waterline extraction analyses [7,22,64]. Our results indicate a uniform zero threshold was unable to consistently reproduce the absolute reference waterline position across the five contrasting environments or three water indices we assessed. In an extreme example, a 'zero' threshold for tidal mudflat environments completely failed to differentiate between land and water for both the MNDWI and $AWEI_{ns}$ indices due to the spectral similarity of wet substrate with turbid water in the short-wave infrared satellite bands that were utilised by both indices ([19,20], Figure 4). Although poor modelling performance was observed using the 'zero' threshold for both sub- and whole-pixel waterlines, the narrower distribution of sub-pixel waterline errors meant that this method was more sensitive to poor threshold selection compared to whole-pixel waterlines whose larger error distribution typically overlapped with the true location of the waterline (Table 1, Figure 4).

To address the limitations of a consistent zero threshold, we assessed model performance using water index thresholds optimised using Otsu's method. Otsu has been proposed as an automated method for deriving locally-tuned threshold values for operational waterline analyses, however, the performance of the approach on sub-pixel waterline accuracy has not previously been assessed. Our results indicate that 'Otsu' thresholding significantly improved the consistency of waterlines extracted using the sub-pixel approach, leading to error distributions which were generally narrower than corresponding zero threshold distributions, and more comparable across different environment types (Figure 4). This was particularly the case for wetland vegetation and tidal mudflat environments, which saw improvements in overall RMSE of up to 16.4 m between 'zero' and 'Otsu' scenarios (e.g., wetland vegetation MNDWI; Table 1, Figure 4). Although Otsu typically did not reproduce the absolute or relative accuracy of 'optimal' thresholds, the improvement in results compared to zero threshold scenarios results indicates that automated threshold selection can complement sub-pixel waterline extraction approaches. This is likely to be particularly significant in large-scale analyses covering diverse environment types, where the influence of inappropriate threshold selection may overwhelm increases in accuracy gained from implementing sub-pixel waterline extraction methods.

3.4. *Real-World Application*

By progressively spatially degrading high resolution (2 m) WorldView-2 remote sensing imagery, it was possible to simulate lower resolution satellite imagery and assess the impact of spatial resolution on waterline extraction performance in a more complex real-world scenario. At 30 m resolution (equivalent to Landsat TM, ETM+ and OLI bands), sub-pixel waterlines were 1.9 to 2.3 times more accurate than whole-pixel waterlines both in absolute terms (e.g., RMSE of 3.28–4.52 m compared to 7.46–8.73 m for whole-pixel waterlines), and in their ability to reproduce the relative shape of the reference coastline (e.g., standard deviation of 3.27–4.42 m compared to 7.46–8.70 m; Table 2, Figure 7). The ability of the sub-pixel approach to accurately reproduce the true waterline even at this relatively coarse resolution can be qualitatively observed in Figure 8a, where sub-pixel waterlines (blue) closely follow small undulations in the path of the dashed 2 m high resolution waterline which are considerably smaller in scale than the pixel resolution captured by the equivalent whole-pixel waterlines.

Table 2. Root mean squared error (RMSE in metre units) and standard deviation (in metre units) of sub-pixel and whole-pixel waterline extraction errors compared to the 2 m resolution WorldView-2 reference waterline. Results are shown for optimal NDWI thresholds for two coastal environments (rocky shorelines and sandy beaches) and four key resolutions (4 m as relevant to Planet Labs 3 m PlanetScope and 5 m RapidEye imagery, 10 m and 20 m as equivalent to Sentinel 2 MSI, and 30 m as equivalent to Landsat TM, ETM+, and OLI). Bold cells for the RMSE and standard deviation columns indicate the best modelling performance for each scenario (i.e., sub-pixel or whole-pixel waterlines). For 'zero' and 'Otsu' threshold results, refer to Appendix A.

Environment	Spatial Resolution (m)	RMSE (m)		Standard Deviation (m)	
		Sub-pixel	Whole-pixel	Sub-pixel	Whole-pixel
Rocky shoreline	4	**0.72**	1.17	**0.72**	1.15
Rocky shoreline	10	**1.43**	2.65	**1.43**	2.65
Rocky shoreline	20	**2.41**	5.07	**2.41**	5.07
Rocky shoreline	30	**3.28**	7.46	**3.27**	7.46
Sandy beach	4	**0.63**	1.19	**0.63**	1.17
Sandy beach	10	**1.47**	2.80	**1.47**	2.79
Sandy beach	20	**2.84**	5.68	**2.84**	5.67
Sandy beach	30	**4.52**	8.73	**4.42**	8.70

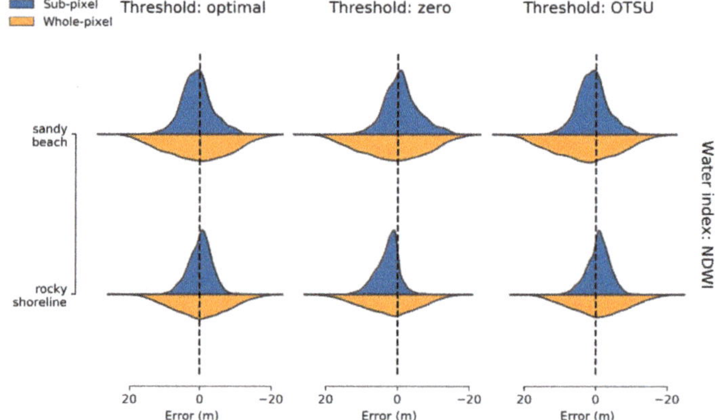

Figure 7. Error distributions compared to the reference waterline for 30 m resolution sub-pixel (blue or dark) and whole-pixel (orange or light) waterline extraction approaches across sandy beach and rocky shoreline environments in the Worldview-2 image. Errors were assessed for the NDWI water index and three thresholding approaches (horizontal panels). Errors distributed around 0 indicate a good reproduction of the absolute waterline position, while tightly distributed errors represent a good reproduction of the relative shape of the waterline (even if waterlines were offset by a constant distance from the reference waterline).

Sub-pixel accuracy increased approximately linearly with increasing resolution, with subpixel shorelines achieving accuracies of up to 2.41 m at 20 m Sentinel 2 resolution equivalent, and 1.43 m at 10 m Sentinel 2 resolution (Figure 9). Waterlines based on the original 2 m Worldview-2 dataset (8515 by 21352 pixels) were extracted in 3.98 seconds (± 115 ms for ten repetitions) for 143 km of coastline on Australia's National Computing Infrastructure's Virtual Desktop infrastructure (8 vCPUs, 32GB RAM). Waterline extractions at 30 m Landsat resolution took 1.93 seconds (± 116 ms) at a rate of 0.013 seconds per kilometer of coastline. This is equivalent to 13 minutes to process the entire ~60,000 km coastline of Australia (the eighth longest coastline in the world), which is considered suitable for large-scale operational analysis. Our results demonstrate that sub-pixel waterlines can be reliably and efficiently extracted across complex heterogeneous coastal environments, and achieve accuracies that consistently exceed the accuracy of traditional waterline extraction approaches regardless of the spatial scale of analysis.

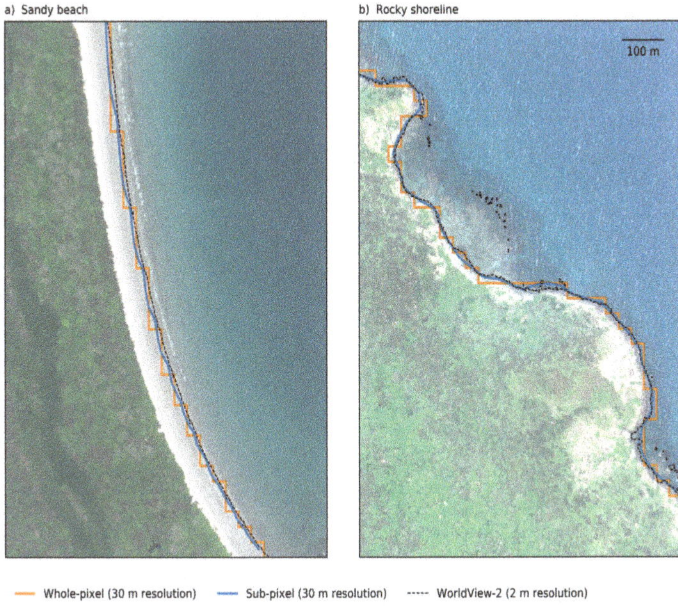

Figure 8. An example of 30 m resolution sub-pixel (blue or dark) and whole-pixel (orange or light) waterlines extracted using 'optimal' threshold selection across two contrasting environments: sandy beaches (left panel) and rocky shorelines (right panel). The reference 2 m resolution WorldView-2 waterline is shown by a black dotted line overlaid over the WorldView-2 image. In high spectral contrast sandy beach environments (**a**), sub-pixel waterlines display subtle undulating artefacts, while in lower contrast rocky shoreline environments (**b**) sub-pixel waterlines closely follow small undulations in the reference waterline position which are considerably smaller in scale than the 30 m pixel resolution.

Comparing two unique coastal environments (sandy beaches and rocky shorelines) allowed us to assess whether we could reproduce the effects of spectral contrast on waterline accuracy that we observed in the simulated landscape analysis. Although the effect was more subtle than in the simulated analysis, Figure 9 highlights that RMSE accuracies for the sandy beach waterlines were routinely lower than rocky shorelines, particularly at coarser resolutions (e.g., 30 m pixels). A qualitative inspection of the resulting shorelines for these two environments revealed that high contrast white sandy beaches were associated with similar repeated undulating artefacts (Figure 8a). While previous studies have shown a relationship between spectral contrast and absolute waterline offsets (e.g., brighter land spectra leading to a land-ward bias in waterline errors, [32]), our results indicate that caution should be used when applying waterline extraction approaches to water indices such as NDWI or MNDWI which react non-linearly to underlying distributions of water and land. This is particularly significant given the majority of waterline extraction analyses to date have focused on analysing sandy beach environments which are likely to be most vulnerable to contrast-driven modelling artefacts. While the limited selection of available WorldView-2 bands prevented us from assessing model performance using an alternative index such as $AWEI_{ns}$, we recommend that future studies evaluate model performance using water indices that respond linearly to the fractional coverage of land and water. Alternatively, sub-pixel waterline extraction could be applied directly to more physically meaningful water index surfaces, such as per-pixel estimates of fractional water coverage derived from spectral unmixing analysis [65–67]. This will ensure that sub-pixel resolution waterline positions can be compared reliably across environments with unique and contrasting spectral characteristics.

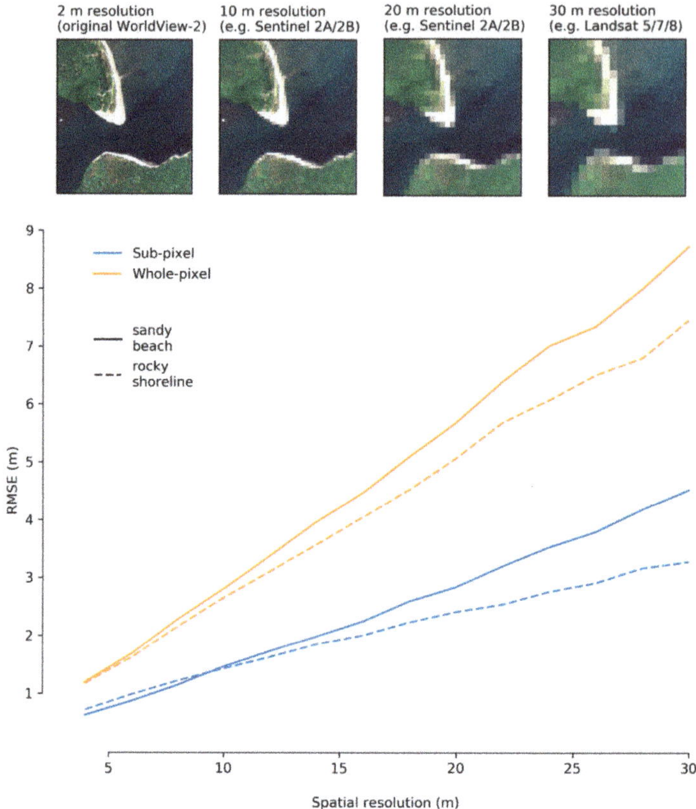

Figure 9. Waterline extraction accuracy (RMSE in metres) of sub-pixel (blue) and whole-pixel (orange) waterlines compared for sandy beach and rocky shoreline environments. The effect of resolution on accuracy was assessed by spatially degrading the 2.0 m resolution WV-2 image from 4 m to 30 m resolution using an 'average' aggregation rule (i.e., the graph's x-axis; see insets for a visual comparison of the aggregated imagery at resolutions corresponding to several common sources of remote sensing data).

4. Conclusions

In this study, we have evaluated a sub-pixel waterline extraction method that shows a strong ability to reproduce both absolute waterline positions and relative shape at a resolution that far exceeds that of traditional whole-pixel thresholding methods, particularly in environments without extreme contrast between water and land. Given the low computational overhead and open source availability of the technique, the approach is likely to be suitable for large-scale waterline extraction analyses as an easy-to-implement substitute for less precise whole-pixel approaches.

Although sub-pixel waterline extraction provides key advantages over whole-pixel approaches, the extra sensitivity of the method makes it vulnerable to two key challenges facing remote sensing-based waterline extraction in general: the selection and thresholding of a reliable water index [1]. Our results indicate that although zero threshold approaches were unable to consistently reproduce absolute waterline positions, sub-pixel waterlines based on a zero threshold were better able to model relative waterline shape compared to whole-pixel waterlines. Simple zero threshold approaches to mapping surface water are therefore likely to be enhanced through integration with sub-pixel waterline delineation, particularly when the application requires the consistent monitoring of relative water line positions through time.

Where absolute accuracy is the priority, our results show that automated index thresholding approaches (e.g., Otsu) can be combined with sub-pixel methods to extract waterlines that approach the accuracy of optimal thresholds. However, while automated thresholding methods may work successfully for small study areas and discrete time-steps, producing seamless and consistent waterline datasets through time and across complex heterogeneous environments remains a challenge. There is a critical need for the development of new continuous along-shore methods for determining thresholds [68,69] and approaches for ensuring the temporal consistency of automated thresholds to support optimising and operationalising the mapping of dynamic waterlines across space and time. Finally, our results also highlight the importance of selecting remote sensing indices which respond linearly to changes in underlying proportions of water and land, and suggest that future work should focus on integrating the technique with more physically meaningful water indices which directly quantify the proportion of water within each pixel.

To support future applications of the approach, we have provided the tools used to extract sub-pixel waterlines as open-source code (Supplementary Materials). These functions can be applied directly to large multi-dimensional satellite datasets, making them suitable for applying to large scale remote sensing workflows such as those enabled by Open Data Cube [3,4] or Google Earth Engine [5]. By enabling the automatic extraction of sub-pixel waterlines consistently across space and time, it is anticipated that these tools may supplement existing open source coastal monitoring software [70] and assist in scaling up current local-scale waterline analyses to provide insights into drivers of environmental change operating at regional, continental or global scales. Future work will focus on applying sub-pixel waterline extraction methods to provide accurate estimates of waterbody area and volume for inland reservoirs and lakes to monitor the impact of drought and water extraction within semi-arid inland Australia, and operationalising the approach to monitor fine-scale erosion and coastal change across complex coastal environments along the entire Australian coastline.

Supplementary Materials: Open-source code for extracting sub-pixel waterlines from multidimensional satellite data arrays are available online: https://github.com/GeoscienceAustralia/dea-notebooks/tree/subpixel_waterlines.

Author Contributions: Conceptualization, R.B.-T., S.S. and L.L.; Formal analysis, R.B.-T.; Investigation, R.B.-T.; Methodology, R.B.-T., S.S. and L.L.; Resources, I.A. and J.S.; Software, R.B.-T., I.A. and J.S.; Validation, R.B.-T.; Visualization, R.B.-T.; Writing – original draft, R.B.-T.; Writing – review & editing, S.S. and L.L.

Funding: This research received no external funding

Acknowledgments: The authors would like to thank T. Dhu, R. Nanson, S. Chua and the three anonymous reviewers for their valuable and constructive comments and suggestions. This research was undertaken with the assistance of resources from the National Computational Infrastructure (NCI) High Performance Data (HPD) platform, which is supported by the Australian Government. This paper is published with the permission of the Chief Executive Officer, Geoscience Australia.

Conflicts of Interest: The authors declare no conflict of interest.

Appendix A

Table A1. Root mean squared error (RMSE in metre units) and standard deviation (in metre units) of sub-pixel and whole-pixel waterline extraction errors compared to the 2 m resolution WorldView-2 reference waterline. Results are shown for two coastal environments (rocky shorelines and sandy beaches), four key resolutions (4 m as relevant to Planet Labs 3 m PlanetScope and 5 m RapidEye imagery, 10 m and 20 m as equivalent to Sentinel 2 MSI, and 30 m as equivalent to Landsat TM, ETM+ and OLI), and three thresholding strategies (optimal thresholding, zero thresholding and automated Otsu thresholding). Bold cells for the RMSE and standard deviation columns indicate the best modelling performance for each scenario (i.e., sub-pixel or whole-pixel waterlines).

			RMSE (m)		Standard Deviation (m)	
Environment	**Spatial Resolution (m)**	**Threshold**	**Sub-pixel**	**Whole-pixel**	**Sub-pixel**	**Whole-pixel**
	4	Optimal	**0.72**	1.17	**0.72**	1.15
Rocky shoreline		Zero	**0.72**	1.19	**0.72**	1.19
		Otsu	**2.10**	2.50	**1.72**	2.24
	10	Optimal	**1.43**	2.65	**1.43**	2.65
		Zero	**1.43**	2.66	**1.43**	2.65
		Otsu	**2.44**	3.36	**1.66**	3.01
	20	Optimal	**2.41**	5.07	**2.41**	5.07
Rocky shoreline		Zero	**2.52**	5.14	**2.35**	5.06
		Otsu	**3.20**	5.58	**2.49**	5.27
	30	Optimal	**3.28**	7.46	**3.27**	7.46
		Zero	**4.23**	7.68	**3.16**	7.44
		Otsu	**3.36**	7.54	**3.26**	7.48
	4	Optimal	**0.63**	1.19	**0.63**	1.17
Sandy beach		Zero	**0.63**	1.25	**0.63**	1.25
		Otsu	2.17	**1.81**	**1.80**	1.43
	10	Optimal	**1.47**	2.80	**1.47**	2.79
		Zero	**1.56**	2.80	**1.51**	2.79
		Otsu	**2.38**	3.23	**1.92**	2.82
	20	Optimal	**2.84**	5.68	**2.84**	5.67
Sandy beach		Zero	**3.19**	5.69	**3.05**	5.65
		Otsu	**2.96**	5.96	**2.84**	5.67
	30	Optimal	**4.52**	8.73	**4.42**	8.70
		Zero	**5.00**	8.79	**4.89**	8.70
		Otsu	**4.52**	9.09	**4.42**	8.66

References

1. Kelly, J.T.; Gontz, A.M. Using GPS-surveyed intertidal zones to determine the validity of shorelines automatically mapped by Landsat water indices. *Int. J. Appl. Earth Obs. Geoinf.* **2018**, *65*, 92–104. [CrossRef]
2. Kopp, S.; Becker, P.; Doshi, A.; Wright, D.J.; Zhang, K.; Xu, H. Achieving the Full Vision of Earth Observation Data Cubes. *Data* **2019**, *4*, 94. [CrossRef]
3. Dhu, T.; Dunn, B.; Lewis, B.; Lymburner, L.; Mueller, N.; Telfer, E.; Lewis, A.; McIntyre, A.; Minchin, S.; Phillips, C. Digital earth Australia—Unlocking new value from earth observation data. *Big Earth Data* **2017**, *1*, 64–74. [CrossRef]
4. Lewis, A.; Oliver, S.; Lymburner, L.; Evans, B.; Wyborn, L.; Mueller, N.; Raevksi, G.; Hooke, J.; Woodcock, R.; Sixsmith, J.; et al. The Australian Geoscience Data Cube—Foundations and lessons learned. *Remote Sens. Environ.* **2017**, *202*, 276–292. [CrossRef]
5. Gorelick, N.; Hancher, M.; Dixon, M.; Ilyushchenko, S.; Thau, D.; Moore, R. Google Earth Engine: Planetary-scale geospatial analysis for everyone. *Remote Sens. Environ.* **2017**, *202*, 18–27. [CrossRef]
6. Luijendijk, A.; Hagenaars, G.; Ranasinghe, R.; Baart, F.; Donchyts, G.; Aarninkhof, S. The state of the world's beaches. *Sci. Rep.* **2018**, *8*, 6641. [CrossRef]
7. Sagar, S.; Roberts, D.; Bala, B.; Lymburner, L. Extracting the intertidal extent and topography of the Australian coastline from a 28 year time series of Landsat observations. *Remote Sens. Environ.* **2017**, *195*, 153–169. [CrossRef]
8. Murray, N.J.; Phinn, S.R.; DeWitt, M.; Ferrari, R.; Johnston, R.; Lyons, M.B.; Clinton, N.; Thau, D.; Fuller, R.A. The global distribution and trajectory of tidal flats. *Nature* **2019**, *565*, 222–225. [CrossRef]
9. Lymburner, L.; Bunting, P.; Lucas, R.; Scarth, P.; Alam, I.; Phillips, C.; Ticehurst, C.; Held, A. Mapping the multi-decadal mangrove dynamics of the Australian coastline. *Remote Sens. Environ.* **2019**. [CrossRef]

10. Bishop-Taylor, R.; Sagar, S.; Lymburner, L.; Beaman, R.J. Between the tides: Modelling the elevation of Australia's exposed intertidal zone at continental scale. *Estuar. Coast. Shelf Sci.* **2019**, *23*, 115–128. [CrossRef]
11. Mueller, N.; Lewis, A.; Roberts, D.; Ring, S.; Melrose, R.; Sixsmith, J.; Lymburner, L.; McIntyre, A.; Tan, P.; Curnow, S.; et al. Water observations from space: Mapping surface water from 25 years of Landsat imagery across Australia. *Remote Sens. Environ.* **2016**, *174*, 341–352. [CrossRef]
12. Pekel, J.-F.; Cottam, A.; Gorelick, N.; Belward, A.S. High-resolution mapping of global surface water and its long-term changes. *Nature* **2016**, *540*, 418. [CrossRef]
13. Vousdoukas, M.I.; Almeida, L.P.M.; Ferreira, Ó. Beach erosion and recovery during consecutive storms at a steep-sloping, meso-tidal beach. *Earth Surf. Process. Landf.* **2012**, *37*, 583–593. [CrossRef]
14. Long, N.; Millescamps, B.; Guillot, B.; Pouget, F.; Bertin, X. Monitoring the Topography of a Dynamic Tidal Inlet Using UAV Imagery. *Remote Sens.* **2016**, *8*, 387. [CrossRef]
15. Almeida, L.P.; Almar, R.; Bergsma, E.W.J.; Berthier, E.; Baptista, P.; Garel, E.; Dada, O.A.; Alves, B. Deriving High Spatial-Resolution Coastal Topography From Sub-meter Satellite Stereo Imagery. *Remote Sens.* **2019**, *11*, 590. [CrossRef]
16. Ogilvie, A.; Belaud, G.; Massuel, S.; Mulligan, M.; Le Goulven, P.; Calvez, R. Surface water monitoring in small water bodies: potential and limits of multi-sensor Landsat time series. *Hydrol. Earth Syst. Sci.* **2018**, *22*, 4349. [CrossRef]
17. Pereira, B.; Medeiros, P.; Francke, T.; Ramalho, G.; Foerster, S.; Araújo, J.C.D. Assessment of the geometry and volumes of small surface water reservoirs by remote sensing in a semi-arid region with high reservoir density. *Hydrol. Sci. J.* **2019**, *64*, 66–79. [CrossRef]
18. Tulbure, M.G.; Broich, M.; Stehman, S.V.; Kommareddy, A. Surface water extent dynamics from three decades of seasonally continuous Landsat time series at subcontinental scale in a semi-arid region. *Remote Sens. Environ.* **2016**, *178*, 142–157. [CrossRef]
19. Ryu, J.-H.; Won, J.-S.; Min, K.D. Waterline extraction from Landsat TM data in a tidal flat: A case study in Gomso Bay, Korea. *Remote Sens. Environ.* **2002**, *83*, 442–456. [CrossRef]
20. Ryu, J.-H.; Kim, C.-H.; Lee, Y.-K.; Won, J.-S.; Chun, S.-S.; Lee, S. Detecting the intertidal morphologic change using satellite data. *Estuar. Coast. Shelf Sci.* **2008**, *78*, 623–632. [CrossRef]
21. Park, W.; Lee, Y.-K.; Shin, J.-S.; Won, J.-S. A tidal correction model for near-infrared (NIR) reflectance over tidal flats. *Remote Sens. Lett.* **2013**, *4*, 833–842. [CrossRef]
22. Xu, H. Modification of normalised difference water index (NDWI) to enhance open water features in remotely sensed imagery. *Int. J. Remote Sens.* **2006**, *27*, 3025–3033. [CrossRef]
23. Bai, J.; Chen, X.; Li, J.; Yang, L.; Fang, H. Changes in the area of inland lakes in arid regions of central Asia during the past 30 years. *Environ. Monit. Assess.* **2011**, *178*, 247–256. [CrossRef] [PubMed]
24. Muala, E.; Mohamed, Y.; Duan, Z.; van der Zaag, P. Estimation of reservoir discharges from Lake Nasser and Roseires Reservoir in the Nile Basin using satellite altimetry and imagery data. *Remote Sens.* **2014**, *6*, 7522–7545. [CrossRef]
25. Rokni, K.; Ahmad, A.; Selamat, A.; Hazini, S. Water Feature Extraction and Change Detection Using Multitemporal Landsat Imagery. *Remote Sens.* **2014**, *6*, 4173–4189. [CrossRef]
26. Yang, X.; Lu, X. Drastic change in China's lakes and reservoirs over the past decades. *Sci. Rep.* **2014**, *4*, 6041. [CrossRef]
27. Schwatke, C.; Scherer, D.; Dettmering, D. Automated Extraction of Consistent Time-Variable Water Surfaces of Lakes and Reservoirs Based on Landsat and Sentinel-2. *Remote Sens.* **2019**, *11*, 1010. [CrossRef]
28. Liu, Q.; Trinder, J.C. Sub-pixel technique for time series analysis of shoreline changes based on multispectral satellite imagery. In *Advanced Remote Sens. Technology for Synthetic Aperture Radar Applications, Tsunami Disasters, and Infrastructure*; IntechOpen: London, UK, 2018.
29. Dai, C.; Howat, I.M.; Larour, E.; Husby, E. Coastline extraction from repeat high resolution satellite imagery. *Remote Sens. Environ.* **2019**, *229*, 260–270. [CrossRef]
30. Li, J.; Knapp, D.E.; Schill, S.R.; Roelfsema, C.; Phinn, S.; Silman, M.; Mascaro, J.; Asner, G.P. Adaptive bathymetry estimation for shallow coastal waters using Planet Dove satellites. *Remote Sens. Environ.* **2019**, *232*. [CrossRef]
31. Woodcock, C.E.; Allen, R.; Anderson, M.; Belward, A.; Bindschadler, R.; Cohen, W.; Gao, F.; Goward, S.N.; Helder, D.; Helmer, E.; et al. Free access to Landsat imagery. *Science* **2008**, *320*, 1011. [CrossRef]

32. Pardo-Pascual, J.E.; Almonacid-Caballer, J.; Ruiz, L.A.; Palomar-Vázquez, J. Automatic extraction of shorelines from Landsat TM and ETM+ multi-temporal images with subpixel precision. *Remote Sens. Environ.* **2012**, *123*, 1–11. [CrossRef]
33. Pardo-Pascual, J.E.; Sánchez-García, E.; Almonacid-Caballer, J.; Palomar-Vázquez, J.M.; Priego de los Santos, E.; Fernández-Sarría, A.; Balaguer-Beser, Á. Assessing the accuracy of automatically extracted shorelines on microtidal beaches from Landsat 7, Landsat 8 and Sentinel-2 imagery. *Remote Sens.* **2018**, *10*, 326. [CrossRef]
34. Almonacid-Caballer, J.; Sánchez-García, E.; Pardo-Pascual, J.E.; Balaguer-Beser, A.A.; Palomar-Vázquez, J. Evaluation of annual mean shoreline position deduced from Landsat imagery as a mid-term coastal evolution indicator. *Mar. Geol.* **2016**, *372*, 79–88. [CrossRef]
35. Foody, G.M.; Muslim, A.M.; Atkinson, P.M. Super-resolution mapping of the shoreline through soft classification analyses. In Proceedings of the IGARSS 2003, 2003 IEEE International Geoscience and Remote Sens. Symposium. Proceedings (IEEE Cat. No.03CH37477), Toulouse, France, 21–25 July 2003; Volume 6, pp. 3429–3431.
36. Foody, G.M.; Muslim, A.M.; Atkinson, P.M. Super-resolution mapping of the waterline from remotely sensed data. *Int. J. Remote Sens.* **2005**, *26*, 5381–5392. [CrossRef]
37. Muslim, A.M.; Foody, G.M.; Atkinson, P.M. Shoreline Mapping from Coarse-Spatial Resolution Remote Sens. Imagery of Seberang Takir, Malaysia. *J. Coast. Res.* **2007**, *23*, 1399–1408. [CrossRef]
38. Liu, Q.; Trinder, J.C.; Turner, I.L. Automatic super-resolution shoreline change monitoring using Landsat archival data: A case study at Narrabeen–Collaroy Beach, Australia. *J. Appl. Remote Sens.* **2017**, *11*. [CrossRef]
39. Vos, K.; Harley, M.D.; Splinter, K.D.; Simmons, J.A.; Turner, I.L. Sub-annual to multi-decadal shoreline variability from publicly available satellite imagery. *Coast. Eng.* **2019**, *150*, 160–174. [CrossRef]
40. Liu, X.; Deng, R.; Xu, J.; Zhang, F. Coupling the modified linear spectral mixture analysis and pixel-swapping methods for improving subpixel water mapping: Application to the Pearl River Delta, China. *Water* **2017**, *9*, 658. [CrossRef]
41. Niroumand-Jadidi, M.; Vitti, A. Reconstruction of river boundaries at sub-pixel resolution: estimation and spatial allocation of water fractions. *ISPRS Int. J. Geo Inf.* **2017**, *6*, 383. [CrossRef]
42. Hagenaars, G.; de Vries, S.; Luijendijk, A.P.; de Boer, W.P.; Reniers, A.J. On the accuracy of automated shoreline detection derived from satellite imagery: A case study of the Sand Motor mega-scale nourishment. *Coast. Eng.* **2018**, *133*, 113–125. [CrossRef]
43. Moreno, L.J.; Kraus, N.C. Equilibrium shape of headland-bay beaches for engineering design. *Proc. Coastal Sediments* **1999**, 860–875.
44. Li, F.; Jupp, D.L.; Thankappan, M.; Lymburner, L.; Mueller, N.; Lewis, A.; Held, A. A physics-based atmospheric and BRDF correction for Landsat data over mountainous terrain. *Remote Sens. Environ.* **2012**, *124*, 756–770. [CrossRef]
45. Wang, X.; Liu, Y.; Ling, F.; Xu, S. Fine spatial resolution coastline extraction from Landsat-8 OLI imagery by integrating downscaling and pansharpening approaches. *Remote Sens. Lett.* **2018**, *9*, 314–323. [CrossRef]
46. Souza Filho, P.W.M.; Farias Martins, E.D.S.; da Costa, F.R. Using mangroves as a geological indicator of coastal changes in the Bragança macrotidal flat, Brazilian Amazon: A remote sensing data approach. *Ocean Coast. Manag.* **2006**, *49*, 462–475. [CrossRef]
47. Nguyen, H.-H.; McAlpine, C.; Pullar, D.; Johansen, K.; Duke, N.C. The relationship of spatial–temporal changes in fringe mangrove extent and adjacent land-use: Case study of Kien Giang coast, Vietnam. *Ocean Coast. Manag.* **2013**, *76*, 12–22. [CrossRef]
48. Nardin, W.; Locatelli, S.; Pasquarella, V.; Rulli, M.C.; Woodcock, C.E.; Fagherazzi, S. Dynamics of a fringe mangrove forest detected by Landsat images in the Mekong River Delta, Vietnam. *Earth Surf. Process. Landf.* **2016**, *41*, 2024–2037. [CrossRef]
49. Zhao, B.; Guo, H.; Yan, Y.; Wang, Q.; Li, B. A simple waterline approach for tidelands using multi-temporal satellite images: A case study in the Yangtze Delta. *Estuar. Coast. Shelf Sci.* **2008**, *77*, 134–142. [CrossRef]
50. McFeeters, S.K. The use of the Normalized Difference Water Index (NDWI) in the delineation of open water features. *Int. J. Remote Sens.* **1996**, *17*, 1425–1432. [CrossRef]
51. Murray, N.J.; Phinn, S.R.; Clemens, R.S.; Roelfsema, C.M.; Fuller, R.A. Continental scale mapping of tidal flats across East Asia using the Landsat archive. *Remote Sens.* **2012**, *4*, 3417–3426. [CrossRef]

52. Feyisa, G.L.; Meilby, H.; Fensholt, R.; Proud, S.R. Automated Water Extraction Index: A new technique for surface water mapping using Landsat imagery. *Remote Sens. Environ.* **2014**, *140*, 23–35. [CrossRef]
53. Cipolletti, M.P.; Delrieux, C.A.; Perillo, G.M.; Piccolo, M.C. Superresolution border segmentation and measurement in remote sensing images. *Comput. Geosci.* **2012**, *40*, 87–96. [CrossRef]
54. van der Walt, S.; Schönberger, J.L.; Nunez-Iglesias, J.; Boulogne, F.; Warner, J.D.; Yager, N.; Gouillart, E.; Yu, T. scikit-image: image processing in Python. *PeerJ* **2014**, *2*, e453. [CrossRef] [PubMed]
55. Otsu, N. A threshold selection method from gray-level histograms. *IEEE Trans. Syst. Man Cybern.* **1979**, *9*, 62–66. [CrossRef]
56. Gilles, S. Shapely: Manipulation and Analysis of Geometric Objects. Available online: https://toblerity.org (accessed on 10 December 2019).
57. Hoyer, S.; Hamman, J. xarray: ND labeled arrays and datasets in Python. *J. Open Res. Software* **2017**, *5*. [CrossRef]
58. Van Der Walt, S.; Colbert, S.C.; Varoquaux, G. The NumPy array: A structure for efficient numerical computation. *Comput. Sci. Eng.* **2011**, *13*, 22. [CrossRef]
59. Jones, E.; Oliphant, T.; Peterson, P. {SciPy}: Open Source Scientific Tools for Python. Available online: http://www.scipy.org/ (accessed on 10 December).
60. McKinney, W. Data structures for statistical computing in Python. In Proceedings of the 9th Python in Science Conference (SciPy 2010), Austin, TX, USA, 28 June–3 July 2010; Volume 445, pp. 51–56.
61. Hunter, J.D. Matplotlib: A 2D graphics environment. *Comput. Sci. Eng.* **2007**, *9*, 90. [CrossRef]
62. Waskom, M. Seaborn: Statistical Data Visualization using Matplotlib. Available online: https://seaborn.pydata.org (accessed on 10 December).
63. Zhou, Y.; Dong, J.; Xiao, X.; Xiao, T.; Yang, Z.; Zhao, G.; Zou, Z.; Qin, Y. Open surface water mapping algorithms: A comparison of water-related spectral indices and sensors. *Water* **2017**, *9*, 256. [CrossRef]
64. Kelly, J.T.; McSweeney, S.; Shulmeister, J.; Gontz, A.M. Bimodal climate control of shoreline change influenced by Interdecadal Pacific Oscillation variability along the Cooloola Sand Mass, Queensland, Australia. *Mar. Geol.* **2019**, *415*, 105971. [CrossRef]
65. Rover, J.; Wylie, B.K.; Ji, L. A self-trained classification technique for producing 30 m percent-water maps from Landsat data. *Int. J. Remote Sens.* **2010**, *31*, 2197–2203. [CrossRef]
66. Xie, H.; Luo, X.; Xu, X.; Pan, H.; Tong, X. Automated subpixel surface water mapping from heterogeneous urban environments using Landsat 8 OLI imagery. *Remote Sens.* **2016**, *8*, 584. [CrossRef]
67. Sun, W.; Du, B.; Xiong, S. Quantifying sub-pixel surface water coverage in urban environments using low-albedo fraction from Landsat imagery. *Remote Sens.* **2017**, *9*, 428. [CrossRef]
68. Sánchez-García, E.; Balaguer-Beser, Á.; Almonacid-Caballer, J.; Pardo-Pascual, J.E. A new adaptive image interpolation method to define the shoreline at sub-pixel level. *Remote Sens.* **2019**, *11*, 1880. [CrossRef]
69. Song, Y.; Liu, F.; Ling, F.; Yue, L. Automatic semi-global artificial shoreline subpixel localization algorithm for Landsat imagery. *Remote Sens.* **2019**, *11*, 1779. [CrossRef]
70. Vos, K.; Splinter, K.D.; Harley, M.D.; Simmons, J.A.; Turner, I.L. CoastSat: A Google Earth Engine-enabled Python toolkit to extract shorelines from publicly available satellite imagery. *Environ. Model. Softw.* **2019**, *122*. [CrossRef]

Article

Satellite Observations of Wind Wake and Associated Oceanic Thermal Responses: A Case Study of Hainan Island Wind Wake

Jin Sha [1,2,*], Xiao-Ming Li [1,2,3], Xue'en Chen [4] and Tianyu Zhang [5]

1 Key Laboratory of Digital Earth Science, Institute of Remote Sensing and Digital Earth, Chinese Academy of Sciences, Beijing 100094, China; lixm@radi.ac.cn
2 Hainan Key Laboratory of Earth Observation, Sanya 572029, China
3 Laboratory for Regional Oceanography and Numerical Modeling, Qingdao National Laboratory for Marine Science and Technology, Qingdao 266235, China
4 College of Oceanic and Atmospheric Sciences, Ocean University of China, Qingdao 266100, China; xchen@ouc.edu.cn
5 State Key Laboratory of Remote Sensing Science, College of Global Change and Earth System Science, Beijing Normal University, Beijing 100875, China; zhangty@mail.bnu.edu.cn
* Correspondence: shajin@radi.ac.cn

Received: 31 October 2019; Accepted: 14 December 2019; Published: 16 December 2019

Abstract: The wind wake on the lee side of Hainan Island in the winter covers the southwest entrance of Beibu Gulf (or Gulf of Tonkin) and is essential to regional ocean dynamics. Using multiple satellite observations including advanced synthetic aperture radar (ASAR), we revisited the wake process during the winter of 2011. Asymmetric oceanic thermal responses were found with a warm band expanding northwestwardly while a cold tongue formed to the southeast. Combining satellite observations, model simulations, and reanalysis data, heat advection terms (ADV) are reconstructed and compared to air-sea heat flux terms. The observed thermal evolution process across the wake footprint is closely related to the balanced spatial variability from the Ekman ADV, the barotropic geostrophic ADV, and the latent heat flux (LHF), which are all on the order of 10^{-5} K·m·s^{-1}. Specifically, the Ekman ADV tends to heat the northwestern side of the wake and cool the southeastern side, while the geostrophic ADV compensates with the Ekman ADV across the wake footprint. This study reveals detailed oceanic responses associated with the wind wake and clarifies the contribution of ADV to the asymmetric spatial thermal variabilities. The identified role of heat advection on a sub-seasonal timescale may further benefit the understanding of regional oceanic dynamics.

Keywords: Coastal process; wind wake; heat advection; multi-sensor; ASAR; oceanic thermal response; Hainan Island

1. Introduction

Wind wakes are commonly observed on the lee side of an island with reduced wind speeds due to frictions of land orography [1]. Concerning the oceanic response, wind wakes are documented to generate regional thermal variability by different mechanisms depending on detailed hydrological conditions [2,3], which could be further complicated by circulations and associated eddies [4,5]. The wind wake of Hainan Island (HIWW) has been observed to the southwest coast of the island with interactions between the land elevated orography and the Asian monsoon, collocating with a warm sea surface temperature band [6]. A weak wind wake is also observed to the northwest of the Vietnamese coast resulting from an orographic blockage from June to October [7], and it is proposed that the corresponding warm water is generated by thermal processes (e.g., latent heat flux) on the seasonal time scale, while dominated by wind mixing processes on the diurnal time scale.

Hainan Island is located in the northwest of the South China Sea (SCS) and is close to the entrance of the semi-enclosed shallow ocean, the Beibu Gulf (or Gulf of Tonkin). The major circulation of the SCS is driven by the monsoon wind [8], with a southward coastal current along the western boundary in the winter and a northward coastal current in the summer [9,10]. Concerning regions around Hainan Island, many previous studies have focused on the summertime upwelling systems off the east coast (e.g., [11–13]). For example, the work of Su and Pohlmann [13] revealed that the southwesterly and southerly winds in the summer are responsible for the formation of the cold water centers off the eastern Hainan coast, while the combined effects of wind and topography lead to the uneven distribution of the upwelling centers. For the west coast, tidal currents are considered as an important factor of local fronts [14,15]. There are also upwellings observed off the west coast during summer despite the downwelling favorable winds, and these are attributed to the joint effects of tidal mixing and background stratification [16,17]. Li et al. [14] also found that the interannual variability of the summer upwellings is related to the combined effects of along-shore wind, tidal mixing, and boundary currents. Few studies have focused on the southwest coast of Hainan Island during winter, except for the work of Li et al. [6] who observed the HIWW.

The steady and strong wind conditions in the boreal winter makes HIWW an ideal case for wind wake studies. An important mechanism forming the warm band in SST is the reduced latent heat flux associated with the decreased wind speed in the wake region [6]. However, it is not clear yet how the heat advection term influences the regional thermal evolution process. The competition between heat advection and the air-sea heat flux in regional heat variabilities has been long discussed world-wide and may vary depending on the time scale and spatial distributions of interest [18–21]. Meanwhile, the seasonal variation of the seawater temperature is influenced by the air-sea heat flux [22], and temperature variabilities on time scales from days to weeks over shallow water could be mainly attributed to heat advection [23]. Hence, the role of heat advection could be essential to understanding the regional thermal evolution process and should be taken into consideration in a detailed investigation of regional seawater variabilities.

In this study, with support from multi-sensor satellite observations including Advanced Synthetic Aperture Radar (ASAR) data and the derived sea surface wind field, sea surface temperature (SST) and merged wind products from microwave and optical sensors, as well as simultaneous model simulations, the characteristics of HIWW and associated oceanic responses are investigated in detail. Our analysis is organized as follows. Firstly, the datasets and the methods adopted are described in Section 2. Observed wind wake patterns and corresponding sea surface thermal responses are introduced in Section 3, followed by an analysis of the roles of heat advection and air-sea heat flux terms. In Sections 4 and 5, the regional thermal evolution mechanism is summarized and discussed.

2. Data and Methods

2.1. Study Area

To study the oceanic spatial variability, a transect is made across the wind wake, starting at 18.8°N,107.5°E and ending at 17.0°N, 109.4°E in the northwestern-southeastern direction (Figure 1). There is a total of 568 sampling points with a spacing interval of 500 m along the transect, spanning over a distance of approximately 283 km.

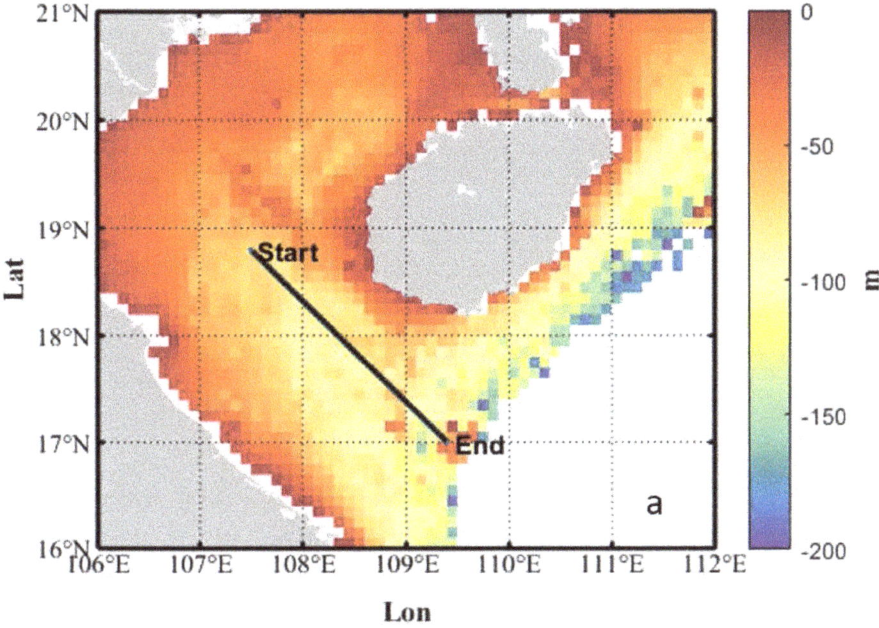

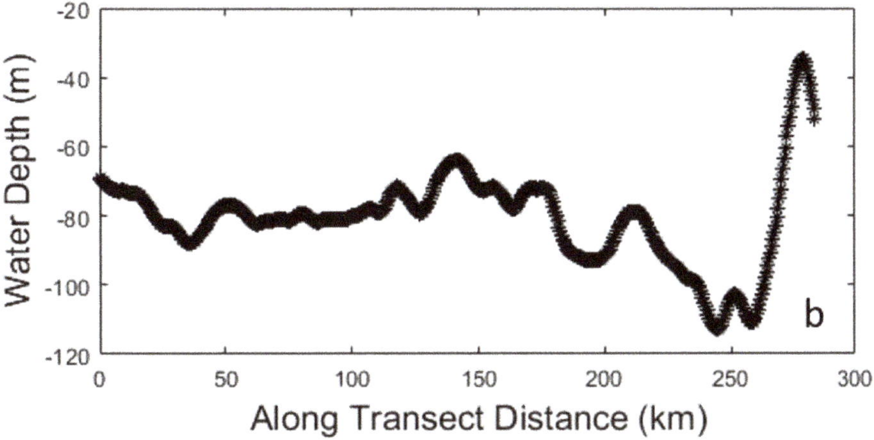

Figure 1. The bathymetry of the study area. (**a**) The bathymetry around Hainan Island, in the northwest of the South China Sea. The color shading is the water depth. Areas with bathymetry deeper than 200 m are not plotted. The black line is a transect made across the wind wake. (**b**) The bathymetry along the transect.

2.2. ASAR Data and Processing

The Envisat/ASAR data (02:47 UTC of December 10th, 2011) in wide swath mode (WSM) was used to retrieve the sea surface wind field. ASAR WSM data were not highly acquired within our study area during winter. The presented case is a good enough example to show a clear wind wake pattern downstream of the island and is thus used for further analysis in this study. The ASAR is one of 10 instruments onboard the Envisat satellite, operating at C-band in a sun-synchronous orbit at an altitude of 800 km. The WSM is one of ASAR's five imaging modes, and the swath width is approximately 400 km with the pixel size of 75 m [24].

To retrieve the sea surface wind field from ASAR vertical-vertical (VV) polarization data, the C-band geophysical model function (GMF) CMOD5.N [25] was applied. As one of geophysical model functions (GMFs), CMOD5.N depicts the relationship between the normalized radar cross section (NRCS) and wind vectors at 10 m height above the sea surface and radar incidence angles, which has been confirmed as a mature method for the retrieval equivalent neutral winds from scatterometer and synthetic aperture radar (SAR) measurements. The difference between the equivalent neutral wind and the real wind is found to be within 0.1~0.2 m/s [26]. The wind field can be retrieved at a resolution of 1 km [27,28], while we use a 5 km grid size [29] so as to keep consistent with model simulations as introduced in Section 2.3. Thus, U and V components were further interpolated to the grid size in the retrieval. Sea surface wind directions from the European Centre for Medium-Range Weather Forecasts (ECMWF) ERA-Interim wind data (0.125° by 0.125°) were used as references. In a comparison [30] between the ASAR-retrieved wind fields and on-land meteorological stations of Hainan Island, the standard deviation, mean error, and the correlation coefficient are 2.09 m/s, 0.27 m/s, and 0.75, respectively.

2.3. Other Satellite Observation and Reanalysis Datasets

SST used in this study comes from daily optimal interpolated maps (OISST) [31] from remote sensing system (RSS) merging both microwave and infrared data at 9 km resolution, which combines the cloud-free capacity of microwave data and high resolution of infrared data of coastal oceans based on the optimal interpolation method. To fully investigate the character of the wind wake, gridded wind vectors from the cross-calibrated multi-platform (CCMP, v2.0) [32] from RSS are also used as a reference in the present study, with a 25 km spatial resolution and 6-h temporal resolution. The sea surface latent heat flux (LHF), sensible heat flux (SHF), longwave thermal radiation, and surface net solar radiation are used as components of the air-sea heat flux, and are available from ERA5 reanalysis [33] of Copernicus Climate Change Service (C3S) (2017) on a grid size of 0.25° with a temporal resolution of 6 h. Senafi et al. [34] evaluated the performance of surface heat flux components of ERA5 by comparing with in-situ measurements and MERRA2/NASA reanalysis at the Arabian Gulf and the Red Sea, showing a bias of 4.5 W/m^2 and 1.59 W/m^2 respectively. Wind vectors from ERA5 03:00 UTC of December 10th, 2011 are also examined qualitatively to make sure the wind wake feature is captured, but no further analysis or calculation is conducted with ERA5 wind fields. To derive the surface absolute geostrophic velocities, we use the absolute dynamic topography (ADT) (version: SSALTO/DUACS Delayed-Time Level-4), which is available from the Copernicus Marine Environment Monitoring Service (CMEMS) containing multi-satellite merged daily maps from 1993 to present on a 0.25° Cartesian grid with tidal and inverse barometer corrections. The time-varying sea level is generally considered with an error of 2~3 cm [19,35]. Different datasets (Table 1) are harmonized by interpolation on a uniform grid with a spatial resolution of 0.1°, a temporal resolution of 1 day, and a vertical resolution of 1 m (if necessary).

Table 1. Observational and Reanalysis Datasets Summary.

Products	Variable	Time	Resolution
ASAR/Envisat/ESA	Normalized radar cross section	2011/12/10 02:47 UTC	150 m
OISST/RSS	Sea surface temperature	2011/01/01 –2014/12/31	9km, daily
CCMP/RSS	10 m wind vector	2011/01/01 –2014/12/31	0.25°, 6-hourly

Table 1. *Cont.*

Products	Variable	Time	Resolution
ERA-Interim/ECMWF	10 m wind vector	2011/01/01 –2014/12/31	0.125°, 3-hourly
ERA5/ECMWF	Latent heat flux, sensible heat flux, surface thermal radiation, surface net solar radiation, 10m wind vector	2011/10/01 –2012/01/31	0.25°, hourly
SSALTO/DUACS Sea Surface Height	Absolute dynamic topography	2011/01/01 –2014/12/31	0.25°, daily

2.4. WRF Model Simulations

The weather research and forecasting (WRF) model with advanced research WRF (ARW) dynamic core is used in this study to get a high-resolution wind field [36]. The WRF method has been validated over the Beibu Gulf [36], showing a standard deviation of 1.41m/s, a mean difference of 0.83m/s, and correlation coefficients squared of 0.67 by comparing offshore winds from WRF and SSM/I. WRF-ARW is a numerical weather prediction and atmospheric simulation system with fully compressible and non-hydrostatic equations, and provides several options to parameterize sub-grid scale physical processes and assimilate data. We use time-updating sea surface temperature (SST) with 0.5° × 0.5° from the National Centers for Environmental Prediction/Marine Modeling and Analysis Branch (NCEP / MMAB) and spectral nudging in order to make sure the lower boundary conditions are updated, and keep the large-scale circulation patterns of simulations close to the forcing field [37–39]. The spectral nudging method is applied to temperature, geopotential height, and wind vector over 700 hPa level in all runs. The double and two-way nested scheme has been adopted in the model.

The model domain is centered at 18.03150°N and 109.6105°E, with a fine grid space of 10 km (dimensions of 135 × 115). The fine domain comprises the whole Hainan Island and Beibu Gulf. The Mercator projection is chosen due to the low-latitude simulated area. The main parameterization schemes used are Yonsei University planetary boundary layer (YSU PBL) scheme [40], Kain-Fritsch Cumulus scheme [41], Lin et al. microphysics scheme [42], the rapid and accurate radiative transfer model (RRTM) longwave radiation scheme [43], Dudhia shortwave radiation scheme [44] and the Noah land surface model [45]. These schemes have proven to be appropriate for sea wind simulation [46,47]. Menendez et al. [46] conducted a sensitivity test to demonstrate that the YSU PBL scheme is slightly more suitable than other PBL schemes in simulations of the sea wind over the Mediterranean Sea. We chose ERA-Interim reanalysis as the driving field and re-initialized the integral approach to conduct the dynamical downscaling spanning from January 1, 2011 to December 31, 2014. Then, the 6-hourly simulation results were interpolated linearly onto the standard grid with the spatial resolution of 0.1° and temporal resolution of one day.

Concerning the variation of the wind speed across the wind wake, curves from ASAR, WRF, CCMP, and ERA5 along the transect (as in Section 2.1) are shown (Figure 2). All datasets could capture the low-speed feature across the wake consistently, with the minimum wind speed occurs between 150~200 km along the transect. The curve along the transect from ASAR-derived wind shows the finest features, while the WRF simulation tends to overestimate the wind speed shear, and the CCMP wind speed shows weakened variation.

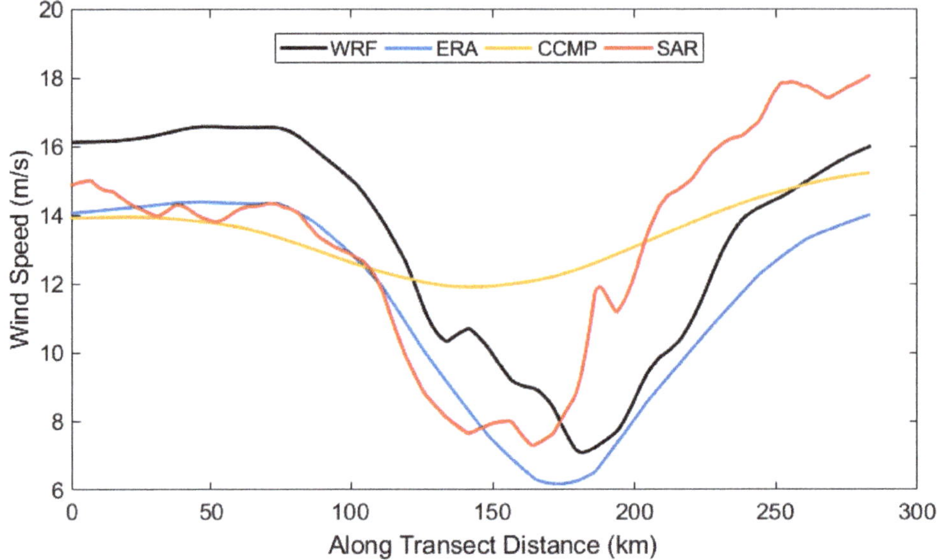

Figure 2. The comparison of sea surface wind speed from different datasets used in this study. The wind speed comes from ASAR, WRF, ERA-5, and CCMP on December 10th, 2011, respectively. The WRF and CCMP data are the averaged value between 0:00 UTC and 6:00 UTC. The ERA-5 data is at 03:00 UTC. The ASAR data is at 02:47 UTC.

To simply provide an uncertainty bound, wind speeds from ASAR observation, WRF model simulation, and CCMP merged observation are cross-compared over the study domain with land area excluded. The ASAR data is at 02:47 UTC. The WRF and CCMP data are the averaged value between 0:00 UTC and 6:00 UTC on Dec 10th, 2011. ERA5 is not included in the comparison since ERA-interim is already used in WRF simulation, and ERA5 also assimilates observational datasets. In the first step, all datasets are interpolated onto the 0.1° grid as mentioned in the above section. There are totally 5485 collocated triplets for ASAR, WRF, and CCMP respectively. We here introduce the triple collocation analysis, which is a data comparison method using three independent products to assess the error magnitude of each product separately and has been popular in the wind and soil moisture products applications [48–51]. Following the calibration model of Stoffelen [48] and the error assessment formula of Pan et al. [50], the estimated error magnitudes between the products and a deterministic "truth" value are approximately 2.30 m/s, 1.34 m/s, and 2.14 m/s for ASAR, WRF, and CCMP, respectively, which is consistent with previous literatures [30,36]. No comparison with in-situ measurement was performed due to a lack of offshore meteorological buoy data.

2.5. Estimation of Wind-induced Currents

Wind time series used to calculate horizontal heat advection are directly from the WRF high-resolution simulation. Based on the conventional Ekman theory [52], the depth-averaged wind current (u_e, v_e) within the Ekman layer can be approximated as:

$$\vec{u}_e = \frac{1}{\rho_0 f D_e} \vec{\tau} \times \vec{k} \tag{1}$$

where ρ_0 is constant seawater density, f the Coriolis parameter, $\vec{k}$ the vertical unit vector, D_e is the Ekman depth, and $\vec{\tau} = (\tau_x, \tau_y)$ is the wind stress, which can be derived from wind vectors. The vertical velocities driven by the wind stress curl is given as:

$$w_e = \frac{1}{\rho_0 f} \nabla \times \vec{\tau} \tag{2}$$

However, the conventional Ekman theory is under the assumption of infinite water depth, which is obviously not the case in coastal oceans. Welander [53] extended the equations to varying water depths by introducing structure functions, and the depth-integrated flow is expressed as

$$\begin{aligned} U &= \frac{1}{\rho_0 f}(C_1\tau_x - C_2\tau_y) + \frac{gh}{f}(C_3\frac{\partial\eta}{\partial x} - C_4\frac{\partial\eta}{\partial y}) \\ V &= \frac{1}{\rho_0 f}(C_2\tau_x + C_1\tau_y) + \frac{gh}{f}(C_4\frac{\partial\eta}{\partial x} + C_3\frac{\partial\eta}{\partial y}) \end{aligned} \tag{3}$$

with structure functions

$$\begin{aligned} C_1 &= \frac{2\sin R \sinh R}{\cos 2R + \cosh 2R} \\ C_2 &= \frac{2\cos R \cosh R}{\cos 2R + \cosh 2R} - 1 \\ C_3 &= \frac{1}{2R}\frac{\sin 2R - \sinh 2R}{\cos 2R + \cosh 2R} \\ C_4 &= -\frac{1}{2R}\frac{\sin 2R + \sinh 2R}{\cos 2R + \cosh 2R} + 1 \end{aligned} \tag{4}$$

where η is the sea level variation. $R = \pi \cdot h \cdot D_e^{-1}$ indicates the ratio of water depth h to the Ekman depth. Currents in Equation (3) consist of contribution from the wind stress, as well as the sea level pressure gradient adjusted to wind and topography [54]. If we focus on the wind-induced drift components by eliminating the sea level variation parts and assume that $D_e \approx 0.4 \cdot f^{-1} \cdot \sqrt{\left|\vec{\tau}\right| \cdot \rho_0{}^{-1}}$ [55,56], the depth-averaged Ekman currents over coastal oceans could then be estimated as

$$\begin{aligned} u_w &= \frac{1}{\rho_0 f h'}(C_1\tau_x - C_2\tau_y) \\ v_w &= \frac{1}{\rho_0 f h'}(C_2\tau_x + C_1\tau_y) \\ h' &= \begin{cases} h & D_e > h \\ D_e & D_e \le h \end{cases} \end{aligned} \tag{5}$$

For infinite depth ($R \rightarrow +\infty$), $C_1 \rightarrow 0$ and $C_2 \rightarrow -1$, and Equation (5) will reduce to Equation (1).

2.6. Heat Advection Estimation

Regarding the temperature variability over the continental shelf, a simplified heat budget [57] over the whole water column is

$$\int_{-H}^{0} \frac{\partial T}{\partial t} dz + \int_{-H}^{0} \vec{u} \cdot \nabla T dz = \int_{-H}^{0} \frac{1}{\rho_0 C_p} \frac{\partial q}{\partial z} dz + Rs \tag{6}$$

where T is the temperature (unit: K), $\vec{u}$ is the water currents (unit: m/s), H is the water depth (unit: m), and q is the vertical heat flux (unit: $W \cdot m^{-2}$). $C_p = 3990$ J kg^{-1} K^{-1} is the seawater heat capacity [55] and ρ_0 is the seawater density. Rs is the residual term representing processes not included, such as diffusions.

We assume that the temperature is vertically uniform, which may not be true for stratified oceans, but is useful and valid for the well-mixed coastal water of Hainan Island during the winter [6]. Then, SST could be used to approximate the vertical temperature, and Equation (6) becomes

$$H \cdot \frac{\partial T_{SST}}{\partial t} \approx -\vec{U}_H \cdot \nabla T_{SST} + \frac{Q}{\rho_0 C_p} \tag{7}$$

where $\vec{U}_H$ is the vertically integrated horizontal transport (unit: $m^2 \cdot s^{-1}$) from both the Ekman drift and barotropic geostrophic currents. Q is the air-sea net heat flux, being the sum of the solar radiation

Q_S, the longwave radiation Q_b, the SHF and the LHF, which can be obtained from the ERA5 reanalysis. The term on the left-hand side represents the local temporal variability of SST. The first term on the right-hand side is the horizontal heat advection (ADV), and the second term corresponds to the air-sea heat flux (ASF). All three terms in Equation (7) have the unit of $K \cdot m \cdot s^{-1}$, and thus a meaningful comparison can be made between the ADV and ASF. Since our intent is to clarify the contribution of the heat advection, no attempt was made to construct a closed heat budget in this study. A weekly moving average is applied on the ADV and ASF terms hereafter when plotting to remove high-frequency variabilities.

3. Wind Wake Characters and Oceanic Responses

3.1. Wind Wake Observed by Spaceborne SAR

A high-resolution ASAR image captured at 02:47 (UTC) on December 10, 2011 and the retrieved sea surface wind field present the capability of capturing sub-mesoscale features of the wind wake. Due to the incidence angle, the whole image shows a light-to-dark trend from the lower right corner to the top left corner. The wind wake is indicated as a darker-than-ambient footprint (Figure 3a) to the southwest side of the island, extending about 200 km from the island to the coast of Vietnam, covering most of the southern entrance of Beibu Gulf. The dark area (low NRCS) corresponds to reduced sea surface wind speeds over the wake footprint, which is about 10 m/s lower than ambient waters (Figure 3b), with the minimum wind speed very close to the coast. Enhanced positive/negative wind stress curl is also found close to the footprint boundary in the retrieved wind stress curl map (Figure 3c), forming sub-mesoscale wind curl stripes extending in the offshore direction.

3.2. Seasonality of the Wake Distribution

The Asian monsoon is known to prevail over the South China Sea. However, a detailed description of local wind condition is necessary, which closely relates to the spatial and temporal distribution of HIWW. The seasonal wind rose diagrams (Figure 4) using the daily WRF wind vectors from 2011 to 2014 represent the characters of local wind. There is almost no southeasterly wind in the region. Northeasterly wind prevails during Autumn and Winter, while southwesterly wind prevails during Summer. The dominant wind intervals, i.e., taking the largest portion during each season, for the winter, spring, summer and autumn are 10~12 m/s (22.9%), 4~6 m/s (31.9%), 4~6 m/s (25.5%), and 6~8 m/s (22.1%), respectively.

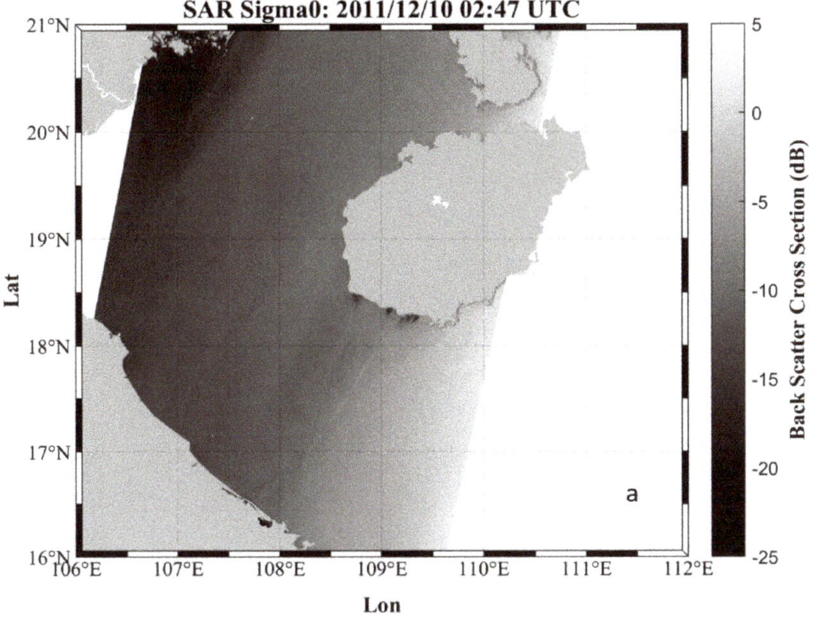

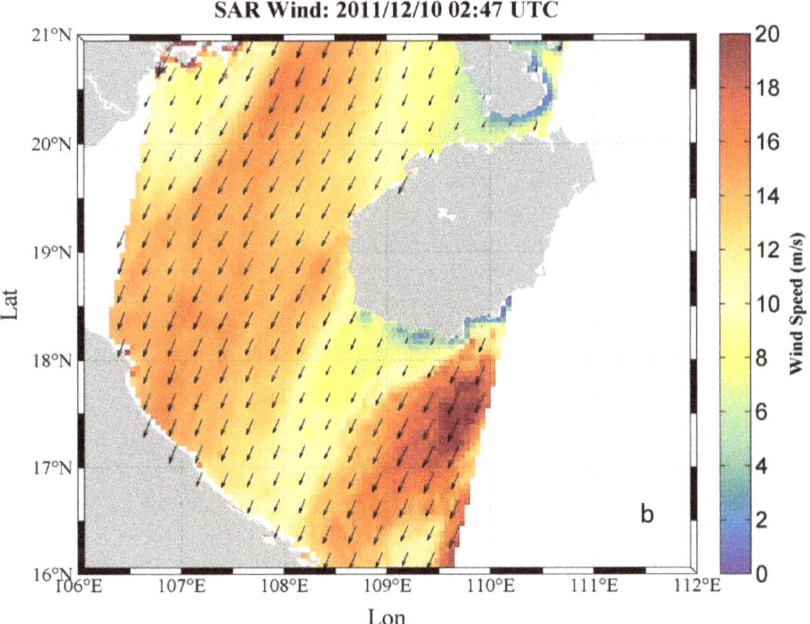

Figure 3. *Cont.*

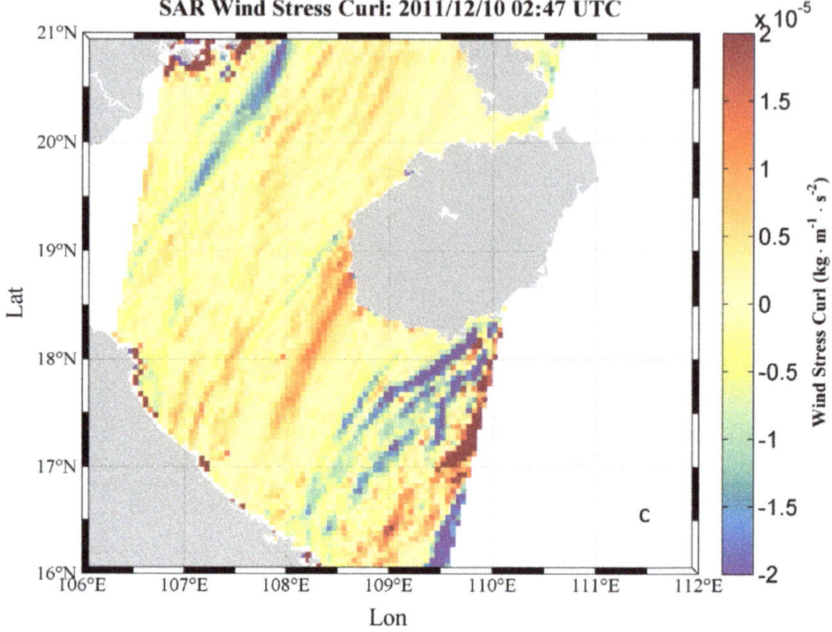

Figure 3. (**a**) An example of the ASAR WSM image acquired at 02:47 UTC on 10 December 2011 showing a typical wind wake in the lee side of Hainan Island. (**b**) The retrieved sea surface wind speed from the ASAR data and (**c**) the calculated wind stress curl based on the ASAR wind field.

HIWW is expected to show a controlled distribution corresponding to the varying wind, which can be represented by the occurrence of anomalous wind speeds (Figure 5) 2 m/s lower than the regional average. The thresholds of other values less than 5 m/s hold consistent distribution patterns. Three areas with high occurrences of anomalous wind are found in the study area: one locates on the southwest coast of Hainan Island in winter, corresponding to the known wind wake footprint. Another one is located on the northwest coast of Hainan, due to the southerly wind in summer. Also, there is one close to the Vietnam coast at the lower-left corner (17°N, 107°E), which is likely due to the interaction between the southwesterly wind and the Vietnamese orography [7]. No occurrence was observed to the east of 110.5°E. In comparison to the northwest Hainan coast, the high occurrence in the southwest coast covers a larger area and larger magnitude, being consistent with the regional monsoon character that the northeasterly wind is strongest and steady from October to January (Figure 4).

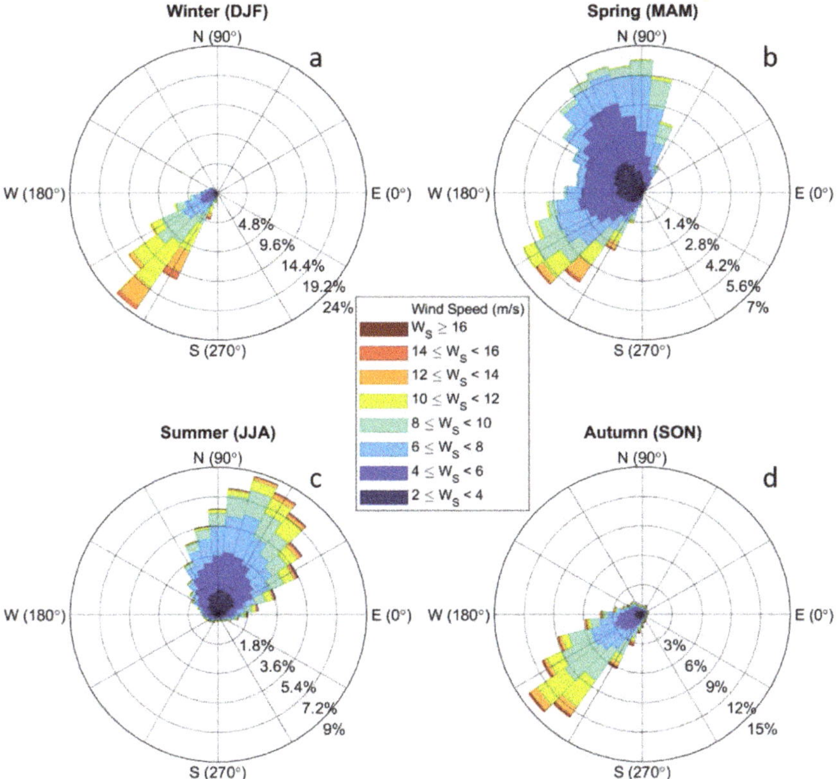

Figure 4. Seasonal wind rose diagrams with WRF wind vectors from 2011 to 2014. Results are for (**a**) winter, (**b**) spring, (**c**) summer, and (**d**) autumn respectively. The wind direction is defined as the direction of the wind vector pointing towards and departs from the east (0°) cyclonically. Land areas are excluded. Color shading indicates the wind speed, with a speed interval of 2 m/s.

3.3. Spatial Variability of Oceanic Thermal Response

To clarify the oceanic thermal response over the wind wake footprint, both the sea surface wind speed anomaly and SST are investigated along the transect across the wind wake. The study period from October 2011 to January 2012 was chosen since this is a period when the regional wind speed is the strongest in the annual cycle and wind direction is steady towards the southwest as mentioned in the above section. No steady wind wakes were observed within the transect before October or from the end of January. On the contrary, a lower-than-ambient wind speed band is clearly shown between distance 100 km and 200 km, together with the local speed minima identified (Figure 6a,b). The daily regional means have been removed to show the spatial variability without the influence of the temporal trend. The temporal-mean wind speed trough is around 5.10 m/s, with a local maximum value of 9~10 m/s on both ambient sides. To simplify the notation, hereafter we divide the transect into two parts, namely the northwestern side (NWS) with a transect distance smaller than the wind speed minima, and the southeastern side (SES) with a transect distance larger than the minima. We also noticed that the wind wake location oscillates during November and from the middle of January.

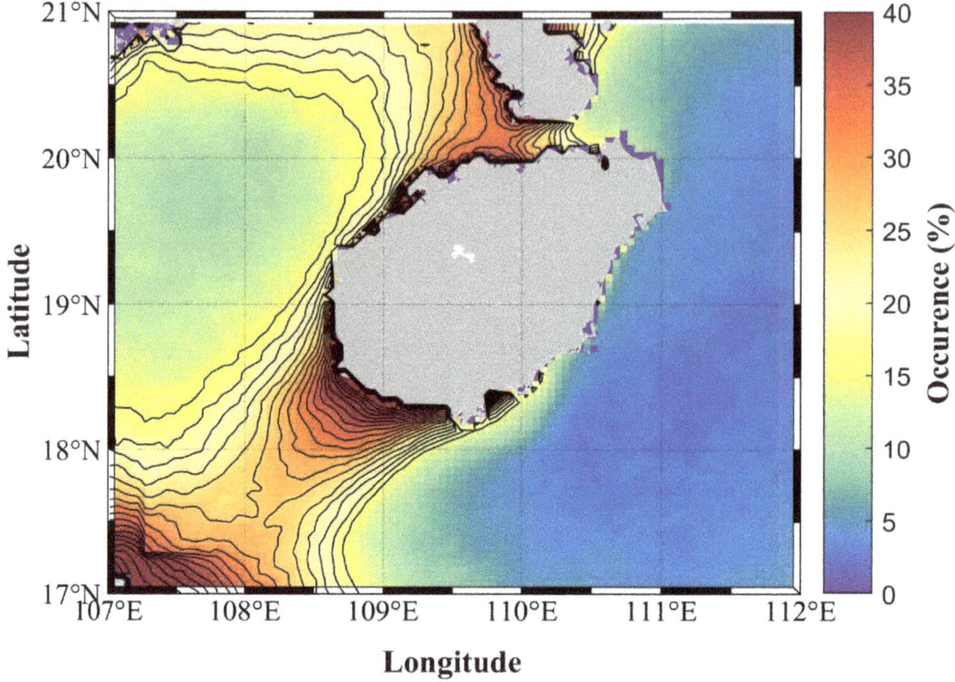

Figure 5. The occurrence of anomalous wind speeds over the study area based on the WRF simulation. The "anomalous" wind speed here represents those wind speeds that is 2 m/s lower than the daily mean over the study area. Then the percentage of occurrence is calculated from 2011 to 2014. The land area is excluded from the calculation. Each contour line indicates a 2% interval from 16% to 40%.

For the SST anomaly (SSTA) along the transect (Figure 6c,d), three features are noted here: firstly, a warm band is observed through the study period collocating with the wind wake pattern. Secondly, the SSTA warm band shows an asymmetric feature comparing with the wake footprint. The warm band covers a larger area than the wind speed band, extending in NWS of the wake. 65% of SST local maxima (red line in Figure 6c) locate within a shorter along transect distance than that of the wind speed minima. Specifically, SSTA with an elevated temperature of at least 0.9 °C is observed on the NWS during December and January (black arrow in Figure 6c). Thirdly, areas in SES (along a transect distance ≥ 200km) keeps reduced temperature steadily during the study period, even forming a cold tongue (magenta arrow in Figure 6c) to the southeast of the warm band from the middle of December until the end of January.

One question then arises naturally as what causes the observed asymmetric SSTA spatial variability. The reduced latent heat flux from the seawater to the atmosphere resulting from the decreased wind speed has been proposed as the forming mechanism of the warming SSTA band [6]. However, the sea surface heat flux solely could not explain the observed warming extension beyond the wind wake. Also, as shown by the retrieved wind stress curl from ASAR data (Figure 3c), a positive wind stress curl occurs on the NWS, which could bring the cooler bottom water, if any, up to the surface [58]. Obviously, the cooling process by upwelling is conflicting with the observed warming extension. The most possible mechanism for this is the redistribution of seawater heat induced by the oceanic heat advection, which is further investigated later on.

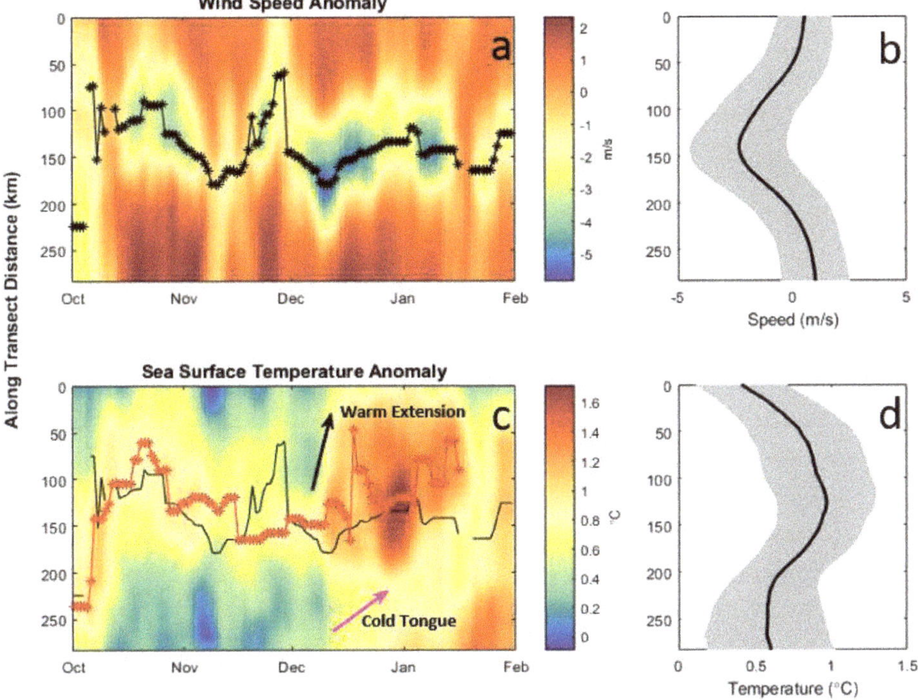

Figure 6. The oceanic variables along the transect. The left panel are the variables (**a**) wind speed anomaly and (**c**) sea surface temperature anomaly varying with time in the winter of 2011. The right panel are the temporal-mean of variables of the left panel. The width of the gray shading area indicates two standard deviations. Blackline in (**a**) and (**c**) is the trough locations (±0.25 km) of wind speed anomaly of each day, the redline in (**c**) is the peak location of the SSTA. The black and magenta arrows in (**c**) indicate the NWS warm band extension and the SES cold tongue respectively. A weekly moving average was applied when plotting.

3.4. The Oceanic Heat Advection

The horizontal heat advection considered here consists of two terms: one is driven by the wind-induced Ekman drift (Figure 7), and the other is driven by the barotropic geostrophic current (Figure 8). Along the transect, the temporal averages of both ADV and LHF terms are in the order of 10^{-5} K·m·s^{-1} (Figure 9), but with different spatial variability across the wind wake, while SHF is one order smaller thus could be eliminated. The spatial difference of LHF along the transect increases from October to December. Focusing on December, we found that collocating with the trough of the wind speed (around 150 km along the transect), reduced LHF is found around 150 km, indicating a reduced heat loss process. Out of the wake footprint, LHF increased in both NWS and SES suggesting an enhanced cooling influence.

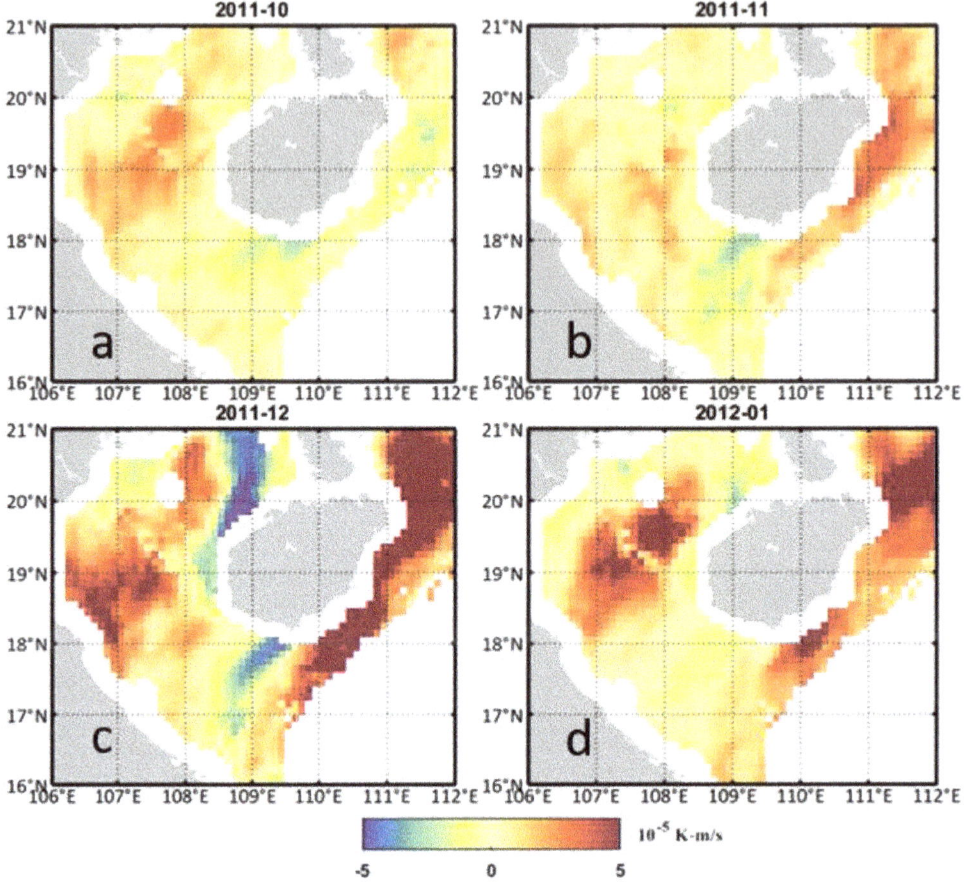

Figure 7. Estimated monthly horizontal heat advection driven by the wind stress along the transect. Monthly averaged values are shown for (**a**) October of 2011, (**b**) November of 2011, (**c**) December of 2011, and (**d**) January of 2012.

Concerning ADV, a much larger temporal variability/uncertainty (yellow error bar in Figure 9) is observed especially in October and January, though the averaged values are still meaningful in terms of the temporal-integrated effects. The source of the temporal uncertainty will be analyzed later in this section. Coherent warming exceeding uncertainty could be found between 50 and 100 km from November to January, while the advective cooling process occurs around 250 km in November and December. In December, both ADV and LHF contribute to a higher heat accumulation process around the wake region than the ambient water (say, the starting and ending point of the transect at 0 km and 283 km) but in an out-of-phase pattern.

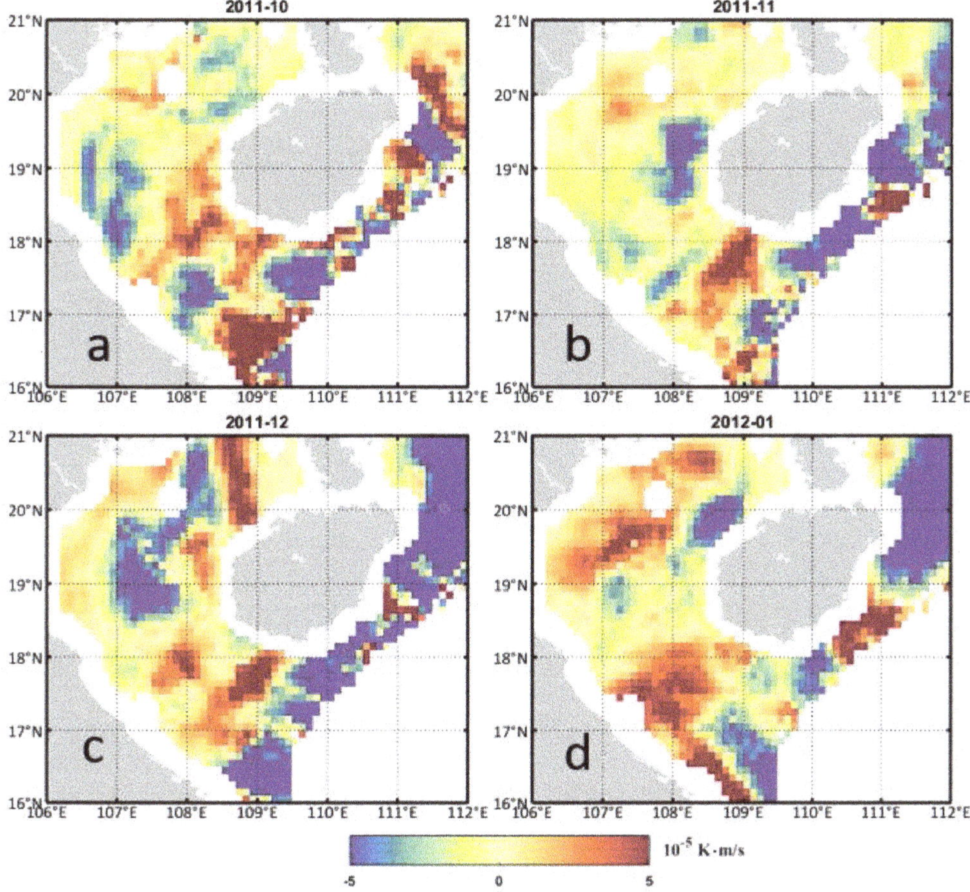

Figure 8. Estimated monthly horizontal heat advection driven by the sea level pressure gradient along the transect. Monthly averaged values are shown for (**a**) October of 2011, (**b**) November of 2011, (**c**) December of 2011, and (**d**) January of 2012.

As mentioned in Section 3.3, a typical warm band extension on NWS (black arrow in Figure 6c) is observed at around December 10th, and a cold tongue forms on SES from the middle of December (magenta arrow in Figure 6c), while wind wake location keeps steady and strong during the same time. Thus, it is necessary to focus on December rather than the whole wake period from October to January to clarify the detailed day-by-day thermal evolution process (Figure 10). The increased ADV values (Figure 10b,c) around December 10th transiting northwestward agree well with the observed SSTA warming extension, and the cold tongue found southeast of the wake in the SST is also consistent with the decreased ADV values (Figure 10e,f).

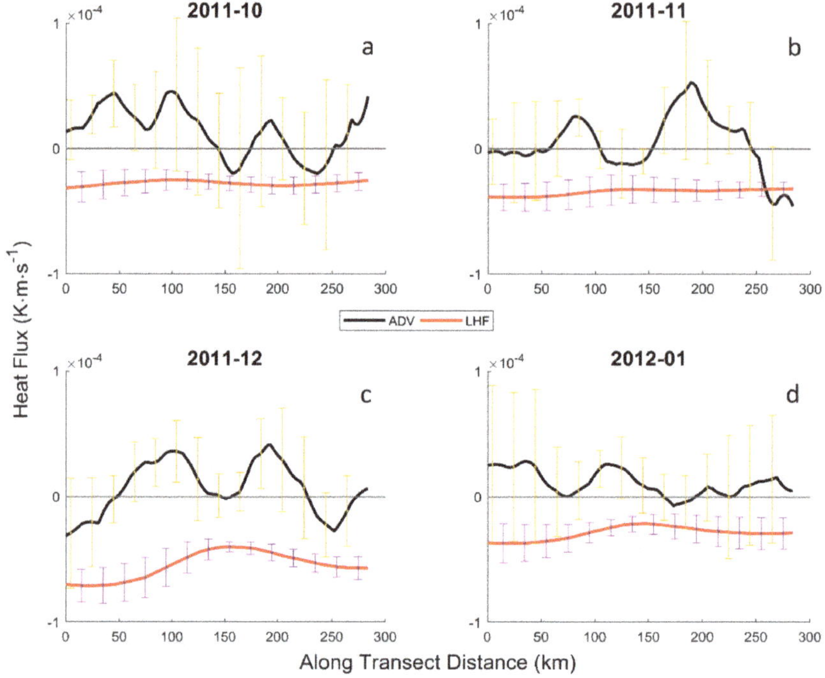

Figure 9. Heat flux terms along the transect averaged over the study period from 2011/10/01 to 2012/01/31. The comparison is made between ADV (black line) and LHF (red line). Uncertainties of two standard deviations within each month are indicated as the length of error bars. The monthly averaged values are shown for (**a**) October of 2011, (**b**) November of 2011, (**c**) December of 2011, and (**d**) January of 2012.

Moreover, it is of interest to clarify the relative contribution of the two driven mechanisms, wind-induced Ekman drift and the geostrophic current, to the total heat advection. Still taking the December of 2011 as an example (Figures 7c and 8c), the wind-driven heat advection tends to warm the water column from the center of the Beibu Gulf seawater to the NWS of the wake at 106°E ~ 108°E, 18°N ~ 20°N, while cool the water on the SES of the wake (approximately -4.23×10^{-5} K·m·s^{-1}) at 109°E, 17.7°N. We also noticed the cooling of the west coast of Hainan Island and warming of the continental shelf, which is beyond our interest in this study though. Meanwhile, the geostrophic heat advection tends to decrease the water temperature in the center of the Beibu Gulf (e.g., -8.25×10^{-5} K·m·s^{-1} at 107.5°E, 18.9°N), and warm both sides of the wake (e.g., 4.53×10^{-5} K·m·s^{-1} at 107.9°E, 18°N, and 9.54×10^{-5} K·m·s^{-1} at 108.9°E, 17.7°N). Along the transect (Figure 11), the Ekman advection shows a consistent distribution through the whole wake period, cooling the NWS and warming the SES, which is probably related to the wind speed distribution associated with the wake. The maximum amplitude occurring in December corresponds to the intensified wind speed during the winter monsoon. On the other hand, the geostrophic advection is subjected to relatively larger temporal variability as represented by the standard deviation (yellow error bar in Figure 11), suggesting the temporal uncertainty observed in the total ADV (Figure 9) mostly comes from the geostrophic advection. Moreover, the geostrophic advection process tends to compensate the Ekman advection in the center of the Beibu Gulf and the SES, especially from October to December.

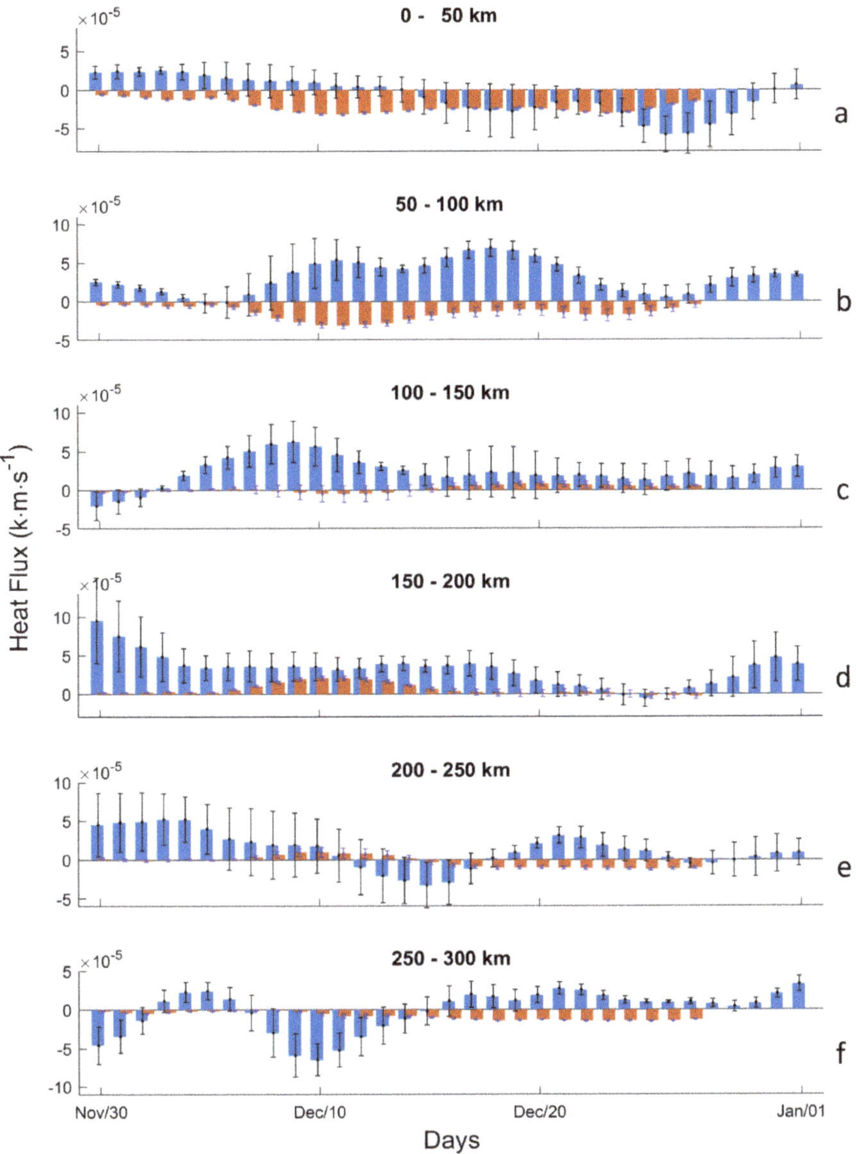

Figure 10. Heat flux anomaly comparison between the heat advection and the latent heat flux in December of 2011. The transect is averaged every 50 km from(**a**) 0~50 km,(**b**) 50~100 km, (**c**) 100~150 km, (**d**) 150~200 km, (**e**) 200~250 km, and (**f**) 250~300 km. Blue bars are the heat advection, and red bars are the latent heat flux. The black and magenta arrows in (**c**) indicate the NWS warm band extension and the SES cold tongue respectively. Uncertainties of two standard deviations are indicated by the length of error bars.

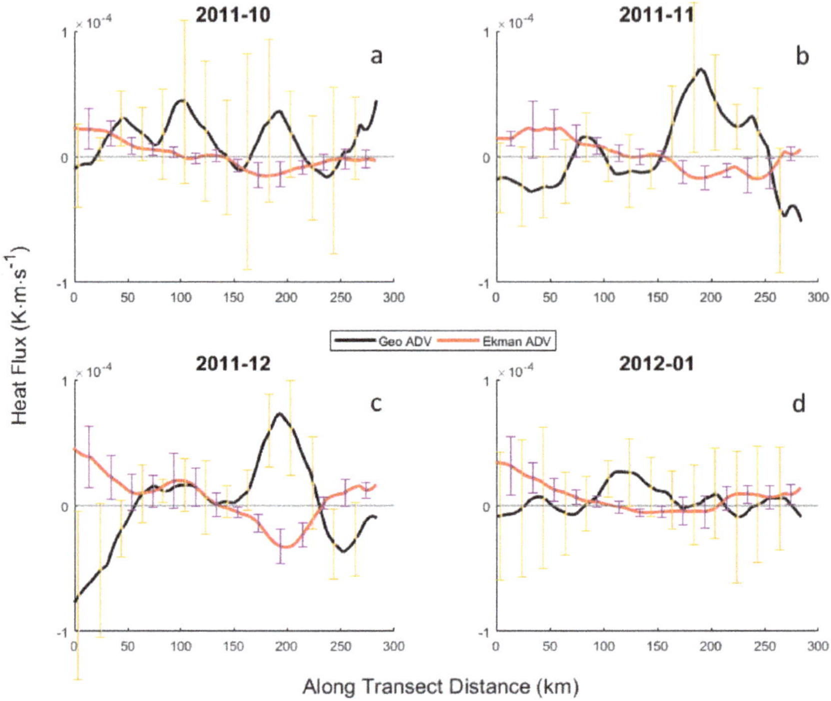

Figure 11. Heat advection terms comparison along the transect averaged over the study period from 2011/10/01 to 2012/01/31. The comparison is made between the geostrophic heat advection (black line) and the wind-driven Ekman heat advection (red line). The monthly averaged values are shown for (**a**) October of 2011, (**b**) November of 2011, (**c**) December of 2011, and (**d**) January of 2012.The error bar along curves represent two times of temporal standard deviation within each month.

4. Discussion

It would be interesting to highlight the role of wind-driven heat advection. The wind advection tends to deplete the heat on the SES of the wake, and accumulate the heat on the NWS, with the sign switched within the wake footprint where the wind speed is the lowest. On the SES, the wind-driven heat advection is the major mechanism of the cooling process, which is enhanced (cooling) by the latent heat flux and compensated (warming) for by the geostrophic advection. The wind stress performs as an important factor driving the spatial thermal variability, especially in shaping the asymmetric warm band corresponding to the wind wake. It should be noted that our estimation in this study is integrated through the whole water column. If considering the surface layer only, the role of the Ekman drifting could even be amplified [52,58] while that of the geostrophic currents are reduced. It is also worth mentioning that the derived wind-driven currents in this study agree well with the reported winter time coastal circulations [59], confirming the influence of the wind in the regional circulation, even taking geostrophic currents into consideration.

To correspond to the snapshot of the ASAR image from December 10, 2011, we conduct a case study of the wind wake from October 2011 to January 2012, when the wind wake location is steady and the wind speed is strong. The similar analyses could also be applied to other periods such as the following winters of 2012, 2013, and 2014 (Figure 12). The NWS warm band extensions beyond the wake could also be observed every winter from 2012 to 2014, and also collocate with similar distributions of the Ekman advection shown in sections above. The Ekman advection always tends to deplete the seawater heat on SES while accumulating the heat on NWS. This again confirms the essential role of the Ekman heat advection associated with the asymmetric thermal distribution across the wind wake. The geostrophic advection is also found to compensate for the Ekman advection during winter seasons somehow but is subjected to large spatial and temporal variability. Moreover, the distribution and the onset time of the wake exhibit interannual variations, which is probably modulated by climate oscillations via wind variabilities [60–62]. The work of Li et al. [6] also reveals that the core temperature of the warm band is significantly related to ENSO events from 1983 to 2011. The detailed climate-modulation process on different heat budget terms across the wake could be investigated in a future work.

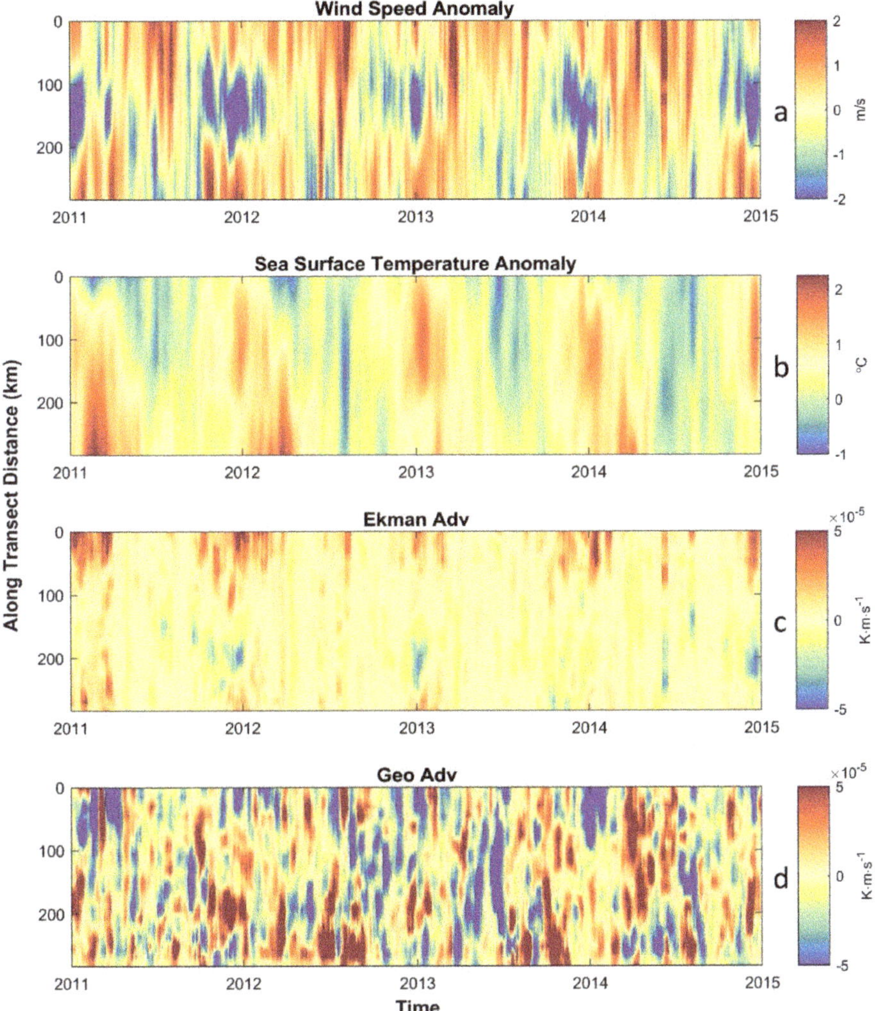

Figure 12. Multi-year time series along the transect from 2011 to 2014. Four variables are plotted including (**a**) wind speed anomaly, (**b**) SSTA, (**c**) Ekman advection, and (**d**) geostrophic advection.

In Section 2.5, our assumption of vertically uniform temperature distribution precludes the heat advection contribution from the Ekman pumping. While from both the ASAR data and the WRF simulation, strong wind stress curls are noticed close to the boundary of the wake, corresponding to a maximum upwelling at an NWS (downwelling at SES) speed of 2.39×10^{-5} m/s. As previously studied, the Ekman pumping velocity of the eastern Hainan coast is in the order of 10^{-4} m/s [13], while the upwelling velocity on the west coast is approximately 2.1×10^{-5} m/s [16]. Obviously, the magnitude of the wake-induced upwelling is comparable to the upwelling systems revealed around Hainan Island. However, the weak background stratification condition could prohibit the wake-induced upwelling to generate strong surface thermal response [17]. In contrast, the wake-induced upwelling could still bring nutrients upward, and might enhance the primary production in the upper ocean [58,63,64]. Thus the identification of the suction/pumping process on the sub-mesoscale is meaningful to regional ecosystem dynamic studies.

A scaling analysis could be further applied considering the warming rate difference ΔT_t between two locations in the transect. Derived from Equation (7), we have $\Delta T_t \sim \Delta ADT/H + \Delta LHF/H$. If we take the starting point at 0 km along the transect as a reference of ambient water, and assume that areas with a transect distance of 50 ± 10 km, 150 ± 10 km, and 250 ± 10 km represent the area of extended warming band out of the wake, the central warming band within the wake, and the cold tongue on the SES, respectively, then the heat flux difference from values at 0km could be estimated (Table 2) As analyzed in previous sections, the averaged values could indicate the temporal-integrated effects, while the uncertainty mostly corresponds to the temporal variations. Quantitatively, we could identify that the geostrophic ADV (Geo. ADV) dominant the warming of area 50 ± 10 km, and Ekman ADV dominant the cooling of area 250 ± 10 km. The ADV terms are important in shaping the boundary of the warm band, though the Ekman ADV and Geo. ADV tend to compensate for each other.

Table 2. Contributions from heat flux/advection terms relative to values at 0 km along the transect.

Along Transect Distance (km)		50 ± 10	150 ± 10	250 ± 10
Heat Flux/Advection Contribution (10^{-5} K·m·s^{-1})	Ekman ADV	−1.4 ± 2.2	−3.3 ± 2.3	−2.9 ± 2.3
	Geo. ADV	2.7 ± 5.1	3.1 ± 5.9	2.2 ± 7.0
	LHF	0.2 ± 0.3	1.4 ± 1.3	0.8 ± 0.8
	Net rate	1.5 ± 4.2	1.2 ± 5.4	−0.2 ± 6.4

Our results are not conflicting with the previously proposed mechanism that LHF could trigger the warm band [6] since we also identified that LHF contributes to the central warming within the wake footprint. However, only if we introduce the ADV terms could the observed spatial variability be explained. On the other side, since $\Delta t \sim \Delta T/(\Delta ADT/H + \Delta LHF/H)$, the net warming rate at 50 ± 10 km (Table 2) corresponds to a time scale of about 30 days to reach a temperature difference of 0.5 K. However, this shorter time scale, rather than the whole study period, indicates there must be other processes, such as tidal currents [14,15,65] or ocean wave processes [66], damping the spatial variability supported by the heat advection and the air-sea heat flux processes, which should be the subject of future works with the support of more in-situ measurements and a high-resolution air-sea coupling model in this region.

Moreover, documented wind wakes are usually presented through microwave instruments on much rougher resolutions [3], or indirectly through optical imagery [1,67]. In this study the high-resolution ASAR data and its derived sea surface wind field show strong observative capacity on wind wake studies, revealing the sub-mesoscale structures within the wake. Furthermore, the combined application of observations from multi-sensor satellites, model simulations, and reanalysis products also support for identifying involved detailed evolution processes and provide complementary observations for model simulation results. This again validated the strong capacity of satellite remote sensing in understanding ocean dynamics.

This study sets up a framework via heat advection reconstruction based on multi-source datasets to clarify the competition process between the air-sea heat flux and the horizontal heat advection terms, which could be applied to similar phenomena as observed in other areas, e.g., in [4,7,68].

The reconstruction then might be confined within the mixed layer in stratified seasons, but could extend to the whole column given support of vertical shear information. The baseline is that the reduced wind speed across the wake modifies the SST distribution not only through the air-sea heat flux (which has been extensively recognized), but also with variations in the horizontal heat advection on sub-seasonal timescale. Although the detailed role of heat advection, especially the wind-driven heat advection, may vary depending on local thermal and circulation conditions [19], or even the interaction with land orography [67], the contribution from heat advection could not be neglected before careful diagnosis.

5. Conclusions

Most of the previous studies around Hainan Island have focused on the upwelling systems on the eastern and western coasts during summer, but only a few have investigated the ocean dynamic to the southwest coast during winter, where HIWW is observed due to the interaction between the winter monsoon and the island's orography. In this study, HIWW occurred during the winter of 2011 is taken as a study case to investigate the characters of the wind wake and associated oceanic thermal responses. Presented by ASAR NRCS and the retrieved sea surface wind field and wind stress curl, a triangular wake footprint with sub-mesoscale features is observed to extend around 200 km between Hainan Island and the Vietnamese coast, covering almost the whole southern entrance of the Beibu Gulf. The identification of strong Ekman suction/pumping bands on the boundary of the wind wake valids the strong capability of SAR sensors on sub-mesoscale study of the wind field. Statistics of the anomalous low wind speed from 2011 to 2014 confirms the monsoon-controlled spatial distribution of the wind wakes off the southwest coast of Hainan Island. It is necessary to further clarify the associated oceanic response to the wake during the wintertime so as to better understand the regional oceanic dynamics.

By a comparison between the wind field and temperature, the collocation of the wind wake footprint and the warm SST band along the transect is noticed. However, the northwest extending of the warm band and the cold tongue appearing to the southeast side could not be explained by the reduced LHF in response to the wind wake. The reconstructed heat advection reveals that the wind-driven horizontal heat advection, the barotropic geostrophic heat advection, and the latent heat flux are the three major factors contributing to the spatial temperature variability and are on the order of 10^{-5} K·m·s^{-1}. The LHF mainly contributes to the central warming across the wind wake footprint. The wind-driven Ekman advection tends to deplete the seawater heat on SES of the wake and increase the heat on the NWS. The largest temporal and spatial variabilities are found within the geostrophic heat advection while compensating for the Ekman advection term somehow. In December 2011, the combined effects of these three terms generated the warming extension on the NWS and the cold tongue on the SES, forming an asymmetric thermal distribution across the wind wake. Our analysis highlights the role of heat advection concerning the thermal evolution process on a sub-seasonal timescale and clarified the regional asymmetric thermal responses to the wind wake in the winter.

Author Contributions: Conceptualization, J.S.; methodology, X.-M.L., X.C. and T.Z.; writing—original draft preparation, J.S.; writing—review and editing, J.S. and X.-M.L.

Funding: This research was supported by Hainan Provincial Natural Science Foundation of China (grant number 417215), National Natural Science Foundation of China (grant number 41706203) and the National Key R&D Program of China (2016YFC1401008).

Acknowledgments: We gratefully acknowledge the Envisat/ASAR data from European Space Agency (ESA), OISST and CCMP data from RSS (http://www.remss.com/), sea surface radiation products from ERA5 of Copernicus Climate Change Service (https://cds.climate.copernicus.eu/cdsapp#!/home), and ADT products from CMEMS (http://marine.copernicus.eu/).The authors would like to thank anonymous reviewers and the editors for suggestions that improved this manuscript. The authors also thank Jianjun Liang and Yili Zhao for providing suggestions and comments.

Conflicts of Interest: The authors declare no conflict of interest.

References

1. Smith, R.B.; Gleason, A.C.; Gluhosky, P.A.; Grubišić, V. The Wake of St. Vincent. *J. Atmos. Sci.* **1997**, *54*, 606–623. [CrossRef]
2. Caldeira, R.M.A.; Groom, S.; Miller, P.; Pilgrim, D.; Nezlin, N.P. Sea-surface signatures of the island mass effect phenomena around Madeira Island, Northeast Atlantic. *Remote Sens. Environ.* **2002**, *80*, 336–360. [CrossRef]
3. Xie, S.-P.; Liu, W.T.; Liu, Q.; Nonaka, M. Far-Reaching Effects of the Hawaiian Islands on the Pacific Ocean-Atmosphere System. *Science* **2001**, *292*, 2057–2060. [CrossRef] [PubMed]
4. Dong, C.; McWilliams, J.C. A numerical study of island wakes in the Southern California Bight. *Cont. Shelf Res.* **2007**, *27*, 1233–1248. [CrossRef]
5. Harlan, J.A.; Swearer, S.E.; Leben, R.R.; Fox, C.A. Surface circulation in a Caribbean island wake. *Cont. Shelf Res.* **2002**, *22*, 417–434. [CrossRef]
6. Li, J.; Wang, G.; Xie, S.-P.; Zhang, R.; Sun, Z. A winter warm pool southwest of Hainan Island due to the orographic wind wake. *J. Geophys. Res.* **2012**, *117*. [CrossRef]
7. Yan, Y.; Chen, C.; Ling, Z. Warm water wake off northeast Vietnam in the South China Sea. *Acta Oceanol. Sin.* **2014**, *33*, 55–63. [CrossRef]
8. Wyrtki, K. *Scientific Results of Marine Investigations of the South China Sea and the Gulf of Thailand 1959–1961*; Naga Report Vol. 2; University of California: San Diego, CA, USA, 1961; pp. 164–169.
9. Shaw, P.-T.; Chao, S.-Y. Surface circulation in the South China Sea. *Deep Sea Res. Part I Oceanogr. Res. Pap.* **1994**, *41*, 1663–1683. [CrossRef]
10. Chu, P.C.; Edmons, N.L.; Fan, C. Dynamical Mechanisms for the South China Sea Seasonal Circulation and Thermohaline Variabilities. *J. Phys. Oceanogr.* **1999**, *29*, 2971–2989. [CrossRef]
11. Xie, L.; Pallàs-Sanz, E.; Zheng, Q.; Zhang, S.; Zong, X.; Yi, X.; Li, M. Diagnosis of 3D Vertical Circulation in the Upwelling and Frontal Zones East of Hainan Island, China. *J. Phys. Oceanogr.* **2017**, *47*, 755–774. [CrossRef]
12. Lin, P.; Cheng, P.; Gan, J.; Hu, J. Dynamics of wind-driven upwelling off the northeastern coast of Hainan Island. *J. Geophys. Res. Oceans* **2016**, *121*, 1160–1173. [CrossRef]
13. Su, J.; Pohlmann, T. Wind and topography influence on an upwelling system at the eastern Hainan coast. *J. Geophys. Res.* **2009**, *114*, C06017. [CrossRef]
14. Minh, N.N.; Patrick, M.; Florent, L.; Sylvain, O.; Gildas, C.; Damien, A.; Van Uu, D. Tidal characteristics of the gulf of Tonkin. *Cont. Shelf Res.* **2014**, *91*, 37–56. [CrossRef]
15. Hu, J.Y.; Kawamura, H.; Tang, D.L. Tidal front around the Hainan Island, northwest of the South China Sea. *J. Geophys. Res.* **2003**, *108*, 3342. [CrossRef]
16. Lü, X.; Qiao, F.; Wang, G.; Xia, C.; Yuan, Y. Upwelling off the west coast of Hainan Island in summer: Its detection and mechanisms. *Geophys. Res. Lett.* **2008**, *35*. [CrossRef]
17. Li, Y.; Peng, S.; Wang, J.; Yan, J.; Huang, H. On the Mechanism of the Generation and Interannual Variations of the Summer Upwellings West and Southwest Off the Hainan Island. *J. Geophys. Res. Oceans* **2018**, *123*, 8247–8263. [CrossRef]
18. Gulev, S.K.; Latif, M.; Keenlyside, N.; Park, W.; Koltermann, K.P. North Atlantic Ocean control on surface heat flux on multidecadal timescales. *Nature* **2013**, *499*, 464–467. [CrossRef]
19. Sha, J.; Yan, X.-H.; Li, X. The horizontal heat advection in the Middle Atlantic Bight and the cross-spectral interactions within the heat advection. *J. Geophys. Res. Oceans* **2017**, *122*, 5652–5665. [CrossRef]
20. Dong, S.; Kelly, K.A. Seasonal and interannual variations in geostrophic velocity in the Middle Atlantic Bight. *J. Geophys. Res.* **2003**, *108*, 3172. [CrossRef]
21. Dong, S.; Kelly, K.A. Heat Budget in the Gulf Stream Region: The Importance of Heat Storage and Advection. *J. Phys. Oceanogr.* **2004**, *34*, 1214–1231. [CrossRef]
22. Mountain, D.G.; Strout, G.A.; Beardsley, R.C. Surface heat flux in the Gulf of Maine. *Deep Sea Res. Part II Top. Stud. Oceanogr.* **1996**, *43*, 1533–1546. [CrossRef]
23. Lentz, S.J.; Shearman, R.K.; Plueddemann, A.J. Heat and salt balances over the New England continental shelf, August 1996 to June 1997. *J. Geophys. Res.* **2010**, *115*. [CrossRef]
24. Closa, J.; Rosich, B.; Monti-Guarnieri, A. The ASAR wide swath mode products. In Proceedings of the IGARSS 2003, 2003 IEEE International Geoscience and Remote Sensing Symposium, Toulouse, France, 21–25 July 2003; Proceedings (IEEE Cat. No.03CH37477). IEEE: New York, NY, USA, 2003; Volume 2, pp. 1118–1120.

25. Hersbach, H. Comparison of C-Band Scatterometer CMOD5.N Equivalent Neutral Winds with ECMWF. *J. Atmos. Ocean. Technol.* **2010**, *27*, 721–736. [CrossRef]
26. Portabella, M.; Stoffelen, A. On Scatterometer Ocean Stress. *J. Atmos. Ocean. Technol.* **2009**, *26*, 368–382. [CrossRef]
27. Horstmann, J.; Koch, W.; Lehner, S. Ocean wind fields retrieved from the advanced synthetic aperture radar aboard ENVISAT. *Ocean Dyn.* **2004**, *54*, 570–576. [CrossRef]
28. Li, X.; Zheng, W.; Yang, X.; Zhang, J.A.; Pichel, W.G.; Li, Z. Coexistence of Atmospheric Gravity Waves and Boundary Layer Rolls Observed by SAR. *J. Atmos. Sci.* **2013**, *70*, 3448–3459. [CrossRef]
29. Horstmann, J.; Koch, W. Measurement of Ocean Surface Winds Using Synthetic Aperture Radars. *IEEE J. Ocean. Eng.* **2005**, *30*, 508–515. [CrossRef]
30. Chang, R.; Zhu, R.; Badger, M.; Hasager, C.; Xing, X.; Jiang, Y. Offshore Wind Resources Assessment from Multiple Satellite Data and WRF Modeling over South China Sea. *Remote Sens.* **2015**, *7*, 467–487. [CrossRef]
31. Reynolds, R.W.; Smith, T.M.; Liu, C.; Chelton, D.B.; Casey, K.S.; Schlax, M.G. Daily High-Resolution-Blended Analyses for Sea Surface Temperature. *J. Clim.* **2007**, *20*, 5473–5496. [CrossRef]
32. Atlas, R.; Hoffman, R.N.; Ardizzone, J.; Leidner, S.M.; Jusem, J.C.; Smith, D.K.; Gombos, D. A Cross-calibrated, Multiplatform Ocean Surface Wind Velocity Product for Meteorological and Oceanographic Applications. *Bull. Am. Meteorol. Soc.* **2010**, *92*, 157–174. [CrossRef]
33. Hersbach, H.; Bell, B.; Berrisford, P.; Horányi, A.; Muñoz, J.; Nicolas, J.; Radu, R.; Schepers, D.; Simmons, A.; Soci, C.; et al. Global reanalysis: Goodbye ERA-Interim, hello ERA5. *ECMWF Newsl.* **2019**, *159*, 17–24.
34. Al Senafi, F.; Anis, A.; Menezes, V. Surface Heat Fluxes over the Northern Arabian Gulf and the Northern Red Sea: Evaluation of ECMWF-ERA5 and NASA-MERRA2 Reanalyses. *Atmosphere* **2019**, *10*, 504. [CrossRef]
35. Cheney, R.; Miller, L.; Agreen, R.; Doyle, N.; Lillibridge, J. TOPEX/POSEIDON: The 2-cm solution. *J. Geophys. Res.* **1994**, *99*, 24555–24563. [CrossRef]
36. Hasager, C.; Astrup, P.; Zhu, R.; Chang, R.; Badger, M.; Hahmann, A. Quarter-Century Offshore Winds from SSM/I and WRF in the North Sea and South China Sea. *Remote Sens.* **2016**, *8*, 769. [CrossRef]
37. Liu, P.; Tsimpidi, A.P.; Hu, Y.; Stone, B.; Russell, A.G.; Nenes, A. Differences between downscaling with spectral and grid nudging using WRF. *Atmos. Chem. Phys.* **2012**, *12*, 3601–3610. [CrossRef]
38. Von Storch, H.; Langenberg, H.; Feser, F. A Spectral Nudging Technique for Dynamical Downscaling Purposes. *Mon. Weather Rev.* **2000**, *128*, 3664–3673. [CrossRef]
39. Waldron, K.M.; Paegle, J.; Horel, J.D. Sensitivity of a Spectrally Filtered and Nudged Limited-Area Model to Outer Model Options. *Mon. Weather Rev.* **1996**, *124*, 529–547. [CrossRef]
40. Hong, S.-Y.; Noh, Y.; Dudhia, J. A New Vertical Diffusion Package with an Explicit Treatment of Entrainment Processes. *Mon. Weather Rev.* **2006**, *134*, 2318–2341. [CrossRef]
41. Kain, J.S.; Fritsch, J.M. A One-Dimensional Entraining/Detraining Plume Model and Its Application in Convective Parameterization. *J. Atmos. Sci.* **1989**, *47*, 2784–2802. [CrossRef]
42. Lin, Y.-L.; Farley, R.D.; Orville, H.D. Bulk Parameterization of the Snow Field in a Cloud Model. *J. Clim. Appl. Meteorol.* **1983**, *22*, 1065–1092. [CrossRef]
43. Mlawer, E.J.; Taubman, S.J.; Brown, P.D.; Iacono, M.J.; Clough, S.A. Radiative transfer for inhomogeneous atmospheres: RRTM, a validated correlated-k model for the longwave. *J. Geophys. Res. Atmos.* **1997**, *102*, 16663–16682. [CrossRef]
44. Dudhia, J. Numerical Study of Convection Observed during the Winter Monsoon Experiment Using a Mesoscale Two-Dimensional Model. *J. Atmos. Sci.* **1989**, *46*, 3077–3107. [CrossRef]
45. Decharme, B. Influence of runoff parameterization on continental hydrology: Comparison between the Noah and the ISBA land surface models. *J. Geophys. Res. Atmos.* **2007**, *112*. [CrossRef]
46. Menendez, M.; García-Díez, M.; Fita, L.; Fernández, J.; Méndez, F.J.; Gutiérrez, J.M. High-resolution sea wind hindcasts over the Mediterranean area. *Clim. Dyn.* **2014**, *42*, 1857–1872. [CrossRef]
47. Wang, G. Characteristics Analysis of High Winds over the Bohai and North Yellow Seas and Numerical Study of a High Wind Event. Master Thesis, Ocean University of China, Qingdao, China, 2013.
48. Stoffelen, A. Toward the true near-surface wind speed: Error modeling and calibration using triple collocation. *J. Geophys. Res.* **1998**, *103*, 7755–7766. [CrossRef]
49. Caires, S. Validation of ocean wind and wave data using triple collocation. *J. Geophys. Res.* **2003**, *108*, 3098. [CrossRef]

50. Pan, M.; Fisher, C.K.; Chaney, N.W.; Zhan, W.; Crow, W.T.; Aires, F.; Entekhabi, D.; Wood, E.F. Triple collocation: Beyond three estimates and separation of structural/non-structural errors. *Remote Sens. Environ.* **2015**, *171*, 299–310. [CrossRef]
51. Yilmaz, M.T.; Crow, W.T. The Optimality of Potential Rescaling Approaches in Land Data Assimilation. *J. Hydrometeorol.* **2013**, *14*, 650–660. [CrossRef]
52. Ekman, V.W. *On the Influence of the Earth's Rotation on Ocean-Currents*; Arkiv För Matematik, Astronomi Och Fysik, Bd. 2, No. 11; Almqvist & Wiksells boktryckeri, A.-B.: Uppsala, Sweden, 1905.
53. Welander, P. Wind Action on a Shallow Sea: Some Generalizations of Ekman's Theory. *Tellus* **1957**, *9*, 45–52. [CrossRef]
54. Estrade, P.; Marchesiello, P.; De Verdière, A.C.; Roy, C. Cross-shelf structure of coastal upwelling: A two —Dimensional extension of Ekman's theory and a mechanism for inner shelf upwelling shut down. *J. Mar. Res.* **2008**, *66*, 589–616. [CrossRef]
55. Cushman-Roisin, B.; Beckers, J.-M. *Introduction to Geophysical Fluid Dynamics: Physical and Numerical Aspects*; Academic Press: Oxford, UK, 2011; Volume 101.
56. Garratt, J.R. *The Atmospheric Boundary Layer*; Cambridge University Press: Cambridge, UK, 1994; ISBN 978-0-521-46745-2.
57. Stevenson, J.W.; Niiler, P.P. Upper Ocean Heat Budget During the Hawaii-to-Tahiti Shuttle Experiment. *J. Phys. Oceanogr.* **1983**, *13*, 1894–1907. [CrossRef]
58. Sha, J.; Jo, Y.-H.; Oliver, M.J.; Kohut, J.T.; Shatley, M.; Liu, W.T.; Yan, X.-H. A case study of large phytoplankton blooms off the New Jersey coast with multi-sensor observations. *Cont. Shelf Res.* **2015**, *107*, 79–91. [CrossRef]
59. Gao, J.; Shi, M.; Chen, B.; Guo, P.; Zhao, D. Responses of the circulation and water mass in the Beibu Gulf to the seasonal forcing regimes. *Acta Oceanol. Sin.* **2014**, *33*, 1–11. [CrossRef]
60. Wang, B.; Huang, F.; Wu, Z.; Yang, J.; Fu, X.; Kikuchi, K. Multi-scale climate variability of the South China Sea monsoon: A review. *Dyn. Atmos. Oceans* **2009**, *47*, 15–37. [CrossRef]
61. Chao, S.-Y.; Shaw, P.-T.; Wu, S.Y. El Niño modulation of the South China Sea circulation. *Prog. Oceanogr.* **1996**, *38*, 51–93. [CrossRef]
62. Zhang, Y.; Sperber, K.R.; Boyle, J.S. Climatology and interannual variation of the East Asian winter monsoon: Results from the 1979-95 NCEP/NCAR reanalysis. *Mon. Weather Rev.* **1997**, *125*, 2605–2619. [CrossRef]
63. Tang, D.; Kawamura, H.; Lee, M.-A.; Van Dien, T. Seasonal and spatial distribution of chlorophyll-a concentrations and water conditions in the Gulf of Tonkin, South China Sea. *Remote Sens. Environ.* **2003**, *85*, 475–483. [CrossRef]
64. Liu, Y.; Tang, D.; Evgeny, M. Chlorophyll Concentration Response to the Typhoon Wind-Pump Induced Upper Ocean Processes Considering Air-Sea Heat Exchange. *Remote Sens.* **2019**, *11*, 1825. [CrossRef]
65. Wilkin, J.L. The Summertime Heat Budget and Circulation of Southeast New England Shelf Waters. *J. Phys. Oceanogr.* **2006**, *36*, 1997–2011. [CrossRef]
66. Shi, Q.; Bourassa, M.A. Coupling Ocean Currents and Waves with Wind Stress over the Gulf Stream. *Remote Sens.* **2019**, *11*, 1476. [CrossRef]
67. Thomson, R.E.; Gower, J.F.R.; Bowker, N.W. Vortex Streets in the Wake of the Aleutian Islands. *Mon. Weather Rev.* **1977**, *105*, 873–884. [CrossRef]
68. Li, J.; Zhang, R.; Ling, Z.; Bo, W.; Liu, Y. Effects of Cardamom Mountains on the formation of the winter warm pool in the gulf of Thailand. *Cont. Shelf Res.* **2014**, *91*, 211–219. [CrossRef]

Article

Comparison of True-Color and Multispectral Unmanned Aerial Systems Imagery for Marine Habitat Mapping Using Object-Based Image Analysis

Apostolos Papakonstantinou *, Chrysa Stamati and Konstantinos Topouzelis

Department of Marine Sciences, Marine Remote Sensing Group, University of the Aegean, University Hill, 81100 Mytilene, Greece; xrysa.stamati93@gmail.com (C.S.); topouzelis@marine.aegean.gr (K.T.)

* Correspondence: apapak@geo.aegean.gr

Received: 18 December 2019; Accepted: 4 February 2020; Published: 7 February 2020

Abstract: The use of unmanned aerial systems (UAS) over the past years has exploded due to their agility and ability to image an area with high-end products. UAS are a low-cost method for close remote sensing, giving scientists high-resolution data with limited deployment time, accessing even the most inaccessible areas. This study aims to produce marine habitat mapping by comparing the results produced from true-color RGB (tc-RGB) and multispectral high-resolution orthomosaics derived from UAS geodata using object-based image analysis (OBIA). The aerial data was acquired using two different types of sensors—one true-color RGB and one multispectral—both attached to a UAS, capturing images simultaneously. Additionally, divers' underwater images and echo sounder measurements were collected as in situ data. The produced orthomosaics were processed using three scenarios by applying different classifiers for the marine habitat classification. In the first and second scenario, the k-nearest neighbor (k-NN) and fuzzy rules were applied as classifiers, respectively. In the third scenario, fuzzy rules were applied in the echo sounder data to create samples for the classification process, and then the k-NN algorithm was used as the classifier. The in situ data collected were used as reference and training data. Additionally, these data were used for the calculation of the overall accuracy of the OBIA process in all scenarios. The classification results of the three scenarios were compared. Using tc-RGB instead of multispectral data provides better accuracy in detecting and classifying marine habitats when applying the k-NN as the classifier. In this case, the overall accuracy was 79%, and the Kappa index of agreement (KIA) was equal to 0.71, which illustrates the effectiveness of the proposed approach. The results showed that sub-decimeter resolution UAS data revealed the sub-bottom complexity to a large extent in relatively shallow areas as they provide accurate information that permits the habitat mapping in extreme detail. The produced habitat datasets are ideal as reference data for studying complex coastal environments using satellite imagery.

Keywords: coastal remote sensing; habitat mapping; unmanned aerial vehicle (UAV); unmanned aircraft system (UAS); drone; object-based image analysis (OBIA); UAS data acquisition

1. Introduction

Coastal zones are among the most populated and most productive areas in the world, offering a variety of habitats and ecosystem services. The European Commission highlights the importance of coastal zone management with the application of different policies and related activities, which were adopted through the joint initiatives of Maritime Spatial Planning and Integrated Coastal Management [1]. The aim is to promote sustainable growth of maritime and coastal activities and to use coastal and marine resources sustainably. Several other environmental policies are included in this initiative, like the Marine Strategy Framework Directive, the Climate Change Adaptation, and the Common Fisheries Policy [1]. Marine habitats have important ecological and regulatory functions and should be monitored in order to detect ecosystem changes [2,3]. Thus, habitat mapping is a prime necessity for environmental planning and management [4,5]. The continued provision of updated habitat maps has a decisive contribution to the design and coordination of relevant actions, the conservation of natural resources, and the monitoring of changes caused by natural disasters or anthropogenic effects [6]. Habitat maps are in critical demand, raising increasing interest among scientists monitoring sensitive coastal areas. The significance of coastal habitat mapping lies in the need to prevent anthropogenic interventions and other factors that affect the coastal environment [7]. Habitat maps are spatial representations of natural discrete seabed areas associated with particular species, communities, or co-occurrences. These maps can reflect the nature, distribution, and extent of disparate natural environments and can predict the species distribution [8].

Remote sensing has long-been identified as a technology capable of supporting the development of coastal zone monitoring and habitat mapping over large areas [1,9]. These processes require multitemporal data, either from satellites or from unmanned aerial systems (UAS). The availability of very high-resolution orthomosaics presents increasing interest, as it provides to scientists and relevant stakeholders detailed information for understanding coastal dynamics and implementing environmental policies [6,10–13]. However, the use of high-resolution orthomosaics created from UAS data is expected to improve mapping accuracy. This improvement is due to the high-spatial resolution of the orthomosaics less than 30 cm.

The increasing demand of detailed maps for monitoring the coastal areas requires automatic algorithms and techniques. Object-based image analysis (OBIA) is an object-based analysis of remote sensing imagery that uses automated methods to partition imagery into meaningful image-objects and generate geographic information in a GIS-ready format, from which new knowledge can be obtained [6,14,15].

In literature, there are several studies presenting the combination of OBIA with UAS imagery in habitat mapping. Husson et al. 2016 demonstrated an automated classification of nonsubmerged aquatic vegetation using OBIA and tested two classification methods (threshold classification and random forest) using eCognition®to true-color UAS images [2]. Furthermore, the produced automated classification results were compared to those of the manual mapping. In another study, Husson et al. 2017 combined height data from a digital surface model (DSM) created from overlapping UAS images with the spectral and textural features from the UAS orthomosaic to test if classification accuracy can be further improved [3]. They proved that the use of DSM-derived height data increased significantly the overall accuracy by 4%–21% for growth forms and 3%–30% for the dominant class. They concluded that height data have a significant potential to efficiently increase the accuracy of the automated classification of nonsubmerged aquatic vegetation.

Ventura et al. 2016 [16] carried out habitat mapping using a low-cost UAS. They tried to locate coastal areas suitable for fish nursery in the study area using UAS data and applying three different classification approaches. UAS data were collected using a video camera, and the acquired video was converted into a photo sequence, resulting in the orthomosaic of the study area. In this study, three classifications were performed using three different methods: (i) maximum likelihood in ArcGIS 10.4, (ii) extraction and classification of homogeneous objects (ECHO) in MultiSpec 3.4, and (iii) OBIA in eCognition Developer 8.7, with an overall accuracy of 78.8%, 80.9%, and 89.01%, respectively [16]. In a subsequent study, Ventura et al. 2018 [12] referred to the island of Giglio in Central Italy, where they carried out habitat mapping in three different coastal environments. Using OBIA and the nearest neighbor algorithm as classifiers, four different classifications were applied to identify Posidonia Oceanica meadows, nurseries for juvenile fish, and biogenic reefs with overall accuracies of 85%, 84%, and 80%, respectively. In another study, Makri D. et al. 2018 [17], a multiscale image analysis methodology was performed at Livadi Beach located on Folegandros Island, Greece. Landsat-8 and Sentinel-2 imagery were georeferenced, and atmospheric and water columns were corrected and analyzed using OBIA. As in situ data, high-resolution UAS data were collected. These data were used, as well in the classification and accuracy assessment. The nearest neighbor algorithm and fuzzy logic rules were used as classifiers. In this study, the overall accuracy was calculated to be 53% for Landsat-8 and 66% for Sentinel- 2 imagery. Duffy J. et al. 2018 [18] studied the creation of seagrass maps in Wales using a light drone for data acquisition. Three different classification methods were examined using the R 3.3.1 software [19]. The first classification performed unsupervised classification to true-color RGB (tc-RGB) data implemented in the 'RStoolbox' package [20,21]. The second classification was realized using tc-RGB data in combination with the texture of the image. Finally, the third classification was based on the object-based image analysis in the Geographic Resources Analysis Support System (GRASS) 7.0 software [22,23]. The accuracy of the classifications was based on the root mean square deflection (RMSD), and the results showed that unsupervised classifications had better accuracy in the seagrass coverage in comparison with the object-based image analysis method. These studies demonstrate that UAS data can provide critical information regarding the OBIA classification process and have the potential to increase classification accuracy in habitat mapping.

This study aimed to investigate the use of an automated classification approach by applying OBIA to high-resolution UAS multispectral and true-color RGB imagery for marine habitat mapping. Based on orthorectified image mosaics (here called UAS orthomosaics), we perform OBIA to map marine habitats in areas with varying levels of habitat complexity on the coastal zone. For the first time, UAS tc-RGB and multispectral orthomosaics were processed following OBIA methodology, and the classification results were compared by applying different classifiers for marine habitat mapping. Furthermore, the performance of the same classifier when applied to different orthomosaics (tc-RGB and multispectral) was examined in terms of accuracy and efficiency in classifications for marine habitat mapping. The object-based image analysis was performed using as classifiers the k-nearest neighbor algorithm and fuzzy logic rules. The validity of the produced maps was estimated using the overall accuracy and the Kappa index coefficient. Furthermore, divers' photos and roughness information derived from echo sounders were used as in situ data to assess the final results. Finally, we compared the results between multispectral and true-color UAS data for the automatic classification habitat mapping and analyzed them concerning the classification accuracy.

2. Materials and Methods

2.1. Study Area

The area used in this study is located in the coastal zone of Pamfila Beach on Lesvos Island, Greece (Figure 1). Lesvos is the third-largest Greek island, having 320 km of coastline, located in the Northeastern Aegean Sea. Pamfila Beach lies in the eastern part of Lesvos Island (39° 9′30.17″ N, 26° 31′53.35″ E), and the islet called Pamfilo is in front of the beach. Furthermore, an olive press and a petroleum storage facility are located close to the beach. This combination creates a unique marine environment; thus, sea meadows mapping is in critical demand for this area. There are four main marine habitats: hard bottom, sand, seagrass, and mix hard substrate. The hard substrate appears in the intertidal and the very shallow zone (0 to 1.5 m). The sand class covers mainly the southern part of the study area, and the mixed hard substrate is mainly located in the center of the study area, at depths of about 2.5 to 6.5 m. The seagrass class (*Posidonia Oceanica*) is dominant in the area and is presented at a depth of about 1.5 to 3.5 m. and 6.5 to 7.9 m.

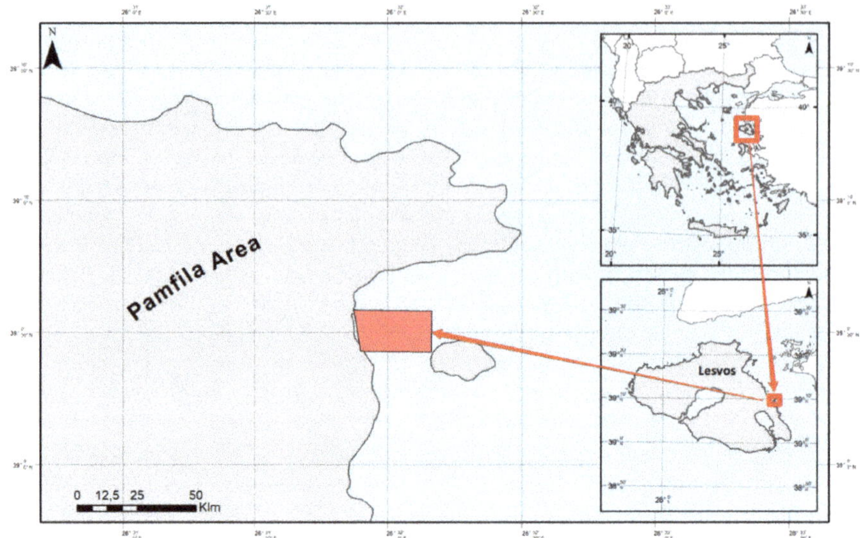

Figure 1. Location map of the Pamfila study area depicted with the red rectangle. Location maps of Greece (top right) and Lesvos Island (bottom right).

2.2. Classification Scheme

Four categories of the European Nature Information System (EUNIS, 2007) habitat classification list were selected for the classification process: (i) hard substrate, (ii) seagrass, (iii) sand, and (iv) mixed hard substrate. Due to the high resolution of orthomosaics (2 to 5 cm), we assume that each pixel is covered 100% by its class; i.e., a pixel is categorized as seagrass when it is covered 100%. The four classes selected as the predominant classes were previously known from local observations and studies. The seagrass class (code A5.535)—namely, *Posidonia Oceanica* beds—is characterized by the presence of the marine seagrass (phanerogam) *Posidonia Oceanica*. This habitat is an endemic Mediterranean species creating natural formations called Posidonia meadows. These meadows are found at depths of 1 to 50 m. The sand class (code A5.235) is encountered in very shallow water where the sea bottom is characterized by fine sand, usually with homogeneous granulometry and of terrigenous origin. The class of hard substrate (code A3.23) is characterized by the presence of many photophilic algae covering hard bottoms in moderately exposed areas. Finally, the mixed hard substrate class is considered as an assemblage of sand, seagrass, and dead seagrass leaves covering a hard substrate.

The depth in the study area was measured using a single-beam echo sounder and had a variation of 0 to 8 m. The depth zones accurately defined and proved very useful in the explanation of the results (Figure 2).

2.3. In Situ Data

In the present study, a combination of photographs taken while snorkeling and measurements derived from a single-beam echo sounder attached to a small inflatable boat were used as in situ data.

2.3.1. Divers' Data Acquisition

The study area was initially investigated using an orthomosaic map from a previous demo flight to create sections and transects that divers would follow to capture underwater images. The selected transect orientation was designed to target all four selected classes equally. Furthermore, reference spots were selected using the abovementioned orthomosaic map to help the divers' orientation during snorkeling. The divers' equipment used for taking photos as in situ data was a GoPro 4 camera. The divers team followed the predetermined transects in the area and captured with an underwater image all the preselected spots having one of the four classes selected for classification. The in situ sampling took place on 06/07/2018, having a shape of a trapezium, and lasted one hour. In total, 125 underwater images were collected from the divers using a GoPro 4 Hero underwater camera. Each one of the underwater images was captured in a way to represent one training class. After quality control, 33 images were properly classified. The number of the images used for the classification was reduced by 92 due to the following reasons: (i) 18 images (14.4%) were not clearly focused on the sea bottom due to the depth; (ii) 21 images (16,8%) were not geolocated due to the repetitive sea bottom pattern (for example, sandy sea bottom); (iii) 25 images (20%) were duplicated, as they were captured in the same position by the divers to assure that the dominant class will be captured; and, finally, (iv) 28 images (23%) were blurry due to the sea state and the water quality. All images were interpreted to define the classes. Additionally, the position where these images were captured was identified by photointerpretation from the divers' team using the tc-RGB orthomosaic.

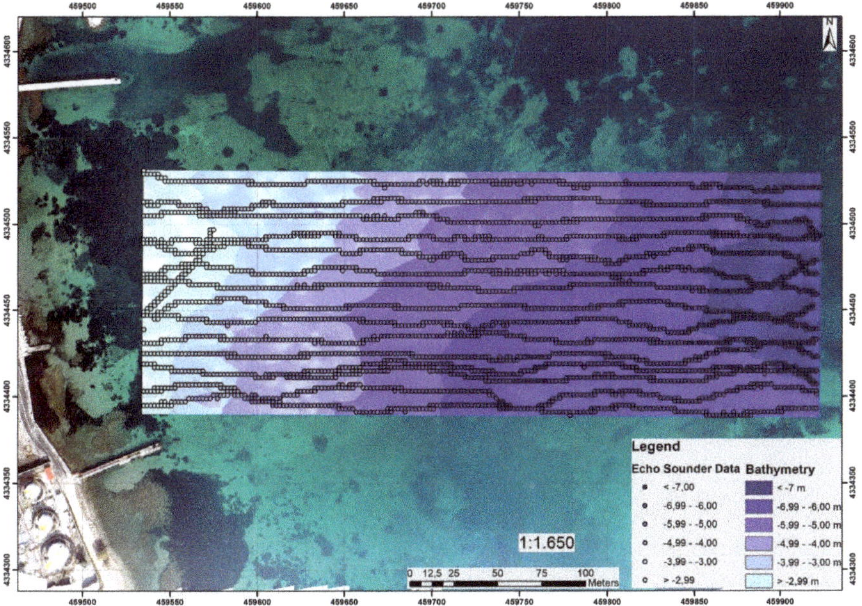

Figure 2. Study area bathymetry and locations where the echo sounder data were collected.

2.3.2. Echo Sounder Data Acquisition

Echo sounder data were collected by SEMANTIC TS personnel using a single-beam sound device attached to a small inflatable vessel on 06/07/2018. The measurements were based on the reflection of the acoustic pulse of the echo sounder device. SEMANTIC TS has developed a signal-processing algorithm based on discriminant analysis to scrutinize the response energy level of the sea bottom pulses [24]. The derived information includes the depth and the substrate roughness, while the precise geographical positions of the acquired data are also recorded. Roughness and bathymetry products of the study area are provided as a raster dataset (Figures 2 and 3). ArcMap 10.3.1 software [25] was used to process the data, and the canvas was converted into a point vector shapefile. The point vector shapefile contained a total of 3.364 points with the roughness and bathymetry info.

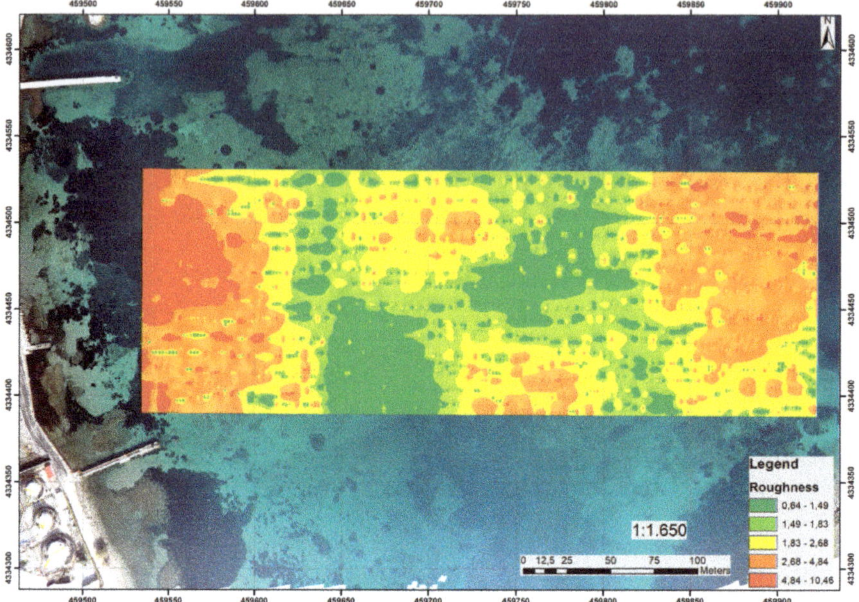

Figure 3. Echo sounder roughness data.

2.4. UAS Data Acquisition

The UAS used for data acquisition was a vertical take-off and landing (VTOL) configuration capable of autonomous flights using preprogrammed flight paths. The configuration was a custom-made airborne system based on the S900 DJI hexacopter airframe, having a 25-min flight time with an attached payload of 1.5 Kg. The payload consisted of two sensors: a multispectral and a tc-RGB. The configuration lies in the Pixhawk autopilot system, which is an open-source UAS [26,27]. For the positioning of the UAS during the flight, a real-time kinematic (RTK) global positioning system was connected to the autopilot.

2.4.1. Air-Born Sensors Used

The true-color RGB sensor used in this study was a Sony A5100 24.3-megapixel camera with interchangeable Sigma ART 19 mm 1:2.8 DN0.2M/0.66Feet lens capable of precise autofocusing in 0.06 sec, capturing high-quality true-color RGB (tc-RGB) images. This sensor was selected because of the lightweight (0.224 kg), manual parameterization and auto-triggering capabilities, using an electronic pulse due to its autopilot. The multispectral camera was a Slantrange 3P (S3P) sensor equipped with an ambient illumination sensor for deriving spectral reflectance-based end-products, an integrated global positioning system, and an inertial measurement unit system. The S3P has a quad-core 2.26 GHz processor and an embedded 2GB RAM for onboard preprocessing. The S3P used is a modified multispectral sensor, having the wavebands adjusted to match the coastal, blue, green, and near-infrared (NIR) wavebands on the Sentinel-2 mission (Table 1). The scope of the waveband modification to the S3P imager aimed at simulating Sentinel-2 data in finer geospatial resolution.

Table 1. Waveband information for the Slantrange 3P imager and Sentinel-2 missions. Coastal (C), blue (B), green (G), and near-infrared (NIR).

	Slantrange 3P		Sentinel-2A		Sentinel-2B	
	Centre (nm)	Bandwidth (nm)	Centre (nm)	Bandwidth (nm)	Centre (nm)	Bandwidth (nm)
C	450	20	442.7	27	442.2	45
B	500	80	492.4	98	492.1	98
G	550	40	559.8	45	559.0	46
NIR	850	100	832.8	145	832.9	133

The open-access Mission Planner v1.25 software was used as a ground station for real-time monitoring of the UAS telemetry and for programming the survey missions [28].

2.4.2. Flight Parameters

Data acquisition took place on the 8th July 2018 in the Pamfila area. Before the data acquisition, the UAS toolbox was used to predict the optimal flight time [29]. The toolbox automates a protocol, which summarizes the parameters that affect the reliability and the quality of the data acquisition process over the marine environment using UAS. Each preprogrammed acquisition flight had the following flight parameters: 85% overlap in-track and 80% side-lap. The UAV was flying at a height of 120 m above sea level (absolute height), having a velocity of 5 m/s. Thus, the tc-RGB sensor was adjusted for capturing a photograph every 3.28 s in the nadir direction and the multispectral every 1 s. The obtained ground resolution of the tc-RGB images acquired from the UAS was 2.15 cm/pixel, and for the multispectral imager was 4.84 cm. Ground sampling resolution varied due to the focal length, pixel pitch, and sensor size of the sensors used. After a quality control inspection, the majority of the images were selected for further processing. The data acquisition information is depicted in the following table (Table 2), followed by the final orthomosaics obtained from the UAV (Figure 4).

Table 2. The number of raw images and spatial resolution acquired using a suite of sensors attached to the unmanned aircraft system (UAS).

Name of Sensor	Number of Images	Sensor Resolution (Pixel)	Flight Height (m)	Ground Resolution (cm/Pixel)
SlantRange 3P	568	1216 × 991	120	4.84
Sony A5100	181	6000 × 4000	120	2.15

Prior to the survey missions, georeferenced targets, designed in a black and white pattern, were used as ground control points (GCP), having dimensions of 40 × 40 cm. In total, 18 targets were placed on the coastal zone of the study area. The GCP's coordinates were measured in the Hellenic geodetic system using a real-time kinematic method yielding a total root mean square error (RMSE) of 0.244 cm.

2.5. Methodological Workflow

An overview of the followed methodological workflow is given below (Figure 5). The methodology is organized into four steps: (i) data acquisition and creation of tc-RGB and multispectral orthomosaics using the UAS-SfM (structure-from-motion) framework [10,30], (ii) orthomosaics preprocessing, (iii) object-based image analysis, and (iv) accuracy assessment. After UAS and in situ data acquisition, the divers' photos passed quality control and were interpreted to define the class that is depicted. Furthermore, in the preprocessing stage, a land mask was applied to both orthomosaics. Then, the OBIA analysis was performed, starting with the objects' segmentation and then performing the classification of benthic substrates of the study area. The classification was implemented following three scenarios. In the first and second scenarios, the k-nearest neighbor and fuzzy rules were applied as classifiers, respectively. In the third scenario, both fuzzy rules and k-NN were applied. The in situ data collected were separated into two nonoverlapping datasets: one for training and one for accuracy assessment.

Figure 4. True-color RGB orthomosaic (top) and multispectral orthomosaic (bottom).

2.6. Orthomosaic Creation

Structure-from-motion (SfM) photogrammetry applied to images captured from UAS platforms is increasingly being used for a wide range of applications. SfM is a photogrammetric technique that creates two and three-dimensional visualizations from two-dimensional image sequences [31,32]. The methodology is one of the most effective methods in the computer vision field, consisting of a series of algorithms that detect common features in images and convert them into three-dimensional information. For the realization of this study, the Agisoft Photoscan 1.4.1 [33] was used, since it automates the SfM process in a user-friendly interface with a concrete workflow [32,34–37].

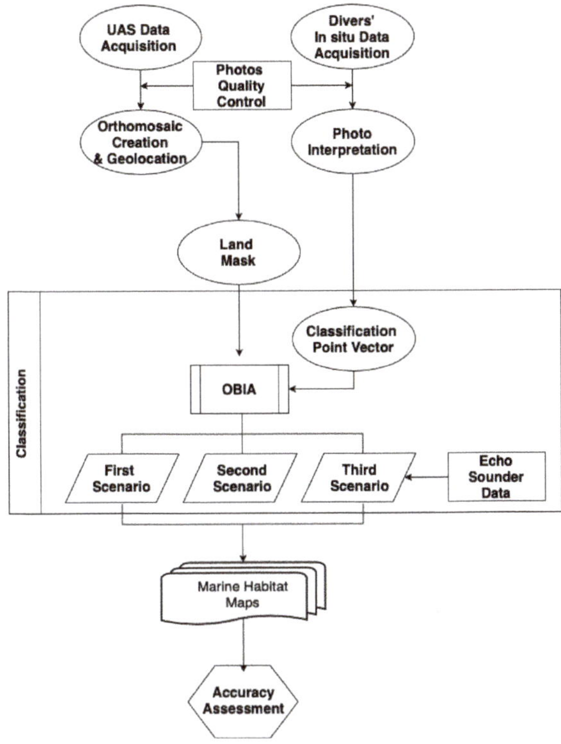

Figure 5. Methodological workflow. UAS: unmanned aircraft system. OBIA: object-based image analysis.

Georeferencing of the tc-RGB orthomosaic was achieved using 18 ground control points (GCPs). The use of GCPs for the geolocation of the tc-RGB orthomosaic had, as a result, an RMSE of 1.56 (cm). The achieved accuracy met the requirements of the authors for creating highly detailed maps. The S3P multispectral orthomosaic was georeferenced using the georeferenced tc-RGB orthomosaic as a base map [38]. An inter-comparison of the georeferenced orthomosaics was implemented to best match common characteristic reference points. Final end-products consisted of georeferenced (i) S3P—coastal (450 nm), blue (500 nm), green (550 nm), and NIR (850 nm) and (ii) Sony A5100 in true-color RGB orthomosaics. The size in pixels for the produced derivatives created from SfM is presented in the following table (Table 3).

Table 3. Size in pixels of produced orthomosaics according to the used sensors: true-color RGB (Sony A5100) and multispectral (Slantrange).

Name of Sensor	Orthomosaic Size (Pixels)
SlantRange 3P	12,122 × 4103
Sony A5100	11,446 × 21,001

Orthomosaic Preprocessing

Before the preprocessing step, the divers' in situ data were interpreted, and the two orthomosaics were initially checked for their geolocation accuracy. Then, the produced orthomosaics were land-masked for extracting information based solely on the pixels of the sea. The land mask was created by editing the coastline as a vector in a shapefile format using ArcMap 10.3.1 software [25].

2.7. Object-Based Image Analysis (OBIA)

For the OBIA, three necessary steps were required: (i) the segmentation procedure; (ii) the definition of the classes that will be later classified; and (iii) the delineation of the classifier, i.e., the classification algorithm defining the class where the segments will be classified.

The first step of the OBIA analysis is to create objects from orthomosaics through the segmentation procedure. The orthomosaics are segmented by a multiresolution segmentation algorithm in eCognition software [39] (Figure 6). The initial outcome of the segmentation is meaningful objects defined from the scale parameters, image layer weights, and composition of the homogeneity criterion [40]. The thresholds used for these parameters were determined empirically, based on the expertise of the image interpreter. For the tc-RGB orthomosaic, the parameters of scale, shape, and compactness were set to 100, 0.1, and 0.9, respectively. For the multispectral orthomosaic, the parameters were set: 15 for the scale, 0.1 for the shape, and 0.9 for the compactness.

Figure 6. Results of the segmentation process for scenarios 1 and 2. The segments depicted are the result of the unification process to form one segment per category: true-color RGB orthomosaic (top) and multispectral orthomosaic (bottom).

In the classification step, two algorithms were selected, the k-nearest neighbor (k-NN) and the fuzzy classification, using eCognition®. The k-nearest neighbor (k-NN) algorithm classifies image objects into a specific feature range and with given samples pertaining to preselected categories. Once a representative set of samples is declared, each object is classified on the resemblance of band values between the training object and the objects to be classified among the k-nearest neighbors. Thus, each segment in the image is denoted with a value equal to 1 or 0, expressing whether the object belongs to a class or not. In fuzzy classification, instead of binary decision-making, the probability of whether an object belongs to a class or not is calculated using membership functions. The limits of each class are no longer restricted using thresholds, but classification functions are used within the dataset, in which every single parameter value will have a chance of being assigned to a class [41,42]. Both classifiers are part of the eCognition®software. The k-NN algorithm was selected as a historical classification algorithm (fast deployment without the need of many samples), and fuzzy logic was selected to add specific knowledge into the classification derived from the training areas.

2.8. Classification Scenarios and Accuracy Assesment

The classification step of the present study was performed in sub-decimeter UAS orthomosaics using three different scenarios for defining the best classifier for marine habitat mapping in complex coastal areas. In the first and second scenarios, the k-nearest neighbor and fuzzy rules were applied as classifiers, respectively. The third scenario was realized by applying fuzzy rules in the echo sounder data for sample creation, and the classification was performed using k-NN.

In the first scenario, in total, 60 segments were selected as training samples (15 samples per class) based on the divers' underwater images. Each segment of the orthomosaic was classified into one of the four predefined classes using the k-NN algorithm as the classifier, and the segments of the same class were merged into one single object. The resulting classification was used for the creation of the final habitat map. This procedure was followed in both tc-RGB and multispectral orthomosaics for the first scenario.

In the second scenario, the same training sample as in the first scenario was used, and the fuzzy rules were defined and applied. The appropriate fuzzy expression for each class was created after examining the relationship between the classes for the mean segment value of the three image bands (tc-RGB). Then, the mean value of the three bands was selected as an input, and the logical rules "AND" and "OR" were used where necessary. The segments were classified using the fuzzy expressions for all classes, and the objects in the same class were merged into one single object. Thus, the results were exported as one polygon vector shapefile. As in the first scenario, the above-presented process was applied to both the tc-RGB and multispectral orthomosaics.

Finally, for the third scenario, the analysis was designed based on the following objectives: (i) examination of usefulness of the roughness echo sounder info to the classification procedure and (ii) comparison of the roughness efficiency against the underwater images photo interpretation for the creation of training samples. More specifically, a new training dataset was created based on the echo sounder's roughness information. In total, 3028 roughness points were derived from the echo sounder dataset, and fuzzy rules were applied to 90% of the points (2724) for the creation of training samples for the classification classes. In the produced segmentation results, the k-NN algorithm was applied as the classifier for the calculation of the final classification for each of the four classes. As the small research vessel was not able to access for safety reasons to depths less than 1.5 m, we manually added samples where necessary.

Accuracy assessment calculates the percentage of the produced map that approaches the actual field reality. In this study, we created a validation dataset that was not overlapped with the calibration dataset. In total, 120 points were generated using the geolocated underwater images (in situ divers' data) and an expert's photointerpretation. Initially, a point vector file was created by locating the 33 underwater images in the tc-RGM orthomosaic. Due to the small number of points produced from in situ data, the dataset was densified via photointerpretation by an expert using the high-resolution tc-RGB orthomosaic. In total, 87 points were added, and, therefore, the final point vector file ended with 120 points (30 per class). In Figure 7, all points are illustrated with colors according to their assigned taxon. In this figure, divers and photointerpretation points are depicted in round and star shapes, respectively. The 120 accuracy assessment points were firstly assigned to their relevant segments (i.e., forming 120 segments) and the total number of pixels forming the segments was used as the accuracy assessment dataset for all classifications. Furthermore, in the third scenario, 10% of the roughness points (304 points) equally distributed to all four classes were used also as accuracy assessment data. The accuracy matrices created for all scenarios provided information regarding the user and producer accuracy, overall accuracy, and the Kappa index coefficient (Table 4, Table 5, Table 6, Table 7, Table 8, and Table 9).

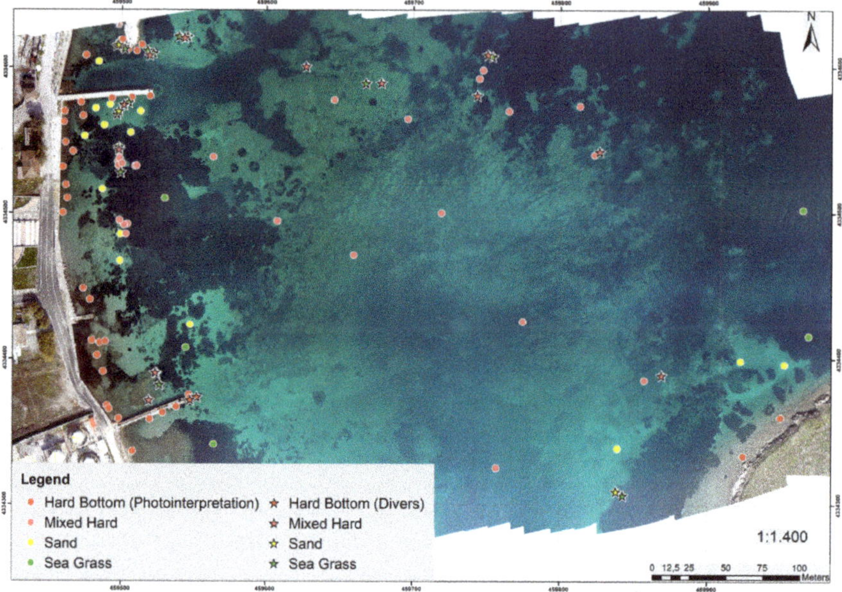

Figure 7. Point vector dataset containing: (i) 33 points (round shape) derived from in situ divers' photos and (ii) 87 points created through photointerpretation (star shape). All points are colored according to their assigned class.

Table 4. Accuracy matrix for the true-color RGB (tc-RGB) orthomosaic, with k-nearest neighbor as the classifier. KIA: Kappa index of agreement.

		Reference Data				
		Sea Grass (Pixels)	**Mixed Hard Substrate (Pixels)**	**Sand (Pixels)**	**Hard Bottom Substrate (Pixels)**	**Sum (Pixels)**
Training Data	**Sea Grass**	37,2384	0	0	4053	37,6437
	Mixed Hard Substrate	18,174	52,3604	164,822	32,669	739,269
	Sand	25,776	85,351	172,498	4167	287,792
	Hard Bottom Substrate	6642	52,941	0	433,426	493,009
	Sum	422,976	661,896	337,320	474,315	
	Producers' accuracy	0.88	0.79	0.51	0.91	
	Users' accuracy	0.98	0.70	0.60	0.88	
	KIA per class	0.85	0.66	0.42	0.88	
		Total accuracy		**0.79**		
		KIA		**0.71**		

Table 5. Accuracy matrix for the multispectral orthomosaic with k-nearest neighbor as the classifier.

		Reference Data				
		Sea Grass (Pixels)	**Mixed Hard Substrate (Pixels)**	**Sand (Pixels)**	**Hard Bottom Substrate (Pixels)**	**Sum (Pixels)**
Training Data	**Seagrass**	13,677	9800	0	2333	25,810
	Mixed Hard Substrate	0	19,428	0	2038	21,466
	Sand	0	0	14,585	1294	15,879
	Hard Bottom Substrate	0	3752	1311	22,312	27,375
	Sum	13,677	32,980	15,896	27,977	
	Producers accuracy	1	0.59	0.92	0.8	
	Users accuracy	0.53	0.91	0.92	0.82	
	KIA per class	1	0.46	0.9	0.71	
		Total accuracy		**0.77**		
		KIA		**0.7**		

Table 6. Accuracy matrix for the tc-RGB orthomosaic applying the fuzzy rules.

		Reference Data				
	Classes	**Sea Grass (Pixels)**	**Mixed Hard Substrate (Pixels)**	**Sand (Pixels)**	**Hard Bottom Substrate (Pixels)**	**Sum (Pixels)**
Training Data	**Sea Grass**	320,025	75,491	21,987	10,838	428,341
	Mixed Hard Substrate	1807	404,691	43,224	4062	453,784
	Sand	101,144	135,194	272,109	73,351	581,798
	Hard Bottom Substrate	0	46,520	0	386,064	432,584
	Sum	422,976	661,896	337,320	474,315	
	Producers accuracy	0.76	0.61	0.81	0.81	
	Users accuracy	0.75	0.89	0.47	0.89	
	KIA per class	0.69	0.49	0.72	0.76	
		Total accuracy		**0.73**		
		KIA		**0.64**		

Table 7. Accuracy matrix for the multispectral orthomosaic applying the fuzzy rules.

		Reference Data				
	Classes	**Sea Grass (Pixels)**	**Mixed Hard Substrate (Pixels)**	**Sand (Pixels)**	**Hard Bottom Substrate (Pixels)**	**Sum (Pixels)**
Training Data	**Sea Grass**	13,677	5761	3333	0	22,771
	Mixed Hard Substrate	0	19,998	0	4371	24,369
	Sand	0	2786	12,563	16,633	31,982
	Hard Bottom Substrate	0	4435	0	6973	11,408
	Sum	13,677	32,980	15,896	27,977	
	Producers accuracy	1	0.6	0.79	0.25	
	Users accuracy	0.6	0.82	0.39	0.61	
	KIA per class	1	0.46	0.68	0.14	
		Total accuracy		**0.59**		
		KIA		**0.46**		

Table 8. The tc-RGB data classification accuracy matrix for the multispectral orthomosaic.

	Classes	Reference Data				
		Sea Grass (Pixels)	**Mixed Hard Substrate (Pixels)**	**Sand (Pixels)**	**Hard Bottom Substrate (Pixels)**	**Sum (Pixels)**
Training Data	**Sea Grass**	379,034	84,089	14,876	28,568	506,567
	Mixed Hard Substrate	40,103	276,346	108,280	8800	433,529
	Sand	3442	259,310	214,157	36,108	513,017
	Hard Bottom Substrate	394	42,069	7	400,810	443,280
	Unclassified	3	82	0	29	114
	Sum	422,976	661,896	337,320	474,315	
	Producers accuracy	0.9	0.42	0.63	0.85	
	Users accuracy	0.75	0.64	0.42	0.9	
	KIA per class	0.86	0.24	0.5	0.8	
		Total accuracy		**0.67**		
		KIA		**0.56**		

Table 9. Echo sounder data classification accuracy matrix for the multispectral orthomosaic.

	Classes	Reference Data				
		Sea Grass (Pixels)	**Mixed Hard Substrate (Pixels)**	**Sand (Pixels)**	**Hard Bottom Substrate (Pixels)**	**Sum (Pixels)**
Training Data	**Sea Grass**	11,858	7811	4530	1987	26,186
	Mixed Hard Substrate	1681	13,615	2156	2936	20,388
	Sand	133	7042	4747	6685	18,607
	Hard Bottom Substrate	2	4504	4457	16,357	25,320
	Unclassified	3	8	6	12	29
	Sum	13,677	32,980	15,896	27,977	
	Producers accuracy	0.87	0.41	0.3	0.58	
	Users accuracy	0.45	0.67	0.26	0.65	
	KIA per class	0.81	0.24	0.12	0.42	
		Total accuracy		**0.51**		
		KIA		**0.35**		

3. Results

In this section, the classification results of the three scenarios are presented for both the tc-RGB and multispectral data. The classification was performed using four habitat classes: (i) hard substrate, (ii) seagrass, (iii) sand, and (iv) mixed hard substrate.

3.1. Scenario 1: k-Nearest Neighbor as Classifier

In the first scenario, k-nearest neighbor (k-NN) was used as the classifier and, when it was applied to the tc-RGB orthomosaic, it resulted in the classification map depicted in Figure 8A. According to the bathymetry, the hard substrate appears in the intertidal and the very shallow zone at depths of 0 to 1.5 m. The sand class covers mainly the southern part of the study area, and the mixed hard substrate is located mainly in the middle of the study area, at depths of 2.5 to 6.5 m. The seagrass (*Posidonia Oceanica*) class is divided into two parts: one on the west side of the beach (right next to the hard substrate) at a depth of about 1.5 to 3.5 m and one on the eastern part of the beach where the depths vary approximately 6.5 to 7.9 m. The percentage of the segments that belong to each of the classes is 28.9%, 11.5%, 34.9%, and 24.8% for hard substrate, sand, mixed hard substrate, and seagrass, respectively. Additionally, the percent areal coverages in square meters for each class were calculated for this scenario. The area covered from the tc-RGB is 178,386 square meters, and the sand class is presented as occupying 16.29% of the total area mapped. The other classes' percentage areal coverage are 33.04% for seagrass, 44.50% for mixed hard substrate, and 6.17% for the hard bottom.

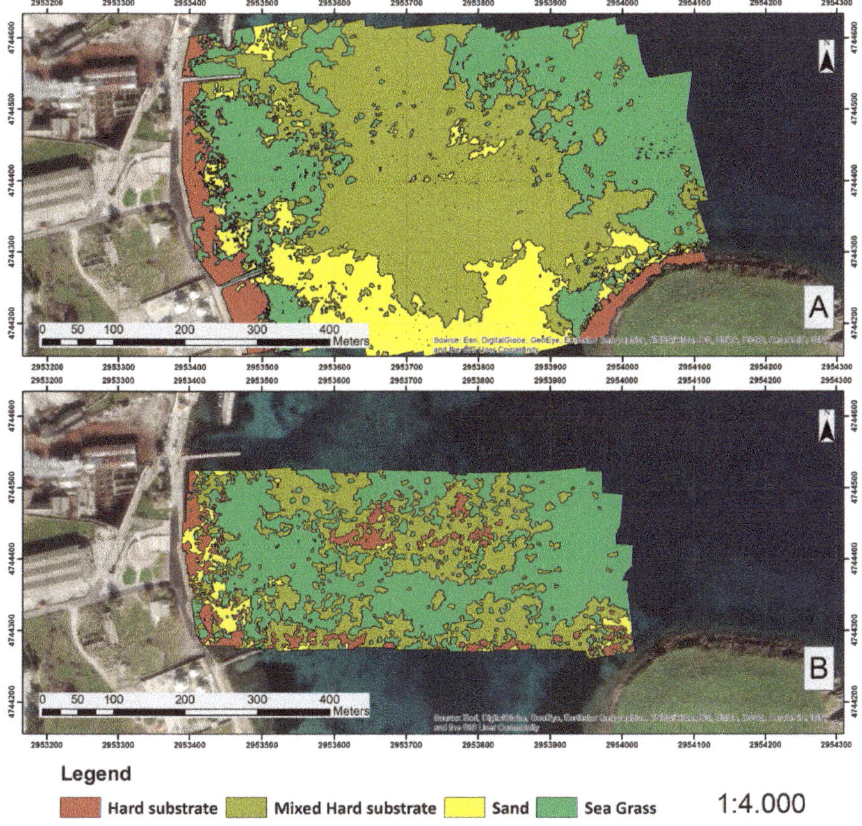

Figure 8. Classified maps for k-nearest neighbor as the classifier for the study area: (**A**) true-color RGB orthomosaic map and (**B**) multispectral orthomosaic map.

The overall accuracy of the classification was 79%, and the Kappa index of agreement (KIA) was 0.71 (Table 4). The sand class presented a lower accuracy compared to the other classes. Several objects in this class have been incorrectly classified as mixed hard substrate.

The classification results of the multispectral orthomosaic are illustrated in Figure 8B. It can be noted that the hard substrate appears mainly in the intertidal and the shallow zone at depths of 0 to 3.5 m, having been assigned 31.9% of the total objects. The sand class covers small areas of the western part of the study area, and the mixed hard substrate is located mainly in the middle of the study area, at depths of about 2.5 to 9.9 m. The seagrass is scattered almost throughout the study area, covering a wide depth range of about 0 to 9.9 m. The classified object percentages for the sand, mixed hard substrate, and seagrass classes are 20.8%, 27.8%, and 19.5%, respectively. In an area of 90,857 square meters covered by multispectral orthomosaic, the percent areal coverages by class are 3.78%, 51.55%, 36.54%, and 8.13% for sand, sea grass, mixed hard substrate, and hard substrate, respectively.

The achieved overall classification accuracy was 77%, and the KIA coefficient was equal to 0.70 (Table 5). In this case, the mixed hard substrate class presented lower accuracy in comparison with the rest of the classes, since several objects have been incorrectly classified as seagrass and mixed hard substrate.

3.2. Scenario 2: Fuzzy Rules as Classifier

In the second scenario, fuzzy rules were used as the classifier, and the four classes created from the tc-RGB and multispectral orthomosaics are illustrated in Figure 9A,B, respectively. According to the classification results for the tc-RGB orthomosaic, the percentage of the assigned objects for the hard substrate, sand, mixed hard substrate, and seagrass classes are 27.9%, 20%, 29.9%, and 23.1%, respectively. For this scenario, the percentage areal coverage was calculated to be 23.35% sand, 35.35% seagrass, 37.27% mixed hard substrate, and 4.02% hard substrate.

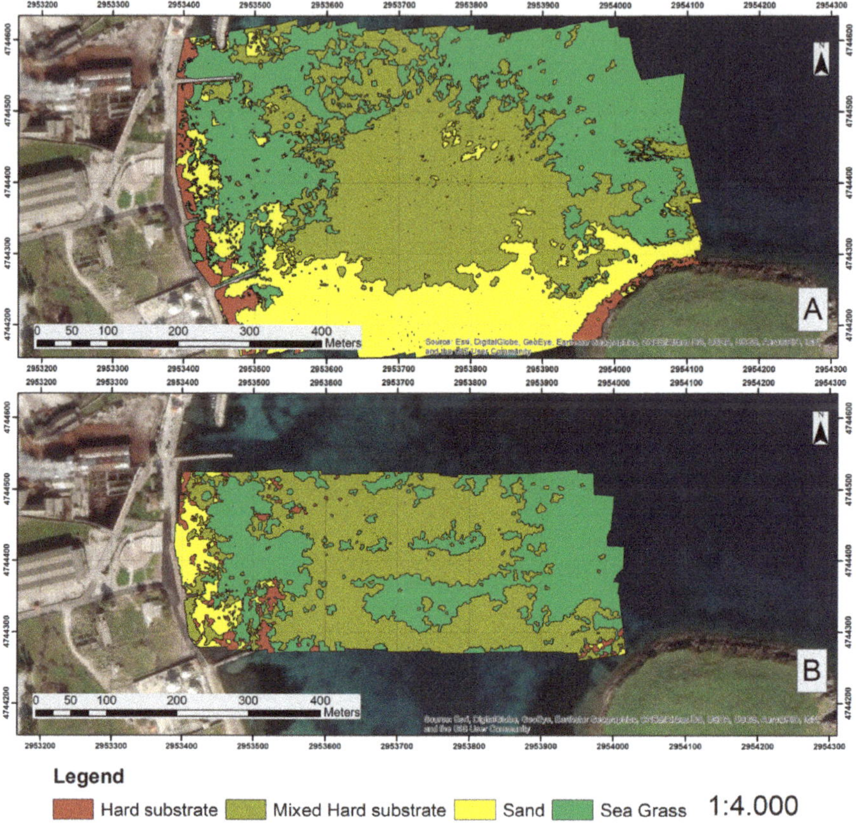

Figure 9. Classification maps for the orthomosaic of the study area applying the fuzzy rules: (**A**) tc-RGB orthomosaic map and (**B**) multispectral orthomosaic map.

When fuzzy rules were used as the classifier, the overall accuracy was 73%, and the coefficient KIA was 0.64 (Table 6). In this scenario, the mixed hard substrate class presented lower accuracy in comparison with the other three classes, since several objects of this class have been incorrectly classified as hard substrate, sand, and seagrass.

From the classification results of the multispectral orthomosaic (Figure 9B), the seagrass class comes into sight, divided into two parts. The first part is on the west side of the beach (right next to the sand class) at depths of about 1.5 to 3.5 m and the second on the eastern part of the beach. In this part, the depths vary approximately from 6.5 to 7.9 m. The percentage of each class object is 13.1%, 23.6%, 37.6%, and 25.7% for hard substrate, sand, mixed hard substrate, and seagrass, respectively. Furthermore, the percentage of areal calculations saw that the sand class covers 4.99%, while the seagrass, mixed hard substrate, and hard bottom are covering 40.25%, 51.50%, and 3.27% of the total classified area, respectively. The overall classification accuracy obtained by the fuzzy rules is 59%, and the coefficient K is equal to 0.46 (Table 7). In this classification scenario, the hard substrate and mixed hard substrate classes present lower accuracy, since several objects in these classes have been incorrectly classified.

3.3. Scenario 3: k-NN and Fuzzy Rules as Classifiers

In the scenario where a combination of the fuzzy rules and k-NN are used as the classifiers, the four classes created from the classification of the tc-RGB orthomosaic are represented in Figure 10A. The percentage of the assigned objects for the four classes is 31.6%, 16.9%, 21.6%, and 29.8% for hard substrate, sand, mixed hard substrate, and seagrass, respectively. The classes created from the tc-RGB orthomosaic present the following results in percent areal coverage. The class sand has 21.66%. The overall classification accuracy is 67%, and the KIA coefficient is equal to 0.56 (Table 8). The mixed hard substrate and sand classes presented lower accuracy compared to the rest of the classes, regarding the tc-RGB orthomosaic in this scenario. Several objects of the mixed hard substrate and sand classes were incorrectly classified into other classes.

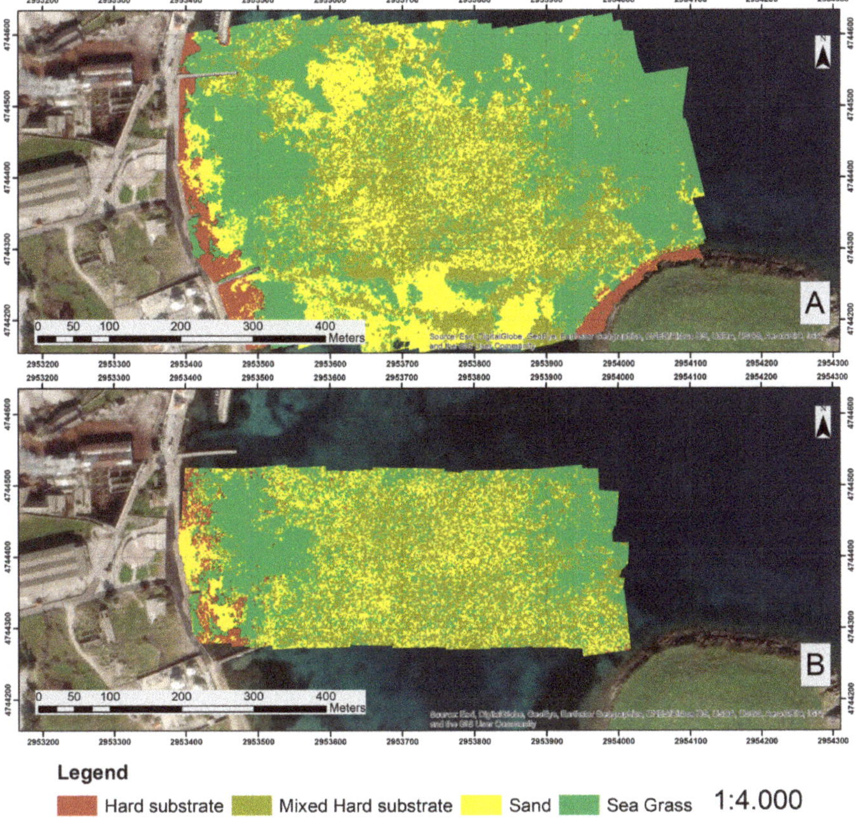

Figure 10. Classification maps for the orthomosaic of the study area, applying the k-NN and fuzzy rules: (**A**) the tc-RGB orthomosaic map and (**B**) multispectral orthomosaic map.

The multispectral orthomosaic classification results derived from scenario 3 are illustrated in Figure 10B. For each class, the percentage of the assigned objects was 35.1%, 10.2%, 29.3%, and 25.5% for hard substrate, sand, mixed hard substrate, and seagrass, respectively. The areal percentage coverage calculation results for the multispectral orthomosaic in scenario 3 were 21.72% coverage for sand, 33.62% for seagrass, 42.94% for mixed hard substrate, and, finally, 2.72% for hard substrate. In this scenario, the overall classification accuracy was relatively low (51%), and the coefficient KIA was very low (0.35) (Table 9). The hard substrate, mixed hard substrate, and sand classes presented lower accuracy compared to the seagrass class using the k-NN and fuzzy rules as classifiers. Several objects in the above-mentioned classes have been incorrectly classified.

The third scenario presents scattered areas of mixed hard substrate, in contrast to scenarios 1 and 2. The main difference is located in the center of the scene where the mixed classes are. On the contrary, all scenarios depict nicely the seagrass class in both shallow and deep waters. Hard bottom is well-defined in all scenarios. The multispectral dataset seems not to classify correctly the mix hard substrate and the sand classes.

Table 10 summarizes the overall accuracy of the three scenarios. In all scenarios, the tc-RGB orthomosaic responded better than the multispectral orthomosaic. The best performance (79%) was given in the first scenario for the k-NN classifier applied to the tc-RGB orthomosaic. The use of echo sounder data as training data did not increase the quality of the final classification maps, as the authors expected. On the contrary, the accuracy was worse when echo sounder data were used as training data, compared to those where solely underwater images were applied.

Table 10. Brief table of total accuracies for the tc-RGB and multispectral orthomosaics of the three scenarios of the study.

Scenario No.	Classifier	Training Data	Orthomosaic	Total Accuracy	Kappa Index
Scenario 1	k-Nearest Neighbor	Underwater images	tc-RGB	79%	0.71
		Underwater images	Multispectral	77%	0.7
Scenario 2	Fuzzy Rules	Underwater images	tc-RGB	73%	0.64
		Underwater images	Multispectral	59%	0.46
Scenario 3	k-Nearest Neighbor & Fuzzy Rules	Echo Sounder roughness	tc-RGB	67%	0.56
			Multispectral	51%	0.35

4. Discussion and Conclusions

In this work, we have shown that the utilization of UAS high in resolution and accuracy aerial photographs, in conjunction with the OBIA, can create quality habitat mapping. We demonstrated that habitat mapping information could be automatically extracted from sub-decimeter spatial true-color RGB images acquired from UAS. High-resolution classification maps produced from UAS orthomosaic using the OBIA approach enables the identification and measurement of habitat classifications (sand, hard bottom, seagrass, and mixed hard substrate) in the coastal zone over the total extent of the mapped area. The detailed geoinformation produced provides scientists with valuable information regarding the current state of the habitat species, i.e., the environmental state of sensitive coastal areas. Moreover, the derived data products enable in-depth analysis and, therefore, the identification of change detections caused by anthropogenic interventions and other factors.

The purpose of this paper was twofold: (a) to compare two types of orthomosaics, the tc-RGB and the multispectral, captured over a coastal area using OBIA with different classifiers to map the selected classes and (b) to examine the usefulness of the bathymetry and the roughness information derived from the echo sounder as training data to the UAS-OBIA methodology for marine habitat mapping.

From the comparison of the classification results, it can be concluded that the tc-RGB orthomosaic produces more valuable and robust results than the multispectral one. This is caused due to the multispectral imager specifications. More specifically, the multispectral sensor receives data from four discrete bands, and a global shutter is used. As a result, the sensor captures photos in a shorter time compared to the tc-RGB camera. Thus, the exposure time is shorter in the multispectral (SlantRange modified imager) than in the true-color RGB (Sony A5100) sensor, causing less light energy. In addition, during the process of the multispectral orthomosaic creation due to the data quality of the initial images, the final derivative was not uniform, thus presenting discrete irregularities. These anomalies appeared due to different illumination conditions and the sun glint; therefore, the multispectral orthomosaic classification presents lower accuracy values. It is crucial to follow a specific UAS flight protocol before each data acquisition, as presented by Doukari et al. in [29,43], for eliminating these anomalies. It should be mentioned that the UAS data acquisition procedure works fine over land, presenting a high accuracy level. However, it is not performed adequately over moving water bodies, especially when having a large extent; thus, no land is presented in the data.

Three scenarios were examined for the classification of the marine habitats using different classifiers. The k-nearest neighbor and fuzzy rules were applied in the first and second scenarios, respectively, and a combination of the fuzzy rules and k-NN algorithm in the third scenario. From the evaluation of the three scenarios' classification results, it can be concluded that the use of the tc-RGB instead of the multispectral data provides better accuracy in detecting and classifying marine habitats by applying the k-NN as the classifier. The overall accuracy using the K-NN classifier was 79%, and the Kappa index (KIA) was equal to 0.71. The results illustrate the effectiveness of the proposed approach when applied to sub-decimeter resolution UAS data for marine habitat mapping in complex coastal areas.

Furthermore, this study demonstrated that the roughness information derived from the echo sounder did not increase the final classification accuracy. Although the echo sounder roughness can be used to discriminate classes and produce maps of the substrates, it cannot be used directly as training data for classifying UAS aerial data. Based on the echo sounder signal, a proper roughness analysis should be initially performed to produce habitat maps, which in later stages could be used as training and validation data for the UAS data.

The results showed that UAS data revealed the sub-bottom complexity to a large extent in relatively shallow areas, providing accurate information and high spatial resolution, which permits habitat mapping with extreme detail. The produced habitat vectors are ideal as reference data for studies with satellite data of lower spatial resolution. Since UAS sub-decimeter spatial resolution imagery will be increasingly available in the future, it could play an important role in habitat mapping, as it serves the needs of various studies in the coastal environment. Finally, the combination of OBIA classification with UAS sub-decimeter orthomosaics implements a very accurate methodology for ecological applications. This approach is capable of recording the high spatiotemporal variability needed for habitat mapping, which has turned into a prime necessity for environmental planning and management.

UAS are increasingly used in habitat mapping [7,12,16–18,44], since they provide high-resolution data to inaccessible areas at a low cost and with high temporal repeatability [10,45,46]. The use of a multispectral camera with similar wavelengths to the Sentinel-2 satellite wavelengths was examined for the first time in the present study. Results indicated that the tc-RGB and multispectral orthomosaic perform similarly, and there is no significant advantage of the multispectral camera. This can be explained twofold: (a) due to the fact that the multispectral camera is designed for land measurements and due to the inherited optical properties that cannot distinguish small radiometric differences in water, and (b) the multispectral orthomosaic was problematic due to the large differences in actual multispectral images as a result of large overlaps between them. The multispectral imager over sea areas should contain small overlaps and should gain data in short shutter speeds, i.e., with larger acquisition times. Additionally, results show that echo sounder roughness should not be used for training classification algorithms. The total accuracy of the third scenario in both orthomosaics clearly indicates the inadequacy of bottom roughness for training datasets.

Moving forward, the authors believe that the rapidly developing field of lightweight drones and the miniaturization and the rapid advance of true-color RGB, multispectral, and hyperspectral sensors for close remote sensing will soon allow a more detailed mapping of marine habitats based on spectral signatures.

Author Contributions: Conceptualization, C.S., A.P., and K.T.; methodology, C.S., A.P., and K.T.; UAS data acquisition, A.P. and K.T.; orthomosaics creation, A.P.; data analysis, C.S. and K.T.; writing—original draft preparation, C.S., A.P., and K.T.; writing—review and editing, A.P. and K.T.; visualization, A.P.; and supervision, A.P. and K.T. All authors have read and agreed to the published version of the manuscript.

Funding: This research received no external funding.

Acknowledgments: The authors would like to acknowledge the SEMANTIC TS company for providing echo sounder data and Michaela Doukari, who helped in the orthomosaic creation process.

Conflicts of Interest: The authors declare no conflict of interest.

References

1. Ouellette, W.; Getinet, W. Remote sensing for Marine Spatial Planning and Integrated Coastal Areas Management: Achievements, challenges, opportunities and future prospects. *Remote Sens. Appl. Soc. Environ.* **2016**, *4*, 138–157. [CrossRef]
2. Husson, E.; Ecke, F.; Reese, H. Comparison of Manual Mapping and Automated Object-Based Image Analysis of Non-Submerged Aquatic Vegetation from Very-High-Resolution UAS Images. *Remote Sens.* **2016**, *8*, 724. [CrossRef]
3. Husson, E.; Reese, H.; Ecke, F. Combining Spectral Data and a DSM from UAS-Images for Improved Classification of Non-Submerged Aquatic Vegetation. *Remote Sens.* **2017**, *9*, 247. [CrossRef]
4. Cicin-Sain, B.; Belfiore, S. Linking marine protected areas to integrated coastal and ocean management: A review of theory and practice. *Ocean Coast. Manag.* **2005**, *48*, 847–868. [CrossRef]
5. Cogan, C.B.; Todd, B.J.; Lawton, P.; Noji, T.T. The role of marine habitat mapping in ecosystem-based management. *ICES J. Mar. Sci.* **2009**, *66*, 2033–2042. [CrossRef]
6. Papakonstantinou, A.; Doukari, M.; Stamatis, P.; Topouzelis, K. Coastal Management Using UAS and High-Resolution Satellite Images for Touristic Areas. *Int. J. Appl. Geospatial Res.* **2019**, *10*, 54–72. [CrossRef]
7. Topouzelis, K.; Doukari, M.; Papakonstantinou, A.; Stamatis, P.; Makri, D.; Katsanevakis, S. Coastal Habitat Mapping in the Aegean Sea Using High Resolution Orthophoto Maps. In Proceedings of the Fifth International Conference on Remote Sensing and Geoinformation of the Environment (RSCy2017), Paphos, Cyprus, 6 September 2017; Volume 1, p. 52.
8. Harris, P.T.; Baker, E.K. GeoHab Atlas of Seafloor Geomorphic Features and Benthic Habitats. In *Seafloor Geomorphology as Benthic Habitat*; Elsevier: Amsterdam, The Netherlands, 2012; pp. 871–890. ISBN 978-0-12-385140-6.
9. McDermid, G.J.; Franklin, S.E.; LeDrew, E.F. Remote sensing for large-area habitat mapping. *Prog. Phys. Geogr.* **2005**, *29*, 449–474. [CrossRef]
10. Papakonstantinou, A.; Topouzelis, K.; Doukari, M. UAS close range remote sensing for mapping coastal environments. In Proceedings of the Fifth International Conference on Remote Sensing and Geoinformation of the Environment (RSCy2017), Paphos, Cyprus, 6 September 2017; Volume 10444, p. 35.
11. Su, L.; Gibeaut, J. Using UAS hyperspatial RGB imagery for identifying beach zones along the South Texas Coast. *Remote Sens.* **2017**, *9*, 159. [CrossRef]
12. Ventura, D.; Bonifazi, A.; Gravina, M.F.; Belluscio, A.; Ardizzone, G. Mapping and Classification of Ecologically Sensitive Marine Habitats Using Unmanned Aerial Vehicle (UAV) Imagery and Object-Based Image Analysis (OBIA). *Remote Sens.* **2018**, *10*, 1331. [CrossRef]
13. Ventura, D.; Bonifazi, A.; Gravina, M.F.; Ardizzone, G.D. Unmanned Aerial Systems (UASs) for Environmental Monitoring: A Review with Applications in Coastal Habitats. In *Aerial Robots—Aerodynamics, Control and Applications*; IntechOpen: Rijeka, Croatia, 2017.
14. Hay, G.J.; Castilla, G. Geographic Object-Based Image Analysis (GEOBIA): A New Name for a New Discipline. In *Lecture Notes in Geoinformation and Cartography*; Blaschke, T., Lang, S., Hay, G.J., Eds.; Springer: Berlin/Heidelberg, Germany, 2008; pp. 75–89.
15. Blaschke, T. Object based image analysis for remote sensing. *ISPRS J. Photogramm. Remote Sens.* **2010**, *65*, 2–16. [CrossRef]
16. Ventura, D.; Bruno, M.; Jona Lasinio, G.; Belluscio, A.; Ardizzone, G. A low-cost drone based application for identifying and mapping of coastal fish nursery grounds. *Estuar. Coast. Shelf Sci.* **2016**, *171*, 85–98. [CrossRef]

17. Makri, D.; Stamatis, P.; Doukari, M.; Papakonstantinou, A.; Vasilakos, C.; Topouzelis, K. Multi-Scale Seagrass Mapping in Satellite Data and the Use of UAS in Accuracy Assessment. In Proceedings of the Sixth International Conference on Remote Sensing and Geoinformation of the Environment (RSCy2018), Paphos, Cyprus, 6 August 2018; Volume 10773, p. 33.
18. Duffy, J.P.; Pratt, L.; Anderson, K.; Land, P.E.; Shutler, J.D. Spatial assessment of intertidal seagrass meadows using optical imaging systems and a lightweight drone. *Estuar. Coast. Shelf Sci.* **2018**, *200*, 169–180. [CrossRef]
19. R Core Development Team. *A Language and Environment for Statistical Computing*; R Core Development Team: Vienna, Austria, 2019; Available online: https://www.R-project.org/ (accessed on 6 February 2020).
20. Leutner, B.; Horning, N. *RStoolbox: Tools for Remote Sensing Data Analysis R Package, Version 0.2.4*; R Core Development Team: Vienna, Austria, 2017.
21. Leutner, B.; Horning, N. *Package 'RStoolbox,' R Foundation for Statistical Computing, Version 0.1*; R Core Development Team: Vienna, Austria, 2017.
22. Neteler, M.; Bowman, M.H.; Landa, M.; Metz, M. GRASS GIS: A multi-purpose open source GIS. *Environ. Model. Softw.* **2012**, *31*, 124–130. [CrossRef]
23. Team, G.D. Geographic Resources Analysis Support System (GRASS GIS) Software, Version 7.2. 2017. Available online: https://grass.osgeo.org (accessed on 6 February 2020).
24. Noel, C.; Perrot, T.; Coquet, M.; Zerr, B.; Viala, C. Acoustic data fusion devoted to underwater vegetation mapping. *J. Acoust. Soc. Am.* **2008**, *123*, 3951. [CrossRef]
25. *ArcGIS Desktop Release 10.3.1*; Environmental Systems Research Institute: Redlands, CA, USA, 2013.
26. Meier, L.; Tanskanen, P.; Fraundorfer, F.; Pollefeys, M. PIXHAWK: A System for Autonomous Flight using Onboard Computer Vision. In Proceedings of the 2011 IEEE International Conference on Robotics and Automation, Shanghai, China, 9–13 May 2011; pp. 2992–2997.
27. Meier, L.; Tanskanen, P.; Fraundorfer, F.; Pollefeys, M. The Pixhawk Open-Source Computer Vision Framework for Mavs. *ISPRS-Int. Arch. Photogramm. Remote Sens. Spat. Inf. Sci.* **2011**, *XXXVIII-1/C22*, 13–18. [CrossRef]
28. ArduPilot, Mission Planner, What Is Mission Planner. 2017. Available online: http://ardupilot.org/planner/docs/mission-planner-overview.html (accessed on 19 December 2019).
29. Doukari, M.; Batsaris, M.; Papakonstantinou, A.; Topouzelis, K. A Protocol for Aerial Survey in Coastal Areas Using UAS. *Remote Sens.* **2019**, *11*, 1913. [CrossRef]
30. Sturdivant, E. Sturdivant, E.J.; Lentz, E.E.; Thieler, E.R.; Farris, A.S.; Weber, K.M.; Remsen, D.P.; Miner, S.; Henderson, R.E. UAS-SfM for Coastal Research: Geomorphic Feature Extraction and Land Cover Classification from High-Resolution Elevation and Optical Imagery. *Remote Sens.* **2017**, *9*, 1020. [CrossRef]
31. Torres, J.C.; Arroyo, G.; Romo, C.; De Haro, J. 3D Digitization using Structure from Motion. In Proceedings of the CEIG-Spanish Computer Graphics Conference, Jaén, Spain, 12–14 September 2012; pp. 1–10.
32. Westoby, M.J.J.; Brasington, J.; Glasser, N.F.F.; Hambrey, M.J.J.; Reynolds, J.M.M. Structure-from-Motion photogrammetry: A low-cost, effective tool for geoscience applications. *Geomorphology* **2012**, *179*, 300–314. [CrossRef]
33. Agisoft, L.L.C. Agisoft PhotoScan, Professional Edition, Version 1.4.1. 2018. Available online: https://www.agisoft.com/pdf/photoscan-pro_1_4_en.pdf (accessed on 6 February 2020).
34. Mathews, A.J.; Jensen, J.L.R. Visualizing and Quantifying Vineyard Canopy LAI Using an Unmanned Aerial Vehicle (UAV) Collected High Density Structure from Motion Point Cloud. *Remote Sens.* **2013**, *5*, 2164–2183. [CrossRef]
35. Dellaert, F.; Seitz, S.M.; Thorpe, C.E.; Thrun, S. Structure from motion without correspondence. In Proceedings of the IEEE Conference on Computer Vision and Pattern Recognition, CVPR 2000 (Cat. No. PR00662), Hilton Head Island, SC, USA, 15 June 2000; Volume 2, pp. 557–564.
36. Nex, F.; Remondino, F. UAV for 3D mapping applications: A. review. *Appl. Geomatics* **2013**, *6*, 1–15. [CrossRef]
37. Lowe, D.G. Distinctive Image Features from Scale-Invariant Keypoints. *Int. J. Comput. Vis.* **2004**, *60*, 91–110. [CrossRef]
38. Topouzelis, K.; Papakonstantinou, A.; Garaba, S.P. Detection of floating plastics from satellite and unmanned aerial systems (Plastic Litter Project 2018). *Int. J. Appl. Earth Obs. Geoinf.* **2019**, *79*, 175–183. [CrossRef]
39. *Trimble eCognition®Reference Book*; Trimble Germany GmbH: Munich, Germany, 2014.

40. Dimitrakopoulos, K.; Gitas, I.Z.; Polychronaki, A.; Katagis, T.; Minakou, C. Land Cover/Use Mapping Using Object Based Classification of SPOT Imagery. In *Proceedings of EARSeL Remote Sensing for Science, Education, and Natural and Cultural Heritage*; Reuter, R., Ed.; University of Oldenburg: Oldenburg, Germany, 2010; pp. 263–272.
41. Jabari, S.; Zhang, Y. Very high resolution satellite image classification using fuzzy rule-based systems. *Algorithms* **2013**, *6*, 762–781. [CrossRef]
42. Shani, A. Landsat Image Classification Using Fuzzy Sets Rule Base Theory. Master's Thesis, San Jose State University, San Jose, CA, USA, 2006; p. 57.
43. Doukari, M.; Papakonstantinou, A.; Topouzelis, K. Fighting the Sunglint Removal in UAV Images. In Proceedings of the 11th International Conference of the Hellenic Geographical Society (ICHGS-2018), Lavrion, Greece, 12–15 April 2018.
44. Hodgson, A.; Kelly, N.; Peel, D. Unmanned aerial vehicles (UAVs) for surveying Marine Fauna: A dugong case study. *PLoS ONE* **2013**, *8*, e79556. [CrossRef] [PubMed]
45. Topouzelis, K.; Papakonstantinou, A.; Doukari, M. Coastline Change Detection Using Unmanned Aerial Vehicles and Image Processing. *Fresenius Environ. Bull.* **2017**, *26*, 5564–5571.
46. Papakonstantinou, A.; Topouzelis, K.; Pavlogeorgatos, G. Coastline Zones Identification and 3D Coastal Mapping Using UAV Spatial Data. *ISPRS Int. J. Geo-Inf.* **2016**, *5*, 75. [CrossRef]

Article

Spatial and Temporal Variability of Open-Ocean Barrier Islands along the Indus Delta Region

Muhammad Waqas [1], Majid Nazeer [1,2,*], Muhammad Imran Shahzad [1,*] and Ibrahim Zia [3,4]

1. Earth & Atmospheric Remote Sensing Lab (EARL), Department of Meteorology, COMSATS University Islamabad, Islamabad 45550, Pakistan; waqas.butt07.vb@gmail.com
2. Key Laboratory of Digital Land and Resources, East China University of Technology, Nanchang 330013, Jiangxi, China
3. National Institute of Oceanography, Karachi 75600, Pakistan; ibrahimzia@hotmail.com
4. Department of Geography, University of Karachi, Karachi 75270, Pakistan

* Correspondence: majid.nazeer@comsats.edu.pk (M.N); imran.shahzad@comsats.edu.pk (M.I.S.); Tel.: +92-51-9247000 (M.N)

Received: 16 January 2019; Accepted: 17 February 2019; Published: 20 February 2019

Abstract: Barrier islands (BIs) have been designated as the first line of defense for coastal human assets against rising sea level. Global mean sea level may rise from 0.21 to 0.83 m by the end of 21st century as predicted by the Intergovernmental Panel on Climate Change (IPCC). Although the Indus Delta covers an area of 41,440 km^2 surrounded by a chain of BIs, this may result in an encroachment area of 3750 km^2 in Indus Delta with each 1 m rise of sea level. This study has used a long-term (1976 to 2017) satellite data record to study the development, movement and dynamics of BIs located along the Indus Delta. For this purpose, imagery from Landsat Multispectral Scanner (MSS), Thematic Mapper (TM), Enhanced Thematic Mapper Plus (ETM+), and Operational Land Imager (OLI) sensors was used. From all these sensors, the Near Infrared (NIR) band (0.7–0.9 μm) was used for the delineation and extraction of the boundaries of 18 BIs. It was found that the area and magnitude of these BIs is so dynamic, and their movement is so great that changes in their positions and land areas have continuously been changing. Among these BIs, 38% were found to be vulnerable to oceanic factors, 37% were found to be partially vulnerable, 17% remained partially sustainable, and only 8% of these BIs sustained against the ocean controlling factors. The dramatic gain and loss in area of BIs is due to variant sediment budget transportation through number of floods in the Indus Delta and sea-level rise. Coastal protection and management along the Indus Delta should be adopted to defend against the erosive action of the ocean.

Keywords: satellite remote sensing; Landsat; coastline; barrier island; morphological change; coastal ocean

---

1. Introduction

Global warming, being caused by the increase in temperature and atmospheric CO_2, acts as a catalyst in the melting of glaciers, expansion of ocean water, increase in sea surface temperature, rise in sea-level, and increase in tropical storms intensity [1–6]. Accelerated sea level rise (SLR) threatens human settlements, and environmental and economic assets which have been tremendously developed in the coastal zones over the last five decades. SLR also shows alarming situation to the low lying sandy beaches and barrier islands (BIs) and intensifies erosion along the coastal areas. It has been predicted that, globally, up-to one meter rise in sea-level for the next hundred years can severely increase salinity of estuarine water, disturb coastal sediment budget supply, disrupt marine life, damage sub-surface and surface fresh water supplies, and damage agricultural and industrial areas [7–9].

Coastal topography may constantly change with numerous oceanic and physical processes including tidal flooding, SLR, and tsunami effect. This may cause sinking of low-land areas, and erosion of sediments which is a great concern for coastal researchers and investigators [10–14]. Further sediments excavation, river modification and construction along coastal areas also play an important role in coastal areas including the BIs [15–19].

BIs are generally defined as the long offshore sand deposits which are 3 to 160 km in length and 1 to 3 km in width, and are separated but parallel to the main coastal area by the water inlets, bays and inland sea [20]. BIs form a first line of defense for the main-land against high-energy tropical cyclones and rise in sea-level, which would otherwise directly transfer their full energy against the seashore. Provided coastal sand supplies are plentiful, and the gradient of sea-bed is gentle, the main-land is usually bordered by a chain of BIs. The water in the estuaries and bays is continuously exchanged with the water of the open shelf by flow through tidal inlets, water gaps separating BIs from one another [21]. High wave energy is likely to close the tidal inlets while strong tidal flux keeps them open [22]. A recent survey reveals that 20,783 km of coastline is occupied by 2149 BIs, worldwide [23]. Therefore, rise in sea-level poses a considerable risk of 10% to the world's total population (7.6 billion) living in lowland coastal areas [24]. Furthermore, tropical cyclones hitting BIs and the coastal systems will cause more devastation to the coastal-urban environment and infrastructure, and also disturb the geometries of the BIs [25–27]. In order to track the changes in BIs shape, area, and their position, several methods have been adopted. For example, the topographic surveys, aerial photogrammetry, and GPS (Global Positioning System) surveys. Recently, the unmanned airial systems (UAS) have revolutionized the science and have proved their practice in remote sensing for short term change detection of coastal land masses, which include both autonomous and remotely piloted aircraft [28]. Although these methods have high spatial and temporal resolution for coastal assessment of short-term changes as well as long-term trends, but they are labor intensive and expensive for non-funded research projects. Thus, the free availability of the medium spatial resolution satellite data (Landsat and Sentinel) in the last decade has enabled the use of multi-sensor, multi-spectral, and multi-temporal satellite imagery of extensive coastal areas to detect and monitor long-term trends of coastal land masses and coastline changes.

1.1. Background of the Study

Pakistan's shoreline extends more than 1000 km along the Arabian Sea which is divided into two coastal areas (i.e., coastal areas of Sindh and Makran). More than 10% of the population of Pakistan is living in the locality of coastal zone, about 20% of the coastal area of Pakistan is comparatively developed, and 40% of the country's industrial areas are located somewhere on or near the shore. The coast of Karachi city is about 70 km long surrounded by a chain of BIs, oriented NW–SE. Currently, 6.8% of the total population of Pakistan (15 million people) is living in the vicinity of the coastal city Karachi of the Indus Delta Region (IDR) which makes it the fifth largest coastal city [29] where a rate of SLR has increased from 1.1 mm/year [30,31] to 1.8 mm/year. IDR is considered as economic hub for Pakistan because Port Qasim and Karachi Port handle about 90% of all external trades. The IDR is stretched more than 200 km and always affected by high southwestern summer monsoon (May–September) winds having average speed of 15 m/s but during the northeastern winter monsoon winds below with an average speed of 5 m/s [32]. Wave measuring at 20 m water depth offshore of Karachi city show a mean significant wave height of 1.8 m during SW monsoon with a mean wave period of 9 s and during the NE monsoon winds, significant wave height is about 1.2 m with a wave period of 6.5 s [33]. The geomorphological changes have occurred at the center of Karachi coast along Bundal and Buddo Islands [34]. IDR is experiencing geomorphological changes in all areas along its major and minor creeks in the form of erosion processes and low deposition rate of sediment budget from Indus River during post-damming era [35]. IDR has been considered as one of the most dynamic deltaic systems which consists of 16 major tidal inlets having 17,000 km^2 area with an active tidal flat area of 10,000 km^2, significant number of BIs [36] and hosts world's largest arid mangrove forest [37].

The IDR is ranked as the fourth highest delta in receiving average high wave energy as compared with other famous deltas around the world (e.g., Nile, Mississippi Niger, Ganges, and Ebro deltas) [38]. The soil of IDR is less resistant to ocean hydro dynamics so the deltaic shoreline recede at average rate of 50 m/y along the central delta coast [35]. SLR followed by sea-water intrusion is claiming more land and changing the geomorphology of the low-lying land masses in the Indus Delta [32,39,40]. A SLR of about 1 meter is expected to sink or sea-encroach an area of 3750 km^2 in the Indus Delta [30]. The global mean sea-level rose at a mean rate of 1.7 mm/year from 1900 to 2010 and at a rate 3.2 mm/year between 1993 and 2010 [41]. Moreover, Intergovernmental Panel on Climate Change (IPCC) f claimed that globally, sea-level will rise from 0.18 m to 0.59 m by the end of 21st century [42] but in the current scenario it will rise to 0.21 m to 0.8 m [43]. Therefore, if the sea-level rose at considerable rate, these narrow and low-lying landforms (i.e., BIs) would be extremely vulnerable so, they either migrate landward; experience erosion or accretion in its land mass in addition to other oceanic factors, but its overall response is unpredictable.

Several studies have investigated the behavior of BIs under different climatic and geographic setting, i.e., at global scale, Stutz and Pilkey (2011) [23] have studied the influence of sediments depositional settings, ocean climatic system and wave-tide regime on open-ocean BIs distribution and morphology. They revealed that 20,783 km of total worldwide coastal shoreline is occupied by 2149 open-ocean BIs. While at local scales, Madurapperuma et al. (2017) [44] used kite aerial photographs for mapping shoreline vulnerabilities at the Oluvil Harbor in Ampara, Sri Lanka and found that a 2 m rise in sea level inundated 90% area of the harbor. Kundu et al. (2014) [45] carried out shoreline mapping of the Sagar Island in West Bengal, India for the period of 1951 to 2011 using geospatial techniques to estimate morphological change and its future prediction. In a study of nearshore and foreshore influence on over-wash of a BI proposed by Matias et al. (2014) [46], identified and described longshore differences in storm impacts along a BI and evaluated the role of sub-aerial and submerged morphological variations in over-wash events on the western segment of Barreta Island, which is part of the Ria Formosa BI chain, in southern Portugal. Morton (2008) [47] has investigated that the Mississippi-Alabama BIs in the north-central Gulf of Mexico are undergoing rapid systematic land loss and translocation because of disruption of the sand-transport system. Moore et al. (2007) [48] has carried out sensitivity analyses on the outer banks of North Carolina and recommended that if sea-level rose by 0.9 m by the end of 21st century, the outer banks may translocate up to 2.5 times more quickly than at the present rate.

1.2. Purpose of the Study

Land reclamation has been a major anthropogenic factor affecting the Pakistan's coast and BIs and there has not been any systematic study for the evaluation. Therefore, this study aims at using the satellite remote sensing as a mean for (i) assessing the development, movement and change (morphology) in position of the BI chain along the Indus Delta from 1976 to 2017 and (ii) assessment of the rate of systematic land-loss due to physical processes.

2. Study Area

IDR is located between the Indian border along 'Sir Creek' on the east to the 'Hub River' on the west having length of 320 km and forming IDR (Figure 1). IDR is located at the head of the Arabian Sea, between Korangi Creek and the Rann of Kutch [30,40]. IDR has 22 tidal creeks which supply the sediment budgets to the BI chain. Main creeks of IDR include Phitti, Waddi Khuddi, Dabbo, Hajambro, Wari, Khar, Keti Bunder, Chann, and Khobar connecting Indus River to Indian Ocean [8].

IDR has 2.6% of the world BI with total length of 567 km [23] (Figure 1). Indus Delta is mixed wave dominated delta [23] where continuous accretion and erosion due to waves and tides up to 3 m in height modify the BIs. Bundal Island is a triangular shaped island located at the eastern side of Karachi harbor and at the intersection of the three major creeks, Gizri creek, Korangi creek, and Phitti creek. Its northern side is wide, and the southern side is narrow. Bundal Island has great social-economic importance in the study area because it protects Phitti creek, a navigational channel of Port Bin Qasim, and the human settlements from the high energy wave.

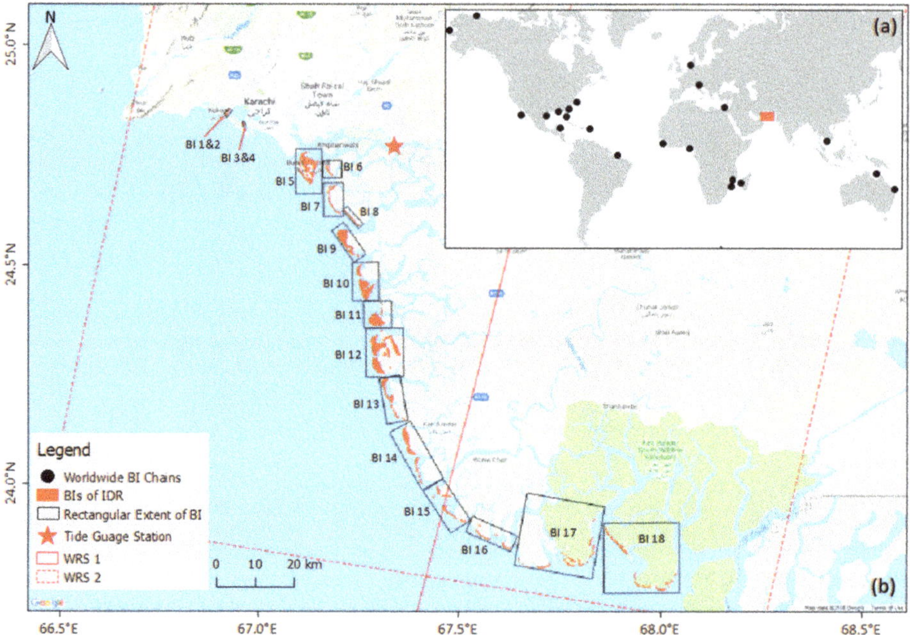

Figure 1. (**a**) Location of study area (filled red rectangle) and worldwide Barrier Island (BI) chains (black dots) derived from [23] and (**b**) barrier islands located within Indus Delta Region (IDR).

3. Methodology

3.1. Satellite Data

This study has used the archived imagery of Landsat-2 (L2) Multispectral Scanner (MSS), Landsat-5 (L5) Thematic Mapper (TM), Landsat-7 (L7) Enhanced Thematic Mapper Plus (ETM+), and Landsat-8 (L8) Operational Land Imager (OLI). L2, L5, L7, and L8 satellites were launched during January 1975, March 1984, April 1999, and February 2013, respectively. L2 and L5 were decommissioned during July 1983 and January 2013, respectively. The MSS sensor had spatial resolution of 80 m with a revisit frequency of 18 days and acquired the imagery on the Worldwide Reference System-1 (WRS-1), while the TM, ETM+ and OLI sensors have spatial and temporal resolutions of 30 m and 16 days, respectively on Worldwide Reference System-2 (WRS-2). All the images acquired from ETM+ after 31 May 2003 suffer from the Scan Line Corrector (SLC) error which causes 22% loss of the data [49]. Therefore, the SLC error was corrected using the "Fill nodata tool" under the raster tools available in QGIS 2.8.8 software.

Overall, this study has used 10 Landsat Collection 1 (C1) images from 1976 to 2017 acquired through L2 MSS, L5 TM, L7 ETM+, and L8 OLI sensors (Table A1). The Level-1 (radiometrically calibrated and orthorectified) MSS image were obtained from the United States Geological Survey (USGS) Earth Explorer website (http://earthexplorer.usgs.gov/) while the atmospherically corrected Level-2 (surface reflectance) images of TM, ETM+, and OLI were obtained from the USGS Earth Resources Observation and Science (EROS) Center Science Processing Architecture (ESPA) On Demand Interface (https://espa.cr.usgs.gov/). Images from different months of the years enabled the investigation of the morphological changes of the BIs over the study area according to the same oceanic and cloud free condition.

3.2. Tidal Data

Tide height was used in the satellite image selection process for the identification and extraction of geometry of the BIs. Ocean tide height along the coast of Sindh, Pakistan is monitored using a fixed tide gauge located at Port Muhammad Bin Qasim (24.7833°N, 67.3500°E) by National Institute of Oceanography (NIO) Karachi, Pakistan (Figure 1b). The maximum and minimum tide heights recorded at this station are 3.8 m and little less than 0.1 m between 1976 to 2017.

3.3. Satellite Data Selection Criteria

Due to seasonal variations in the ocean controlling factors (wave, tides, sea level rise, storms and sediment budget) and suitable atmospheric conditions, selection of satellite images at the same time of the year is important for studying the morphology of BIs [50]. Therefore, only those Landsat (MSS, TM, ETM+, and OLI) images were selected which had 0% cloud cover over the study area duly verified from the metadata file (.MTL) which is provided along each image. The objective for taking only those scenes with 0% cloud cover is due to the limitation of the optical imagery. This criterion can be relaxed to 10% to 20% provided the studied area (or the specific barrier island) is not affected. This limitation can be overcome by using the Synthetic Aperture Radar (SAR) imagery and authors do not exclude this possibility.

Furthermore, the images from non-flooding months (December to April) were selected because tropical storms affect the coastal areas of Pakistan during May to June and September to November. Only those images were selected which had a tide height of <1 m within a time window of ±1 h before or after the image acquisition time. This image selection criteria (Figure 2) resulted in ten images including one image from L2 MSS, two images from L5 TM, six images from L7 ETM+, and one image from L8 OLI sensor from 1976 to 2017. The selected images were obtained as Level-2 surface reflectance products for the Landsat TM, ETM+ and OLI sensors, due to the unavailability of the surface reflectance product Landsat MSS, surface reflectance was estimated in house using the 6S atmospheric correction method [51] (supplementary materials).

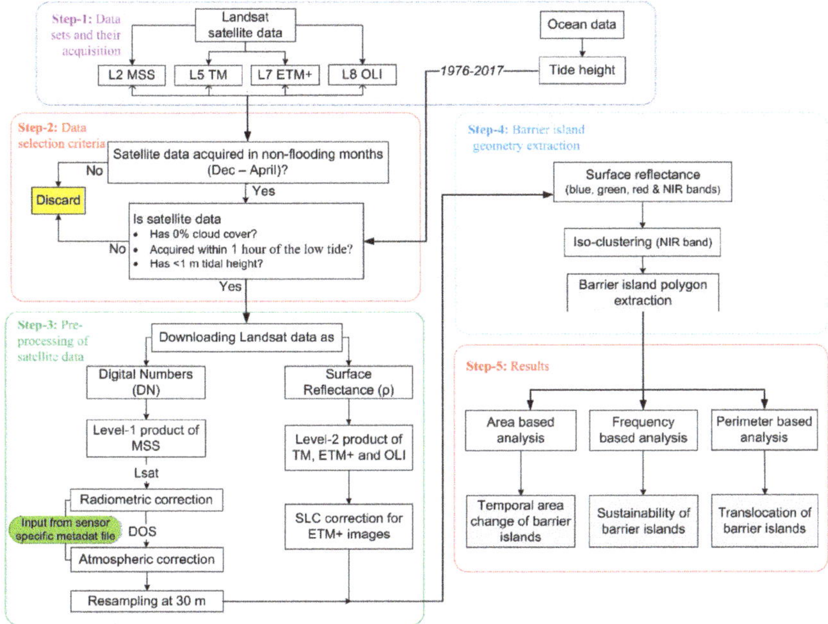

Figure 2. Flowchart of the methodology for the morphological change detection in the IDR barrier islands.

3.4. Barrier Island Identification and Extraction

A BI can be identified based on its association with different features i.e., a BI always has a back-barrier lagoon, a tidal inlet, it faces the shore and it is distant from the land [52]. Taking advantage of the higher reflectivity of land features in the Near Infrared (NIR) wavelength, the NIR band of L2 MSS (0.7–0.8 μm), L5 TM (0.76–0.90 μm), L7 ETM+ (0.77–0.90 μm), and L8 OLI (0.85–0.88 μm) sensors was used for developing water/land mask through Iso-cluster image classification method [53].

The classified images (water/land mask) were then visually examined with their corresponding true color images by displaying at a scale of 1:75,000 for visually identifying the boundary of BIs. Shorelines of the wet marshy sand beaches around the perimeter of BIs are less reliable and vary more with tide level than sandy shorelines of island [54]. So, only elongated sand deposits which fulfill the criteria for BI identification [52] were extracted. The boundaries of the classified BIs were then converted to polygon features by using the raster to vector conversion tool available in QGIS 2.8.8 software. Where necessary, minor edits were made to the boundaries of BIs to extract the more accurate boundary. Small polygons found near the main perimeter of an island were merged to that specific island polygon. Furthermore, polygon areas were calculated and added to the polygon attribute tables. Through this process 18 BIs (Figure 1) were delineated for the whole IDR.

3.5. Pixel by Pixel Frequency of Barrier Island

After finalizing the boundary of each BI from each of the ten selected images (Table A1), raster overlay operation was applied to determine the pixel by pixel frequency of each existing valid pixel of island body. For this purpose, polygon feature of each BI, for each year was converted to a binary raster (with a native grid size of 30 × 30 m) using the vector to raster conversion tool available in QGIS 2.8.8 software where a filled value of 1 was assigned to each pixel of BI, and 0 for each pixel representing any feature(s) other than BI.

The boundary of each BI was different in each of the 10 images but for raster overlay operation, one need to have the same spatial extent for each BI. Therefore, a fixed spatial extent (suitable for all years for each BI) was created by drawing a rectangular polygon to the maximum extent on the east, west, north and south directions of that specific BI (Figure 1—red rectangles). At this point when spatial extent of each BI was same, raster overlay by addition operation was applied, which calculated the frequency value of 1 to 10 for each pixel of the BI. The pixel which shows frequency value of 10, indicates that the island body exists at that pixel (location) in each of the ten rasters (images) while descending number of frequency values (9 to 1) shows that existence of valid island pixels for each image is decreasing.

4. Results

4.1. Boundary Delineation of Barrier Islands

The variant changes observed in the boundary of BIs during study period indicate that generally, there is a retreat of the boundary, and shapes of the BIs have evolved (Figure 3). In Figure 3, the color variations represent the shape of a BI in a specific year. For BIs 1 to 13 the 1976 boundary (filled grey colored polygons) was used as a reference to gauge the movement of the BIs, while it should be noted that for BIs 14 to 18 there is no boundary for the year 1976 which was due to the unavailability of the L2 MSS image satisfying the image selection criteria (Figure 2), hence 1990 boundary (filled grey colored polygons with red outline) was used as a reference. It is evident that, except BIs 1 to 4, 6, and 13, the shape of the BIs has been evolving significantly which can be due to the climatic and oceanic factors. It has been observed that some of the BIs are shrinking while others are expanding in both directions—i.e., oceanside and landside—against the accelerated SLR.

Among all the 18 BIs, BIs 5 and 9 (Figure 3) have lost significant areas and moved towards the land. BI5 also known as Bundal Island (local name) is the largest island located near the most western part of the Indus Delta. In personal communication with the NIO officials we came to know that the island has faced erosion from tidal currents passing through small prolonged tidal inlets on the eastern side and due to the passage of the cargo ships from the southern part of the island at the opening of the Phitti Creek, which produces constant backward and forward ocean currents. The island is abundant in mangroves on the northern side which helps it to retain its northern side. The boundary of the island is shrinking slowly from the eastern and southern sides while expanding from the western side as it is protected from any direct erosive factor (Figure 3). An overall summary of BI wise morphological changes from 1976 to 2017 has been presented in Table 1.

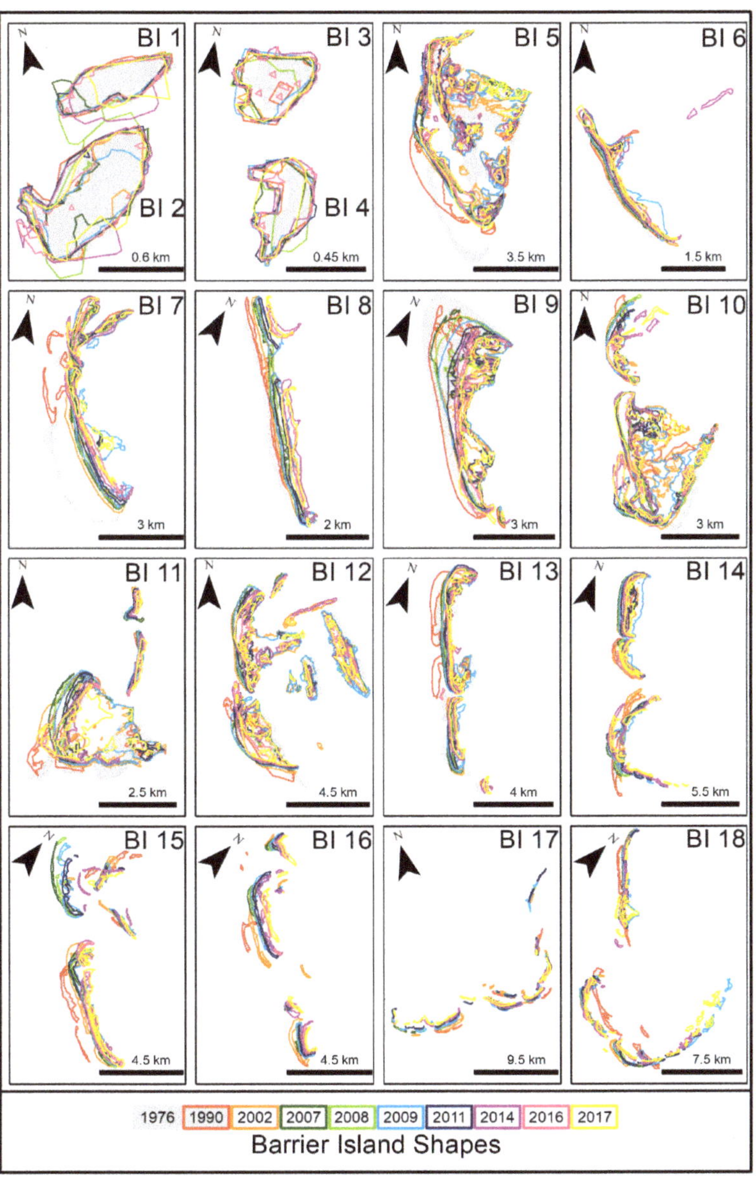

Figure 3. Barrier islands shapes as delineated from 1976 to 2017 from Landsat images.

Table 1. Morphological changes in barrier islands with respect to area and perimeter from 1976 to 2017 in different temporal periods.

	Area Change (%) During Period									Sum	Net Change (%) in	
BI#	1976–1990	1990–2002	2002–2007	2007–2008	2008–2009	2009–2011	2011–2014	2014–2016	2016–2017		Area	Perimeter
1	3	−11	−79	541	−30	1	47	−35	−9	429	−22	−17
2	20	−26	−1	42	−40	52	13	−25	−1	34	−5	−3
3	33	−7	−3	−15	43	0	7	−11	1	48	40	4
4	49	6	5	5	2	8	−1	−9	0	64	71	−1
5	−49	63	−64	29	119	−48	1	−52	82	82	−61	6
6	18	170	−76	84	328	−68	71	−74	76	528	46	63
7	−5	149	−50	114	175	−67	54	−50	33	354	134	142
8	−39	292	−72	52	146	−64	56	−39	49	382	30	5
9	−27	7	−31	−2	54	−48	−27	−45	83	−37	−69	8
10	101	79	−65	8	127	−48	14	−47	92	260	82	49
11	−46	243	−84	13	378	−75	−13	−47	235	604	−37	57
12	−54	308	−55	16	97	−48	5	−55	63	276	−24	54
13	−12	37	−39	19	97	−47	7	−30	10	42	−25	−9
14	-	55	−66	53	148	−48	−8	−49	62	147	−22	44
15	-	113	−55	40	118	−39	4	−45	36	172	40	39
16	-	183	−71	40	255	−62	73	−64	113	467	106	87
17	-	195	−82	76	296	−47	−6	−55	138	514	98	63
18	-	−15	−45	4	275	−56	27	−64	243	370	28	59
Sum	−6	1841	−933	1120	2588	−706	323	−797	1306			

Note: Net area change, and net perimeter change for barrier island 1 to 13 and barrier islands 14 to 18 are with reference to the year 1976 and 1990, respectively.

4.2. Sustainability of Barrier Islands

The vulnerability or sustainability of BI over the study period against the climatic and oceanic factors such as storm erosion, reductions in sediment, longshore drift and SLR shows that the BI are mostly vulnerable to greater extent (Figure 4). The pixel frequency values from 1 to 10 for each BI revels about the existence of the island body covering 10 years of study period at that pixel which is important to assess the sustainability of the BI chain of the Indus Delta. Increasing pixel value from 1 to 10 shows more stable part of the island, such that 10 indicates an area present throughout the study period whereas 1 indicates a part of the island that appears only once in study period (Figure 4).

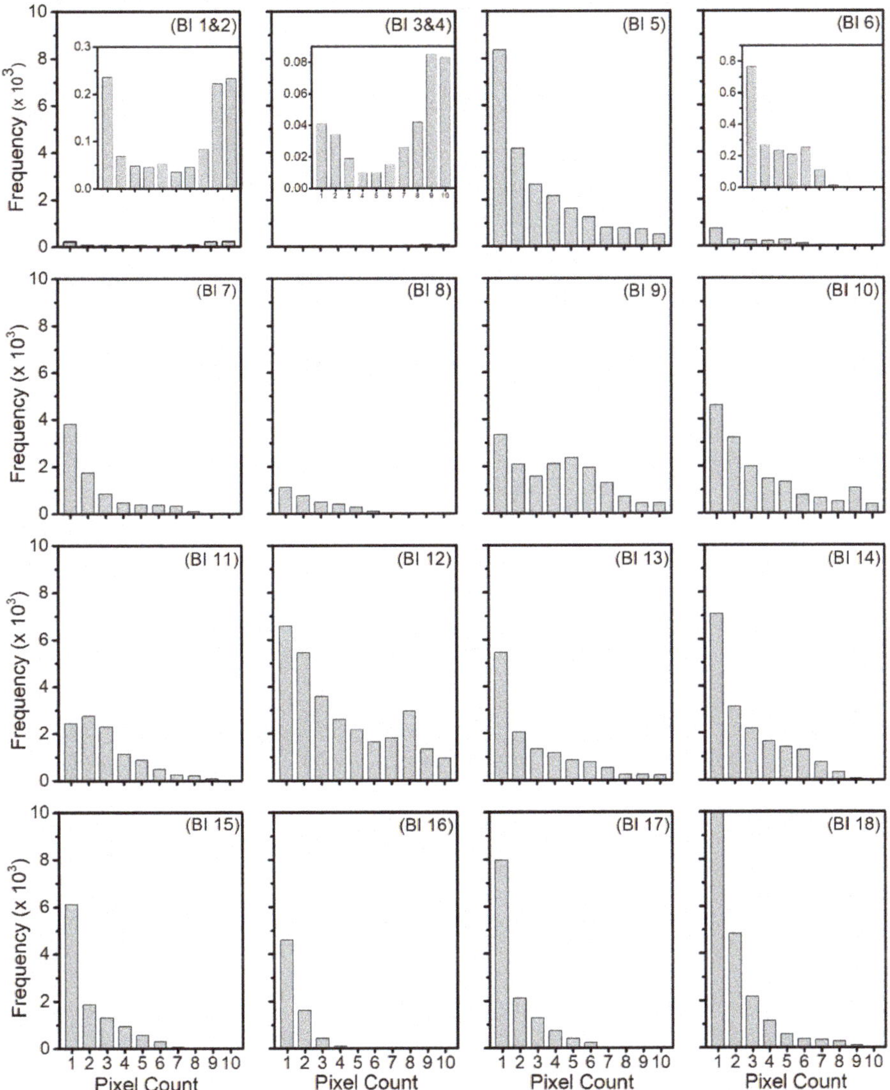

Figure 4. Frequency pixel count of Barrier Island. The inset views for BI 1&2, BI 3&4, and BI 6 show the frequency of the data between 0 and 1×10^3.

4.3. Change in Area of Barrier Islands

The box and whisker plot (Figure 5) show the percent change in area from 1976 to 2017 (for each BI) on the vertical axis, while the horizontal axis shows the BIs from 1 to 18. The red line shows the zero line of the data distribution. The value above the zero line shows the gain in area of a specific BI while the values below show the loss in area. It can be observed that Island 4 sustained, so we can say that the size of the box varies as there is variant change in area of the BI from 1976 to 2017. The Island 11 (Figure 5) shows the most vigorous change above the median value means the variance in data is higher in the 50th to 75th percentile of the data.

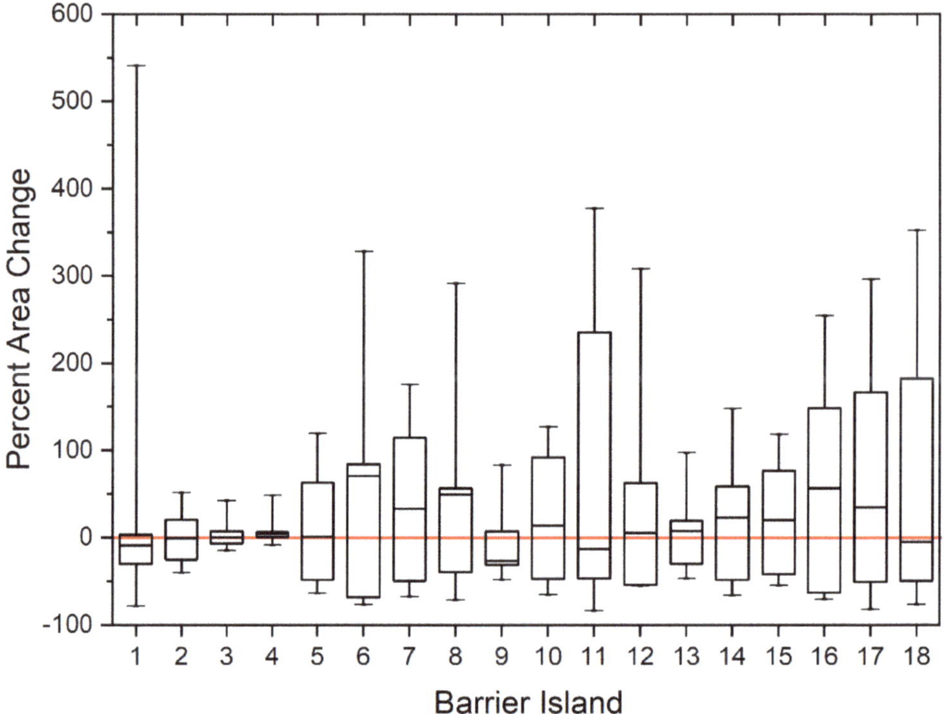

Figure 5. Barrier island percent area change from 1976 to 2017.

4.4. Translocation of Barrier Islands

The coastal erosion over centuries is a result of natural processes and sea level change. The rate of erosion seems to have increased at some points along the BIs of the Indus Delta. Translocation of BIs was estimated based on the movement of that specific BI relative to the reference shape. The reference shape (Figure 6, black colored polygon) for each BI was the shape delineated from the L2 MSS (from 1976) and L5 TM (from 1990) images for BIs 1 to 13 and 14 to 18, respectively. In Figure 6, the shades of blue and red colors represent less sustainable and highly sustainable pixel of the BIs. Due to the existence of BIs 1 to 4 on the back side of the Manora Beach, they were not significantly migrated or translocated (Figure 6). A significant translocation was observed for Bundal Island (BI 5) which migrated landward approximately 1.5 km (Table 2) from the ocean exposing side of the island between 1976 and 2017 with a rate of 38 m per year due to the action of ocean dynamical factors. BIs 6, 7, 8, 15, 16, and 17 have gained a significant area since 1976 and overall, they have started their development towards the land. While BIs 5, 9, 11, and 12 were found to be highly vulnerable to the oceanic conditions and have lost a significant area.

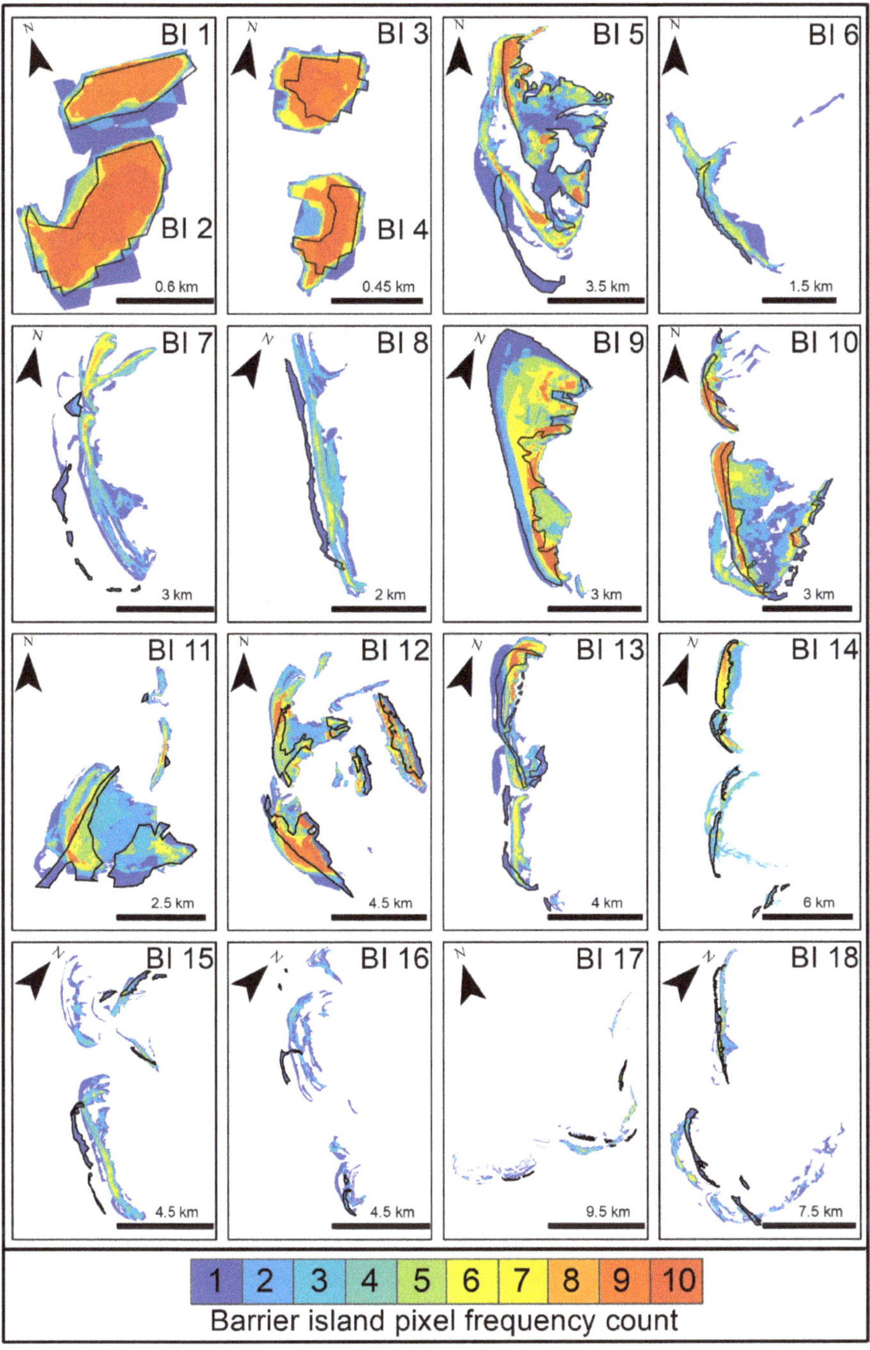

Figure 6. Pixel by pixel frequency. The black colored polygons show the extent of a barrier island in 1976 (for barrier islands 1 to 14) and 1990 (for barrier islands 14 to 18).

A BI wise status of its vulnerability or sustainability has been devised based on the frequency of the pixels—i.e., a BI whose pixel frequency count of 1 is high indicates that specific BI is 'vulnerable' (V) to oceanic factors, frequency pixel counts of 2 to 4, 5 to 8, and 9 to 10 indicate that specific BI is 'partially vulnerable' (PV), 'partially sustainable'(PS), 'and sustainable' (S) during the study period respectively (Table 2). The movement either it is landward, oceanward or in both directions, has also been reported in Table 2 along with the fate of BI whether it is shrinking or expanding. It can be seen that only islands 1 to 4 have more than 40% of their areas which are sustainable, islands 5 to 14 have areas majority of the areas between partially vulnerable to partially sustainable while islands 15 to 18 have more than 50% of their areas which are vulnerable to oceanic factors. Among the 18 BI of the Indus Delta BI chain, majority (11 out of 18) have been found to be moving towards land, a smaller amount (6 out of 18) of BIs has been found which exhibited the mixed behavior of movement (i.e., some parts are moving towards land and some towards ocean) and only 1 out of 18 are found to be moving towards ocean (Table 2). This is an alarming situation and indicates that the sea-level is rising and encroaching the land areas of Indus Delta BI chain. Furthermore, a notable remark is that 10 out of 18 BIs have been found which are expanding while 8 out of 18 BIs are shrinking (Table 2).

Table 2. Net translocation, translocation rate, vulnerability/sustainability status, movement and the fate of barrier islands.

BI#	Net Translocation	Translocation Rate	V	PV	PS	S	Movement	Fate
	(km)	(m/year)	(%)	(%)	(%)	(%)		
1&2	0.06	1.30	22	15	20	43	↕	−
3&4	0.05	1.20	11	17	26	46	↑	+
5	1.56	38.05	36	39	19	6	↑	−
6	0.20	4.88	41	39	20	0	↑	+
7	1.15	28.05	47	38	15	0	↑	+
8	0.51	12.34	36	52	12	0	↑	+
9	1.55	37.80	21	35	39	5	↑	−
10	0.50	12.20	29	42	20	9	↓	+
11	0.80	19.51	23	59	17	1	↕	−
12	0.50	12.20	22	40	30	8	↕	−
13	0.50	12.20	42	35	19	4	↑	−
14	0.40	14.81	40	39	21	0	↑	−
15	0.80	29.63	55	37	8	0	↑	+
16	1.10	40.73	68	32	0	0	↑	+
17	0.90	33.33	62	32	6	0	↕	+
18	1.40	51.85	57	35	7	1	↕	+

Notes: (1) Net translocation and translocation rate for barrier island 1 to 13 and barrier islands 14 to 18 are with reference to the periods 1976 to 2017 and 1990 to 2017, respectively. (2) V, PV, PS, and S represent vulnerable, partially vulnerability, partially sustainability and sustainable status of barrier islands, respectively and the signs '−', '+', '↑', '↓' and '↕' represent shrinkage, expansion, landward movement, oceanward movement, and movement in both directions, respectively.

5. Discussion

This study attempts to study the historical variability of the barrier islands (BIs) located along the Indus Delta Region by employing the Landsat satellite imagery from 1976 to 2017. Overall, 18 BIs were delineated, and their morphological behavior was studied. The spatial resolution is very important in detection and monitoring of the trends of costal land masses and coastline changes. Therefore, taking the advantage of the medium spatial resolution and free availability of the Landsat imagery, this study has used the Landsat TM, ETM+, and OLI imagery at 30 m spatial resolution (except for MSS sensor which has a spatial resolution of 60 m). In recent years, image acquisition systems with high spatial resolution such as spaceborne sensors and unmanned aircraft systems (UAS) or drones have emerged as an important platform for high spatial resolution data collection [28]. The evaluation of detection accuracy is always the focus of discussion. Therefore, results of this study have been verified by authors themselves during frequent field visits with NIO officials and personal communications with the residents, where applicable.

It was found that BIs 5 and 9 faced significant erosion and moved towards the land. BI 5 also known as Bundal Island (local name) is the largest island located near the most western part of the Indus Delta. This island is of high socio-economic importance for the region as it has human settlements from past three decades. Significant net loss of 61% in the area of the Bundal Island and translocation of about 1.56 km of its boundary landward with a rate of 38 m/year shows vulnerability of the socioeconomic coastal environment associated to the island. Historical morphological changes show maximum of 36% of the Bundle Island was vulnerable to oceanic controlling factors with only 6% of the island sustained throughout the study period. Tidal currents contributed to the erosive and accretion action along the Bundal Island. The tidal inlets with narrow opening at the mouth and a large area from inside the Island let the high tide water enters inside and makes island marshy and erosive. Among the other BIS, BIs 6, 7, 8, 15, 16, and 17 had gained a significant area since 1976 and overall, they had started their development towards the land. While BIs 5, 9, 11, and 12 were found to be highly vulnerable to the oceanic conditions and have lost a significant area.

Results suggested that the BIs 1 to 4, 5 to 14, and 15 to 18, are sustainable, partially sustainable and vulnerable to the oceanic factors, respectively. Majority of the Indus Delta BIs have been found to be moving towards land which alerts the local authorities and indicates that the sea-level is rising and encroaching the land areas of the BIs. Climate change and anthropogenic activities (land reclamation and modification) can trigger extreme climatic events such as flooding, tropical cyclone and SLR along the deltaic regions (e.g., Indus Delta Region) leading to an increased vulnerability of coastal landmarks. Cyclones and storms usually develop in the Arabian sea during pre-monsoon (March–May) and post-monsoon (October–November) seasons with favorable month as October and pushed by the prevailing monsoon winds along their direction. Arabian Sea is in northwest of the Indian ocean and share coast among India, Pakistan, Oman, Iran, Siri-Lanka, Maldives, and Somalia.

6. Conclusions

An archive of Landsat imagery from 1976 to 2017, has been used in this study to explore the morphological changes in the barrier island of the Indus Delta Region (IDR). Unlike global scale assessment of the barrier islands, a regional study allows in depth spatial assessment of the island. In general, it has been found that, accretion is more significant in the most parts of the islands, but erosion has also remarkably increased in most of the islands. The relative rates of change of islands morphology and migration are recorded in the IDR barrier islands chain. The natural future trends of change for the IDR barrier islands will be continued in rapid land mass change, change in perimeter, and migration as a result of accelerated sea level rise, frequent intense storms, and sand budget supply. As a result of global warming which causes the sea level rise increases beyond the expected rate will also likely to increase the land loss of the barrier islands regionally and worldwide. The total percent area of barrier islands relatively changed from 1976 to 2017 was found −16.73%. Overall, 44.4% of the total barrier island are shrunk while the rest of them are expanded due to heavy sediments budget supply. The translocation of the barrier island found about of 66.1% of the total islands moved landward, 5.6% along seaward and the rest of the islands moved along both sides. A major part of these barrier islands chain is vulnerable which is about 38.3% and 36.7% is partially vulnerable while 17.4% is partially sustainable and only 7.6% sustained in the study area against the ocean controlling factors.

Most of the barrier islands along the IDR are still undisturbed by the human intervention. The existence of the barrier islands at a place is mostly determined by the sand supply and sea level change. Protection against the sea level rise is an especially important issue which can damage barrier beaches and coastal system. Coastal protection and coastal management along the IDR should be adopted to defend against the erosive action of the ocean. In future, increase in land loss can be mitigated by placing dredged material on the backshore and shore facing side of the island so the island experiences nourishment and rebuilding.

Supplementary Materials: The following are available online at http://www.mdpi.com/2072-4292/11/4/437/s1, Text 1: Radiometric correction, Text 2: Atmospheric correction [49,50,55–59].

Author Contributions: This research was conceived by M.I.S and I.Z.; M.W., M.N., and M.I.S. designed the research experiments, analyzed the results, and contributed to writing; M.W. wrote the initial draft and final edits to the manuscript; I.Z. provided review of the draft and final manuscript.

Funding: This research received no external funding.

Acknowledgments: The authors would like to thank USGS for the distribution of Landsat MSS, TM, ETM+ and OLI data products; the QGIS team for the development and support of the QGIS software; the National Institute of Oceanography, Karachi Pakistan for providing the tide height data; Mr. Ghulam Mustafa (CUI Islamabad) for providing access to his space for running data processing; Sana Almas Khan (CUI Islamabad) for sharing her image processing expertise; and three anonymous reviewers for providing constructive comments which improved the quality of this article greatly.

Conflicts of Interest: The authors declare no conflict of interest. The funders had no role in the design of the study; in the collection, analyses, or interpretation of data; in the writing of the manuscript, or in the decision to publish the results.

Appendix A

Table A1. List of Landsat images used in this study.

Sr. #	Satellite	Sensor	Scene Identifier	Path/Row	Acquisition Date
1	Landsat-7	ETM+	LE07_L1TP_152043_20170219_20170317_01_T1	152/43 [b]	19 Feb 2017
2	Landsat-8	OLI	LC08_L1TP_152043_20160413_20170326_01_T1	152/43 [b]	13 Apr 2016
3	Landsat-7	ETM+	LE07_L1TP_152043_20140126_20161119_01_T1	152/43 [b]	26 Jan 2014
4	Landsat-5	TM	LT05_L1TP_152043_20110211_20161010_01_T1	152/43 [b]	11 Feb 2011
5	Landsat-7	ETM+	LE07_L1TP_152043_20090418_20161222_01_T1	152/43 [b]	18 Apr 2009
6	Landsat-7	ETM+	LE07_L1TP_152043_20080314_20161230_01_T1	152/43 [b]	14 Mar 2008
7	Landsat-7	ETM+	LE07_L1TP_152043_20070224_20170104_01_T1	152/43 [b]	24 Feb 2007
8	Landsat-7	ETM+	LE07_L1TP_152043_20021211_20170127_01_T1	152/43 [b]	11 Dec 2002
9	Landsat-5	TM	LT05_L1TP_152043_19900217_20170131_01_T1	152/43 [b]	17 Feb 1990
10	Landsat-2	MSS	LM02_L1TP_163043_19761214_20180425_01_T2	163/43 [a]	14 Dec 1976

Note: The superscripts a and b represent the path/row for WRS-1 and WRS-2, respectively.

References

1. Emanuel, K.A. The dependence of hurricane intensity on climate. *Nature* **1987**, *326*, 483–485. [CrossRef]
2. Knutson, T.R.; Tuleya, R.E. Impact of CO_2-Induced Warming on Simulated Hurricane Intensity and Precipitation: Sensitivity to the Choice of Climate Model and Convective Parameterization. *J. Clim.* **2004**, *17*, 3477–3495. [CrossRef]
3. Emanuel, K. Increasing destructiveness of tropical cyclones over the past 30 years. *Nature* **2005**, *436*, 686–688. [CrossRef] [PubMed]
4. Webster, P.J.; Holland, G.J.; Curry, J.A.; Chang, H.R. Atmospheric science: Changes in tropical cyclone number, duration, and intensity in a warming environment. *Science* **2005**, *309*, 1844–1846. [CrossRef] [PubMed]
5. Mimura, N.; Nurse, L.; McLean, R.F.; Agard, J.; Briguglio, L.; Lefale, P.; Payet, R.; Sem, G. Small islands. In *Climate Change 2007: Impacts, Adaptation, and Vulnerability. Contribution of Working Group II to the Fourth Assessment Report of the Intergovernmental Panel on Climate Change*; Cambridge University Press: Cambridge, UK, 2007; pp. 687–716.
6. Knutson, T.R.; McBride, J.L.; Chan, J.; Emanuel, K.; Holland, G.; Landsea, C.; Held, I.; Kossin, J.P.; Srivastava, A.K.; Sugi, M. Tropical cyclones and climate change. *Nat. Geosci.* **2010**, *3*, 157–163. [CrossRef]
7. Romine, B.M.; Fletcher, C.H.; Frazer, L.N.; Genz, A.S.; Barbee, M.M.; Lim, S.-C. Historical Shoreline Change, Southeast Oahu, Hawaii; Applying Polynomial Models to Calculate Shoreline Change Rates. *J. Coast. Res.* **2009**, *256*, 1236–1253. [CrossRef]
8. Shahzad, M.I.; Meraj, M.; Nazeer, M.; Zia, I.; Inam, A.; Mehmood, K.; Zafar, H. Empirical estimation of suspended solids concentration in the Indus Delta Region using Landsat-7 ETM+ imagery. *J. Environ. Manag.* **2018**, *209*, 254–261. [CrossRef]

9. Chu, Z.X.; Sun, X.G.; Zhai, S.K.; Xu, K.H. Changing pattern of accretion/erosion of the modern Yellow River (Huanghe) subaerial delta, China: Based on remote sensing images. *Mar. Geol.* **2006**, *227*, 13–30. [CrossRef]
10. Mills, J.P.; Buckley, S.J.; Mitchell, H.L.; Clarke, P.J.; Edwards, S.J. A geomatics data integration technique for coastal change monitoring. *Earth Surf. Process. Landforms* **2005**, *30*, 651–664. [CrossRef]
11. Marfai, M.A.; Almohammad, H.; Dey, S.; Susanto, B.; King, L. Coastal dynamic and shoreline mapping: Multi-sources spatial data analysis in Semarang Indonesia. *Environ. Monit. Assess.* **2008**, *142*, 297–308. [CrossRef]
12. Ryabchuk, D.; Leont'yev, I.; Sergeev, A.; Nesterova, E.; Sukhacheva, L.; Zhamoida, V. The morphology of sand spits and the genesis of longshore sand waves on the coast of the eastern Gulf of Finland. *Baltica* **2011**, *24*, 13–24.
13. Mujabar, P.S.; Chandrasekar, N. Shoreline change analysis along the coast between Kanyakumari and Tuticorin of India using remote sensing and GIS. *Arab. J. Geosci.* **2013**, *6*, 647–664. [CrossRef]
14. Li, L.; Huang, Z.; Qiu, Q.; Natawidjaja, D.H.; Sieh, K. Tsunami-induced coastal change: Scenario studies for Painan, West Sumatra, Indonesia. *Earth Planets Sp.* **2012**, *64*, 799–816. [CrossRef]
15. Morton, R.A.; Sallenger, A.H., Jr. Morphological Impacts of Extreme Storms on Sandy Beaches and Barriers. *J. Coast. Res.* **2003**, *19*, 560–573. [CrossRef]
16. Forbes, D.L.; Parkes, G.S.; Manson, G.K.; Ketch, L.A. Storms and shoreline retreat in the southern Gulf of St. Lawrence. *Mar. Geol.* **2004**, *210*, 169–204. [CrossRef]
17. Hapke, C.J.; Himmelstoss, E.A.; Kratzmann, M.G.; List, J.H.; Thieler, E.R. *National Assessment of Shoreline Change: Historical Shoreline Change along the New England and Mid-Atlantic Coasts*; U.S. Geological Survey Open-File Report; U.S. Geological Survey: Reston, VA, USA, 2010; 57p.
18. Slott, J.M.; Murray, A.B.; Ashton, A.D. Large-scale responses of complex-shaped coastlines to local shoreline stabilization and climate change. *J. Geophys. Res. Atmos.* **2010**, *115*, 1–19. [CrossRef]
19. Hapke, C.J.; Plant, N.G.; Henderson, R.E.; Schwab, W.C.; Nelson, T.R. Decoupling processes and scales of shoreline morphodynamics. *Mar. Geol.* **2016**, *381*, 42–53. [CrossRef]
20. Trujillo, A.P.; Harold, V. *Thurman Essentials of Oceanography*, 12th ed.; Pearson: Dallas, TX, USA, 2016; ISBN 9780134073545.
21. Pinet, P.R. *Invitation to Oceanography*; Jones & Bartlett: Boston, MA, USA, 2009; ISBN 1449667988.
22. Godfrey, P.J. Barrier beaches of the east coast. *Oceanus* **1976**, *19*, 27–40.
23. Stutz, M.L.; Pilkey, O.H. Open-ocean barrier islands: Global influence of climatic, oceanographic, and depositional settings. *J. Coast. Res.* **2011**, *27*, 207–222. [CrossRef]
24. McGranahan, G.; Balk, D.; Anderson, B. The rising tide: Assessing the risks of climate change and human settlements in low elevation coastal zones. *Environ. Urban.* **2007**, *19*, 17–37. [CrossRef]
25. Greening, H.; Doering, P.; Corbett, C. Hurricane impacts on coastal ecosystems. *Estuar. Coasts* **2006**, *29*, 877–879. [CrossRef]
26. Mallin, M.; Corbett, C. How hurricane attributes determine the extent of environmental effects: Multiple hurricanes and different coastal systems. *Estuar. Coasts* **2006**, *29*, 1046–1061. [CrossRef]
27. Sallenger, A.H.; Stockdon, H.F.; Fauver, L.; Hansen, M.; Thompson, D.; Wright, C.W.; Lillycrop, J. Hurricanes 2004: An overview of their characteristics and coastal change. *Estuari. Coasts* **2006**, *29*, 880–888. [CrossRef]
28. Lippitt, C.D.; Zhang, S. The impact of small unmanned airborne platforms on passive optical remote sensing: A conceptual perspective. *Int. J. Remote Sens.* **2018**, *39*, 4852–4868. [CrossRef]
29. Martínez, M.L.; Intralawan, A.; Vázquez, G.; Pérez-Maqueo, O.; Sutton, P.; Landgrave, R. The coasts of our world: Ecological, economic and social importance. *Ecol. Econ.* **2007**, *63*, 254–272. [CrossRef]
30. Rabbani, M.M.; Inam, A.; Tabrez, A.R.; Sayed, N.A.; Tabrez, S.M. The impact of sea level rise on Pakistan's coastal zones—In a climate change scenario. In Proceedings of the 2nd International Maritime Conference, Karachi, Pakistan, 25–27 March 2008.
31. Ali Khan, T.M.; Razzaq, D.A.; Chaudhry, Q.U.Z.; Quadir, D.A.; Kabir, A.; Sarker, M.A. Sea level variations and geomorphological changes in the coastal belt of Pakistan. *Mar. Geod.* **2002**, *25*, 159–174. [CrossRef]
32. Rasul, G.; Mahmood, A.; Sadiq, A.; Khan, S.I. Vulnerability of the Indus Delta to Climate Change in Pakistan. *Pak. J. Meteorol.* **2012**, *8*, 89–107.

33. Rizvi, S.H.; Ali, A.; Naeem, S.A.; Tahir, M.; Baquer, J.; Saleem, M.; Tabrez, S.M. Comparison of the physical properties of seawater offshore the Karachi coast between the northeast and southwest monsoons. In *International Conference of American Institute of Biological Sciences*; M. Thompson, N.M.T., Ed.; Marine Science of the Arabian Sea: Washington, DC, USA, 1988.
34. Siddiqui, M.N.; Maajid, S. Monitoring of geomorphological changes for planning reclamation work in coastal area of Karachi, Pakistan. *Adv. Sp. Res.* **2004**, *33*, 1200–1205. [CrossRef]
35. Giosan, L.; Constantinescu, S.; Clift, P.D.; Tabrez, A.R.; Danish, M.; Inam, A. Recent morphodynamics of the Indus delta shore and shelf. *Cont. Shelf Res.* **2006**, *26*, 1668–1684. [CrossRef]
36. Syvitski, J.P.M.; Kettner, A.J.; Overeem, I.; Giosan, L.; Brakenridge, G.R.; Hannon, M.; Bilham, R. Anthropocene metamorphosis of the Indus Delta and lower floodplain. *Anthropocene* **2013**, *3*, 24–35. [CrossRef]
37. Inam, A.; Clift, P.D.; Giosan, L.; Tabrez, A.R.; Tahir, M.; Rabbani, M.M.; Danish, M. The Geographic, Geological and Oceanographic Setting of the Indus River. In *Large Rivers: Geomorphology and Management*; John Wiley & Sons Ltd: Chichester, England, 2008; pp. 333–346. ISBN 9780470849873.
38. Wells, J.T.; Coleman, J.M. Deltaic morphology and sedimentology, with special reference to the Indus River Delta. *Mar. Geol. Oceanogr. Arab. Sea Coast. Pak.* **1985**, *424*, 85–100.
39. Chandio, N.H.; Anwar, M.M.; Chandio, A.A. Degradation of Indus delta, Removal mangroves forestland Its Causes: A Case study of Indus River delta. *Sindh Univ. Res. J.Sci. Ser.* **2011**, *43*, 67–72.
40. Zia, I.; Zafar, H.; Shahzad, M.I.; Meraj, M.; Kazmi, J.H. Assessment of sea water inundation along Daboo creek area in Indus Delta Region, Pakistan. *J. Ocean Univ. China* **2017**, *16*, 1055–1060. [CrossRef]
41. Hay, C.C.; Morrow, E.; Kopp, R.E.; Mitrovica, J.X. Probabilistic reanalysis of twentieth-century sea-level rise. *Nature* **2015**, *517*, 481–484. [CrossRef] [PubMed]
42. Bindoff, N.L.; Willebrand, J.; Artale, V.; Cazenave, A.; Gregory, J.M.; Gulev, S.; Hanawa, K.; Le Quere, C.; Levitus, S.; Nojiri, Y.; et al. Observations: Oceanic climate change and sea level. *Changes* **2007**, *AR4*, 385–432.
43. Church, J.A.; Clark, P.U.; Cazenave, A.; Gregory, J.M.; Jevrejeva, S.; Levermann, A.; Merrifield, M.A.; Milne, G.A.; Nerem, R.; Nunn, P.D.; et al. Sea level change. In *Climate Change 2013: The Physical Science Basis. Contribution of Working Group I to the Fifth Assessment Report of the Intergovernmental Panel on Climate Change*; Cambridge University Press: Cambridge, UK, 2013; pp. 1137–1216.
44. Madurapperuma, B.D.; Dellysse, J.E.; Iyoob, A.L.; Resources, W.; Secretariat, D.; Lanka, S. Mapping shoreline vulnerabilities using kite aerial photographs at Oluvil harbour in Ampara. In Proceedings of the 7th International Symposium, Oluvil, Sri Lanka, 7–8 December 2017; pp. 197–204.
45. Kundu, S.; Mondal, A.; Khare, D.; Mishra, P.K.; Shukla, R. Shifting shoreline of Sagar Island Delta, India. *J. Maps* **2014**, *10*, 612–619. [CrossRef]
46. Matias, A.; Carrasco, A.R.; Loureiro, C.; Almeida, S.; Ferreira, Ó. Nearshore and foreshore influence on overwash of a barrier island. *J. Coast. Res.* **2014**, *70*, 675–680. [CrossRef]
47. Morton, R.A. Historical Changes in the Mississippi-Alabama Barrier-Island Chain and the Roles of Extreme Storms, Sea Level, and Human Activities. *J. Coast. Res.* **2008**, *246*, 1587–1600. [CrossRef]
48. Moore, L.J.; List, J.H.; Williams, S.J.; Stolper, D. Modelling barrier island response to sea-level rise in the outer banks, North Carolina. In Proceedings of the Sixth International Symposium on Coastal Engineering and Science of Coastal Sediment Process, New Orleans, LA, USA, 13–17 May 2007; pp. 1153–1164.
49. Chander, G.; Markham, B.L.; Helder, D.L. Summary of current radiometric calibration coefficients for Landsat MSS, TM, ETM+, and EO-1 ALI sensors. *Remote Sens. Environ.* **2009**, *113*, 893–903. [CrossRef]
50. Ozturk, D.; Sesli, F.A. Shoreline change analysis of the Kizilirmak Lagoon Series. *Ocean Coast. Manag.* **2015**, *118*, 290–308. [CrossRef]
51. Nazeer, M.; Nichol, J.E.; Yung, Y.K. Evaluation of atmospheric correction models and Landsat surface reflectance product in an urban coastal environment. *Int. J. Remote Sens.* **2014**, *35*, 6271–6291. [CrossRef]
52. Oertel, G.F. The barrier island system. *Marine Geolog.* **1985**, *63*, 1–18. [CrossRef]
53. Armenakis, C.; Leduc, F.; Cyr, I.; Savopol, F.; Cavayas, F. A comparative analysis of scanned maps and imagery for mapping applications. *ISPRS J. Photogramm. Remote Sens.* **2003**, *57*, 304–314. [CrossRef]
54. Guy, B.K.K.; Jewell, S. *Barrier Island Shorelines Extracted from Landsat Imagery*; U.S. Geological Survey: Reston, VA, USA, 2015.

55. Masek, J.G.; Vermote, E.F.; Saleous, N.E.; Wolfe, R.; Hall, F.G.; Huemmrich, K.F.; Gao, F.; Kutler, J.; Lim, T.K. A landsat surface reflectance dataset for North America, 1990-2000. *IEEE Geosci. Remote Sens. Lett.* **2006**, *3*, 68–72. [CrossRef]
56. U.S. Geological Survery. *Landsat 4-7 Surface reflectance (LEDAPS) product guide (version 1.0)*; 2018. Available online: https://landsat.usgs.gov/sites/default/files/documents/ledaps_product_guide.pdf (accessed on 19 February 2019).
57. U.S. Geological Survery. *Landsat 8 Surface Reflectance Code (LaSRC) product guide (version 1.0)*; 2018. Available online: https://landsat.usgs.gov/sites/default/files/documents/lasrc_product_guide.pdf (accessed on 19 February 2019).
58. Chávez, P.S.J. Image-based atmospheric corrections - revisited and improved. *Photogramm. Eng. Remote Sensing* **1996**, *62*, 1025–1036. [CrossRef]
59. Mahiny, A.; Turner, B. A comparison of four common atmospheric correction methods. *Photogramm. Eng. Remote Sens.* **2007**, *73*, 361–368. [CrossRef]

Article

Characterizing and Monitoring Ground Settlement of Marine Reclamation Land of Xiamen New Airport, China with Sentinel-1 SAR Datasets

Xiaojie Liu [1], Chaoying Zhao [1,2,3,*], Qin Zhang [1,2,3], Chengsheng Yang [1,2,3] and Jing Zhang [1,2,3]

1 School of Geology Engineering and Geomatics, Chang'an University, Xi'an 710054, China; 2018026010@chd.edu.cn (X.L.); dczhangq@chd.edu.cn (Q.Z.); yangchengsheng@chd.edu.cn (C.Y.); racheljing@chd.edu.cn (J.Z.)
2 National Administration of Surveying, Mapping and Geoinformation, Engineering Research Center of National Geographic Conditions Monitoring, Xi'an 710054, China
3 State Key Laboratory of Geo-Information Engineering, Xi'an 710054, China
* Correspondence: cyzhao@chd.edu.cn; Tel.: +86-29-8233-9251

Received: 10 February 2019; Accepted: 8 March 2019; Published: 11 March 2019

Abstract: Artificial lands or islands reclaimed from the sea due to their vast land spaces and air are suitable for the construction of airports, harbors, and industrial parks, which are convenient for human and cargo transportation. However, the settlement process of reclamation foundation is a problem of public concern, including soil consolidation and water recharge. Xiamen New Airport, one of the largest international airports in China, has been under construction on marine reclamation land for three years. At present, the airport has reached the second phase of construction, occupying 15.33 km^2. The project will last about twenty years. To investigate the temporal and spatial evolution of ground settlement associated with land reclamation, Sentinel-1 synthetic aperture radar (SAR) data, including intensity images and phase measurements, were considered. A total of 82 SAR images acquired by C-band Sentinel-1 satellite covering the time period from August 2015 to October 2018 were collected. First, the spatial evolution process of land reclamation was analyzed by exploring the time series of SAR image intensity maps. Then, the small baseline subset InSAR (SBAS–InSAR) technique was used to retrieve ground deformation information over the past three years for the first time since land reclamation. Results suggest that the reclaimed land experienced remarkable subsidence, especially after the second phase of land reclamation. Furthermore, 26 ground settlement areas (i.e., 0.015% of the whole area) associated with land reclamation were uncovered over an area of more than 1200 km^2 of the Xiamen coastal area from January 2017 to October 2018. This study offers important guidance for the next phase of land reclamation and the future construction of Xiamen New Airport.

Keywords: ground settlement; marine reclamation land; time series InSAR; Sentinel-1; Xiamen New Airport

1. Introduction

With the rapid development of modern cities, the efficiency of land-resource utilization has significantly risen. To alleviate pressure from dense populations, it is an important way to carry out marine reclamation land projects and expand urban space in coastal cities. Therefore, land reclamation activities have been performed in many countries around the world, including the USA [1], Singapore [2], the Netherlands [3], Japan [4], China [5], and other countries [6]. However, because land reclamation usually involves dumping uncompacted filling materials over unconsolidated marine sediments [7], the settlement process of reclamation foundation has become a problem of great public concern. It can cause severe damage to structures such as harbors, highways, airport runways, and underground facilities. In addition, it greatly threatens the environment and public safety [8]. Therefore, it is necessary to characterize and monitor ground settlement associated with land reclamation to facilitate a better understanding of its temporal and spatial evolution. Accordingly, it can reduce economic loss and guarantee the safety of facility construction.

Traditionally, ground deformation could be quantitatively monitored by employing in-situ measurements such as GNSS, leveling, piezometers, and inclinometers. Although these methods have high accuracy, they still have maintenance problems and high costs due to discrete benchmark installation and campaign/continuous measurements [9]. More recently, repeat-pass spaceborne synthetic aperture radar interferometry (InSAR), which has been used in several regions, has proven to be a powerful geodetic tool for investigating ground settlement associated with land reclamation, with the advantages of wide spatial coverage, fine spatial resolution, and day-and-night and all-weather working capabilities. For instance, Jiang and Lin [10] investigated the long-term reclamation settlement of Hong Kong's Chek Lap Kok Airport by integrating InSAR and geological data; Yu et al. [11] applied COSMO–SkyMed and Sentinel-1 satellite images to investigate ground deformation in ocean-reclaimed areas of Shanghai, China; Liu et al. [12] determined surface deformation associated with land reclamation in Shenzhen, China from 2004 to 2010, and from 2013 to 2017, by combining the Envisat, COSMO–SkyMed, and Sentinel-1 datasets; Aslan et al. [13] investigated the spatial extent and rate of ground deformation related to land reclamation in the megacity of Istanbul, Turkey by using the ERS-1/2, Envisat, and Sentinel-1 datasets. These studies mainly focus on surface deformation investigation several years after the completion of land reclamation. Further, it would be great if the InSAR technique could be used for the real-time monitoring of ground deformation during the land reclamation process, because such a study could play an important decision-making role in the entire land reclamation process.

Compared with in situ measurements, the main concern of the InSAR technique lies in measuring precision. Theoretically, the measurement precision of InSAR largely depends on the coherence of interferograms, which is affected by many factors, including atmospheric artefacts, and temporal and spatial decorrelation [9,14]. Thanks to the successful launch of some new SAR satellites, e.g., TerraSAR-X, ALOS/PALSAR-2, and Sentinel-1, with the characteristics of short revisit period, short spatial baseline, and the provision of various wavelengths. For instance, the Sentinel-1 satellite flies in an orbital tube with a radius of 50 m, thus forming small orbit InSAR baselines in the order of 150 m [15], which significantly improves interferogram coherence and facilitates the improvement of InSAR measurement accuracy. Furthermore, advanced multitemporal InSAR (MT-InSAR) techniques could overcome many intrinsic temporal and spatial decorrelations of traditional InSAR techniques [16]. Currently, two main families of MT-InSAR techniques broadly exist, namely, the small baseline subsets (SBAS) [17–19] and persistent scatterers (PS) [20–22] methods. The former inverts surface deformation evolution through the singular-value decomposition (SVD) method using coherent interferograms.

Xiamen New Airport, as one of the most important international airports in China, four-fifths of which occupy an approximate area of 26 km^2, will be produced by marine reclamation land. Currently, the reclamation area has reached 10.58 km^2, and the construction of the airport will be finished in the next twenty years. Surface deformation occurs in land reclamation areas due to the soil consolidation of the underlying unconsolidated marine sediments [10]. However, there are no bibliographies available for ground deformation analysis on Xiamen New Airport. To bridge this gap, in this study, the SAR images of Sentinel-1 satellite were involved, and the time series InSAR method was employed to explore the spatiotemporal deformation characteristics of Xiamen New Airport after land reclamation. In addition, potential ground settlement areas associated with land reclamation in the coastal area of the city of Xiamen were also mapped and analyzed.

2. Study Area

Xiamen Xiang'an International Airport (Xiamen New Airport hereinafter) is located in the Xiang'an district, Xiamen, China [23], as shown in Figure 1, which is about 25 km to the west of the city center of Xiamen, and 15 km to the south of the county of Jinmen, Taiwan, China. The total planned area of the airport is about 31 km^2, as shown in the red line in Figure 2b, including the island of Xiaodeng, the shoal area between the islands of Dadeng and Xiaodeng, and part of the sea area around the island of Dadeng, as well as part of dry land in the east of Dadeng. Approximately 84% of the total airport area, occupying 26 km^2, will be created through the land reclamation project, which will be carried out in three phases: The first phase is mainly blowing and filling dredged mud, an area of about 3 km^2; the second phase is mainly blowing and filling sands to create land, an area of about 7.58 km^2; and the third phase is about 15 km^2, as shown in Figure 2a. At present, the first and the second phases of the project have been completed.

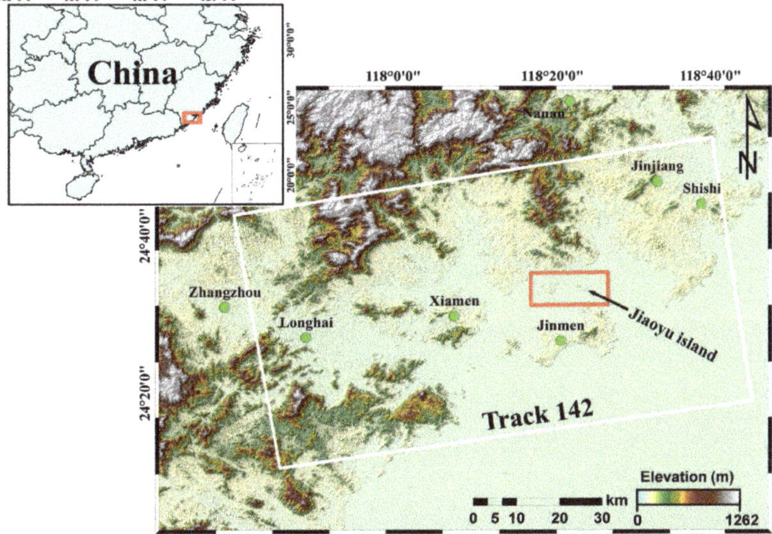

Figure 1. Study area location and synthetic aperture radar (SAR) data coverage. The background is the shaded topography generated from the shuttle radar topography mission digital elevation model (SRTM DEM), where the coverage of Sentinel-1 SAR data is superimposed by the white rectangle, green dots indicate the location of the major cities, and the red rectangle indicates the location of the study area. Inset indicates the study-area location in China.

The function area of the airport mainly includes the flight area, the terminal area, and the supporting area, and the runway is about 3800 m in length [23]. The engineering geology of Xiamen New Airport can be divided into five layers from top to bottom, that is, sand-mixed silt (Q_4^m) with a thickness of around 7.86 m, silt and silt mixed with sand (Q_4^m) with a thickness of around 7 m, silty clay and clay (Q_3^{al}) with a thickness of around 2.1 m, silty sand (Q_3^{al}) with a thickness of around 4.9 m, and residual clay (Q^{el}) with a thickness of around 3.1 m. Two Landsat-8 remote sensing images of the airport were acquired on 26 March 2014 and 13 March 2018, which are shown in Figure 2a,b, respectively. It can be deduced that the airport did not start to reclaim until March 2014. However, the second phase of land reclamation was completed after March 2018.

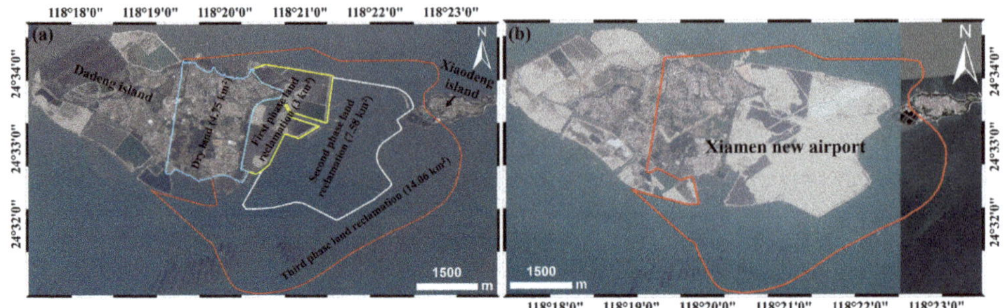

Figure 2. Landsat-8 remote-sensing images of the study area acquired on (**a**) 26 March 2014 and (**b**) 13 March 2018. The blue line represents the dry land area (4.75 km^2) of the airport, the yellow line indicates the first phase (3 km^2), the white line indicates the second phase (7.58 km^2), and the red line indicates the third phase (14.06 km^2) of marine reclamation land.

3. Data and Methodology

3.1. Datasets

A total of 82 ascending Sentinel-1 images acquired from 11 August 2015 to 24 September 2018 were employed to characterize and monitor the ground settlement of marine reclamation land. During the entire SAR data monitoring period, the land reclamation project was still underway, which caused the ground surface to greatly change. Therefore, in order to avoid the effect of SAR image decorrelation caused by ground surface changes, SAR data were divided into two groups for InSAR processing, that is, Group I, from November 2015 to December 2016, and Group II, from January 2017 to October 2018. The shuttle radar topography mission (SRTM) digital elevation model (DEM) with a resolution of 30 m was applied to remove topographic phase contributions. The topographic phase over Xiamen New Airport was negligible due to small elevation changes and the DEM missing during the land reclamation period. A multilooking factor of four in range direction was used in data processing, and the spatial resolution of multilooked SAR images was about 16 m in both range and azimuth directions. SAR images with this spatial resolution can best detect ground deformation.

GAMMA software was used to process the Sentinel-1 datasets [24]. The thresholds of the temporal baseline and the perpendicular baseline were set to 60 days and 150 m, respectively. Therefore, a total of 314 interferograms were generated and, eventually, 208 high-quality interferograms were selected to further calculate ground-surface deformation. The spatiotemporal baseline distributions of the high-quality interferograms are shown in Figure 3a,b, representing the interferograms in Group I and II, respectively.

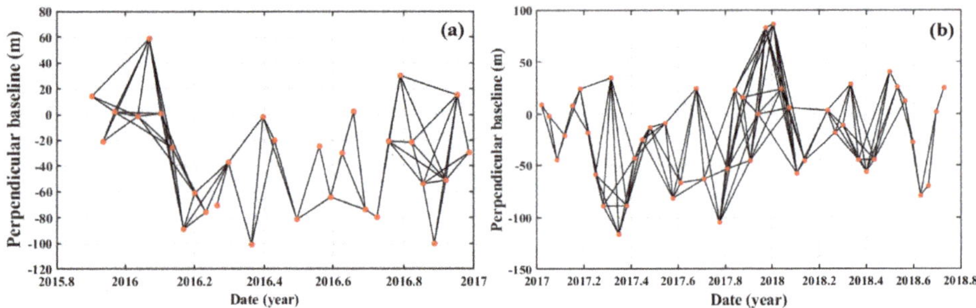

Figure 3. Baseline distribution of high-quality interferometric pairs used in this study. (**a**) Group I of SAR images, acquired from November 2015 to December 2016; (**b**) Group II of SAR images, acquired from January 2017 to October 2018.

3.2. Methodology

The temporal and spatial evolution of land changes and reclaimed land deformation of Xiamen New Airport were jointly investigated with the intensity and phase information of the SAR images.

First, Sentinel-1 images were preprocessed including coregistration, azimuth spectral filtering, image resampling, and data deramping and reramping [15]. With regard to the imaging mode of terrain observation by progressive scans (TOPS) for Sentinel-1 images, the Doppler Centroid rapidly varies along the track [15]. Phase ramps occur in individual bursts if coregistration accuracy is less than one-thousandth of one pixel, which is equivalent to 2 cm in azimuth [25]. Hence, highly precise coregistration, especially in the azimuth direction, is needed. Iterative coregistration refinement, estimated using intensity matching followed by the spectral diversity algorithm [26], was adopted for burst overlap areas to ensure that all SAR images were accurately coregistered.

As for analysis of the land reclamation process, intensity images were considered, which are robustly filtered with homogeneous pixels. First, statistically homogeneous pixels (SHPs) were selected using the fast SHP selection algorithm with the confidence interval for each pixel based on the central limit theorem [27]. Then, the intensity stacks were filtered by using the updated Lee filter algorithm [28] to remove speckle noise. It is worth noting that single-look intensity images and interferograms were used in this processing. As the accurate selection of SHPs is of great significance for the accurate estimation of intensity, SAR images were divided into four different time periods (i.e., from 11 August 2015 to 30 June 2016, from 24 July 2016 to 20 January 2017, from 1 February 2017 to 19 July 2017, and from 31 July 2017 to 24 September 2018) in order to avoid the influence of surface changes caused by land reclamation, and SHPs were estimated separately.

As for ground settlement monitoring, the differential interferograms were generated, filtered [29,30], and unwrapped [31,32] after the accurate coregistration of the SLC stacks. Baseline refinement was conducted to remove the residual orbital ramp phase [33]. The artifacts of atmospheric disturbance were reduced by using a quadratic polynomial model, as follows:

$$w(x,y) = a_0 + a_1 x + a_2 y + a_3 xy + a_4 x^2 + a_5 y^2 \tag{1}$$

where $w(x,y)$ represents the unwrapped phase for a generic pixel (x,y), a_i indicates the unknown coefficients, and i represents the subscript of the unknown coefficient a, corresponding to 1, 2, 3, 4, and 5.

As the study area is located in the coast, and the terrain is relatively flat, there is no need to consider the influence of the stratified atmospheric delay [34]. The average deformation rates and time series of the ground surface were calculated using Equation (2) [24] and the SBAS algorithm [17], respectively.

$$V_{phase} = \frac{\sum_{i=1}^{N} \Delta t_i \varphi_i}{\sum_{i=1}^{N} \Delta t_i^2} \tag{2}$$

where N is the number of interferograms, φ_i represents the unwrapped phase of each interferogram, and Δt_i is the time interval of each interferogram. Finally, the temporal and spatial deformation characteristics of ground settlement after land reclamation were analyzed in depth based on the deformation maps.

The ground deformation rate and time series obtained by Equation (2) and the SBAS algorithm are the sum of the projections of the real three-dimensional (i.e., north–south, east–west, and up–down) ground deformations in the line-of-sight (LOS) direction. For land subsidence caused by land reclamation, the vertical displacement dominates ground deformation [16]. Hence, to analyze the airport's deformation characteristics, the vertical deformations were mainly considered, and LOS displacement was purely back-projected into the vertical direction considering the local incidence angle [35].

4. InSAR Results

The surface deformation rate maps of Xiamen's coastal areas were retrieved from November 2015 to October 2018, as shown in Figure 4, where Figure 4a shows the one from November 2015 to December 2016, and Figure 4b from January 2017 to October 2018. Note that the deformation rate was in the LOS direction, and the positive values (blue color) indicate the motion toward the satellite (uplift), and the negative values (red color) indicate the motion away from the satellite (settlement). Figure 4 shows that most areas were stable (between −10 and 10 mm/year). However, several small-scale land deformation areas (<−10 mm/year) were obviously uncovered along the coastal areas of the city of Xiamen. Quantitatively, 20 land subsidence areas were detected from November 2015 to December 2016, and 26 land subsidence areas were detected from January 2017 to October 2018. The maximum annual subsidence rate from November 2015 to December 2016, and from January 2017 to October 2018 reached −130 mm/year and −126 mm/year, respectively.

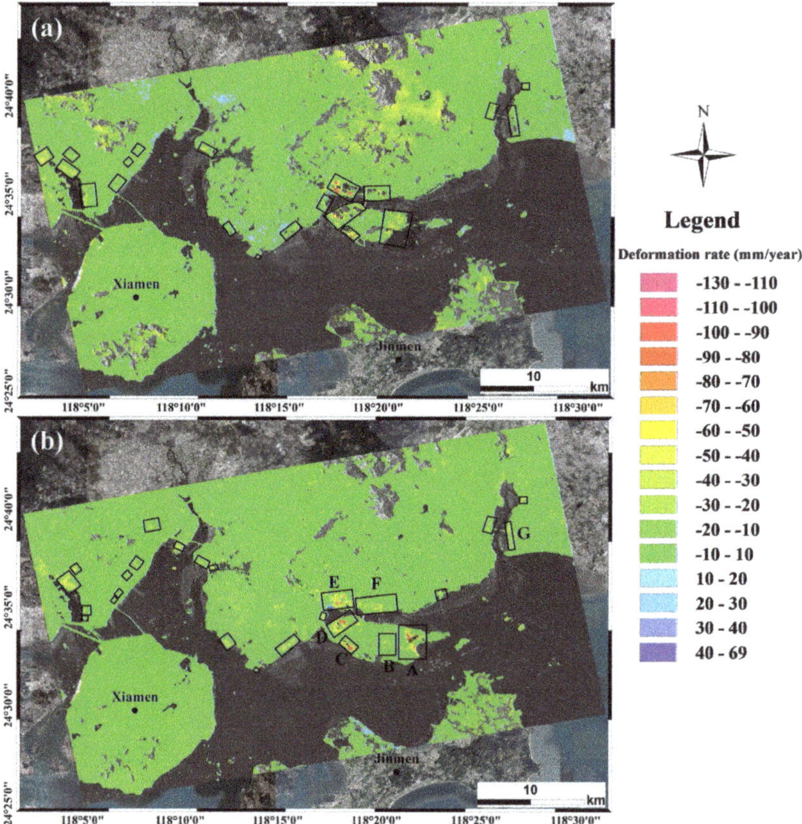

Figure 4. Average deformation rate maps calculated with Sentinel-1 datasets over the whole coastal areas of the city of Xiamen. Black rectangles represent potential deformation areas identified by InSAR, and Regions A–G in Figure 4b are analyzed in Section 5. (**a**) The deformation rate map from November 2015 to December 2016; (**b**) the one from January 2017 to October 2018.

It can also be seen from Figure 4 that moderate (between −30 and −40 mm/year) to strong (<−40 mm/year) land subsidence areas were mainly concentrated in the central coastal areas, i.e., the Xiang'an district of Xiamen. There was only one strong land subsidence area observed from November 2015 to December 2016, with the maximum deformation rate of −130 mm/year. However, it increased to four regions (as shown in Regions A, C, D, and E in Figure 4b) from January 2017 to October 2018, which will be discussed in Section 5. In addition, some slight to moderate subsidence areas were also observed in the eastern (i.e., the eastern part of the Xiang'an district) and western (i.e., Jimei district) coastal areas of Xiamen, which were continuously deformed from November 2015 to October 2018, with the deformation rate ranging from −10 to −30 mm/year.

5. Analysis and Discussion

5.1. Spatial Evolution of Land Reclamation at Xiamen New Airport

Time series SAR intensity images of the Sentinel-1 over Xiamen New Airport from August 2015 to October 2016 were obtained, where four main land reclamation stages could be roughly reflected, as shown in Figure 5a,c,e,g. Four optical remote sensing images with similar acquisition dates are shown in Figure 5b,d,f,h. It can be seen from Figure 5 that the spatial evolution of land reclamation was clearly recorded in the SAR intensity maps including the following four stages, which could be well validated by the optical remote sensing images.

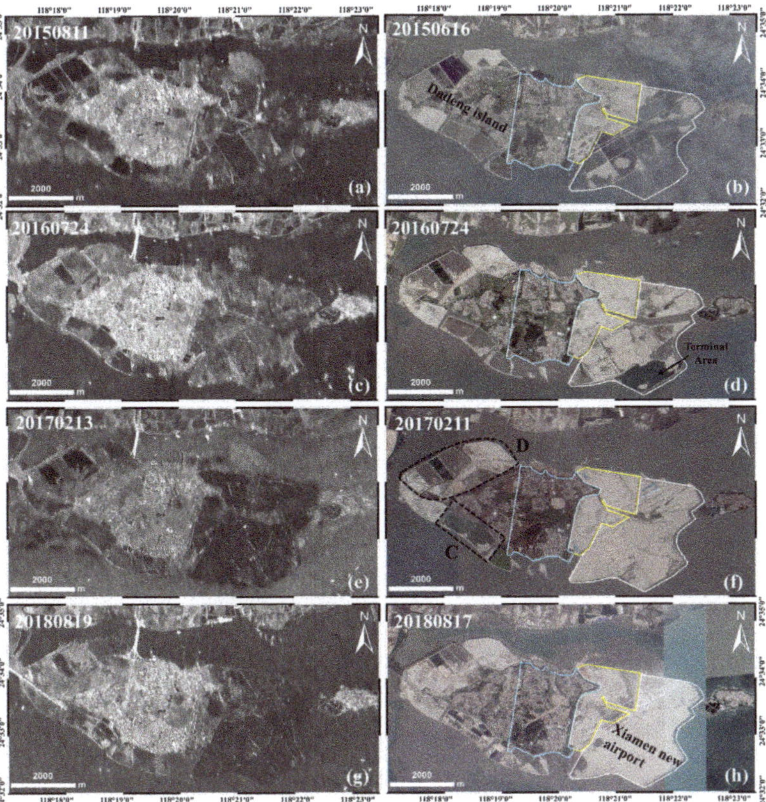

Figure 5. Four SAR intensity images and four remote sensing images of Xiamen New Airport. Intensity images were acquired on (**a**) 11 August 2015, (**c**) 24 July 2016, (**e**) 13 February 2017, and (**g**) 19 August 2018, respectively; remote sensing images were acquired on (**b**) 16 June 2015, (**d**) 24 July 2016, (**f**) 11 February 2017, and (**h**) 17 August 2018, respectively. Solid blue line represents the pre-existing land area of Xiamen New Airport, solid yellow line indicates the area of the first phase of land reclamation, and solid white line indicates the one in the second phase.

(1) The first phase of land reclamation, as shown with the yellow lines in Figure 5, was completed before 16 June 2015 (see Figure 5a,b); (2) except for the southeastern corner of the airport, i.e., the terminal area, other areas of the second phase of land reclamation were reclaimed before 24 July 2016 (see Figure 5c,d); (3) the second stage of land reclamation was fully completed on 13 February 2017, as shown with white lines in Figure 5e,f. However, the operation of soil compaction was not carried out, so water seepage can still be observed in local areas, e.g., in the central and southeastern parts of the airport (see Figure 5e,f); (4) the whole land reclamation area of the second phase was compacted on 19 August 2018 (see Figure 5g,h).

In addition, it can be seen from Figure 5 that the land reclamation project was conducted in Region C on 16 June 2015 (see Figure 5a,b) but was subsequently covered by seawater (see Figure 5c,d,e,f), and land reclamation was restarted on 19 August 2018 (see Figure 5g,h). For Region D, it is evident that land reclamation began on 24 July 2016 (see Figure 5a–d). By 19 August 2018, except for a small part of the central section, the land reclamation project in other areas was basically completed (see Figure 5g,h). However, water seepage could also be observed in the northeastern part of the region (see Figure 5g,h).

5.2. Spatiotemporal Deformation Patterns of Xiamen New Airport

To reveal the deformation characteristics and patterns of Xiamen New Airport after land reclamation, the deformation rate from January 2017 to October 2018 was enlarged in Figure 8a. Two remote sensing images, acquired on 22 January 2017 and 13 March 2018, were segmented into three sections to aid ground settlement analysis, as shown in Figure 8b,b′,c,c′,d,d′. It can be seen from Figure 8 that the pre-existing land is quite stable. However, four obvious subsidence areas associated with land reclamation could be successfully detected, two of which are located in the area of the first and second phases of the airport's land reclamation (i.e., Regions A and B), and the other two are located in the western and southwestern parts of the island of Dadeng (i.e., Regions C and D). Severe land subsidence occurred at Regions A, C, and D from January 2017 to October 2018, and the maximum deformation rate reached −126 mm/year in the LOS direction, while moderate land subsidence was observed in Region B, with a maximum deformation rate of around −48 mm/year. As mentioned in Section 5.1, the land reclamation date of Region B is much earlier than those of the other three regions, which indicates that the severe land subsidence occurred over newly reclaimed areas. An unusual phenomenon was observed in the terminal area (see Figure 5d), which showed to be quite stable even it was reclaimed at the latest date, which may be correlated with the reclamation methods and materials in the later stage. A similar phenomenon was observed in Chek Lap Kok Airport, Hong Kong [10].

Figure 6 shows the six cross-sections of vertical deformation rates in different areas to quantitatively analyze the correlation between the reclamation phases and ground surface settlement, where different colors indicate the different times of completed land reclamation. Ground settlement funnels can be clearly observed in all six profiles. However, the deformation patterns of settlement funnels in different reclamation areas show inconsistencies, i.e., the magnitudes of deformation vary in different areas. On the basis of Section 5.1 and Figure 6, it is evident that settlement funnels along the profiles of AA′ and DD′ experienced the largest deformation (<−130 mm/year; see Figure 6a,d), where the land reclamation project was completed on 13 February 2017. Deformation rates in the vertical direction were over −140 mm/year from January 2017 to October 2018. The settlement funnels along the profiles of CC′, EE′, and FF′ experienced moderately large deformation (between −130 and −100 mm/year; see Figure 6c,e,f), deformation rates in the vertical direction were −111, −125, and −124 mm/year, respectively. It can be seen from Figures 5 and 8 that land reclamation of the maximum deformation area along profile CC′ was completed on 13 February 2017. The settlement funnel along profile EE′ was reclaimed on 16 June 2015 (see Figure 5b), but it was subsequently covered by seawater (see Figure 5d,f), and the land reclamation project was conducted again on 17 August 2018 (see Figure 5h). The settlement funnel along profile FF′ was reclaimed on 24 July 2016, and water seepage could still be observed on 17 August 2018. The settlement funnel along the profile BB′ experienced relatively less deformation (>−100 mm/year) compared to former profiles, where land reclamation projects were completed on 15 June 2015, with a deformation rate as large as −93 mm/year in the vertical direction. Therefore, it can be concluded that the magnitude of ground deformation after land reclamation at Xiamen New Airport is closely correlated to the completion time of land reclamation projects, i.e., significant deformation occurred in newly reclaimed areas, and the deformation rate decreased as time went by. Such a deformation pattern is in good agreement with that of other land reclamation areas, such as Lingang New City in Shanghai, China [7]. In addition, the results reveal that most areas (e.g., Region A in Figure 8a) of the second phase of land reclamation are in a state of severe deformation, with the deformation rate in the vertical direction greater than −93 mm/year from January 2017 to October 2018.

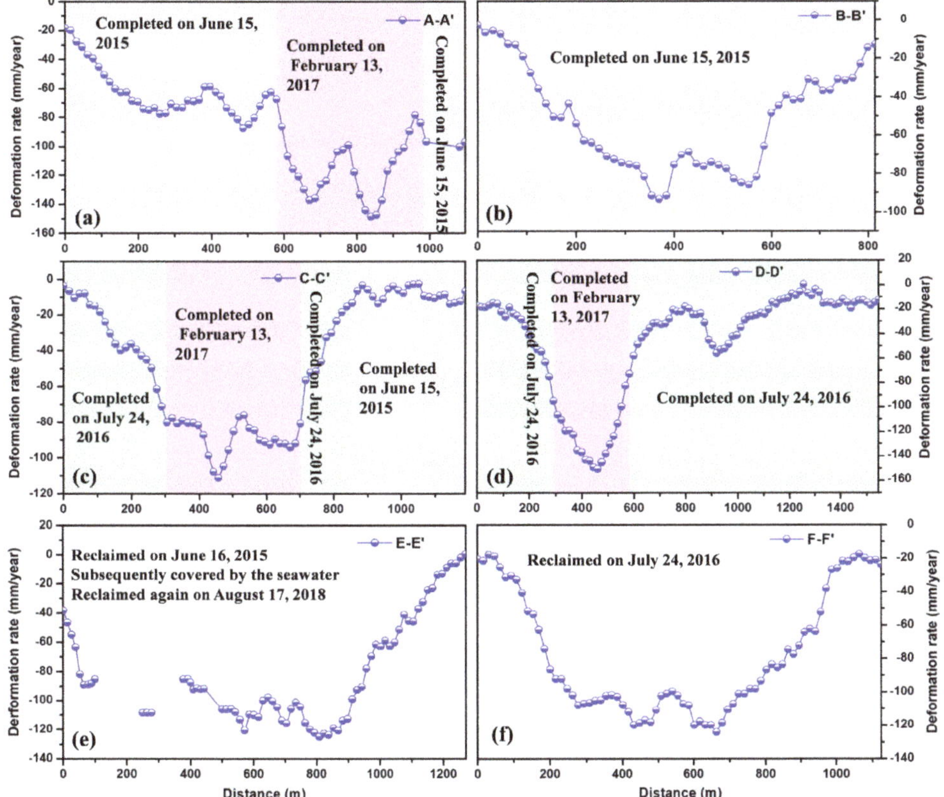

Figure 6. Cross-sections of average vertical deformation rates of Xiamen New Airport from January 2017 to October 2018 along six profiles, whose positions are marked in Figure 8. (**a**) Profile A–A′; (**b**) profile B–B′; (**c**) profile C–C′; (**d**) profile D–D′; (**e**) profile E–E′; (**f**) profile F–F′.

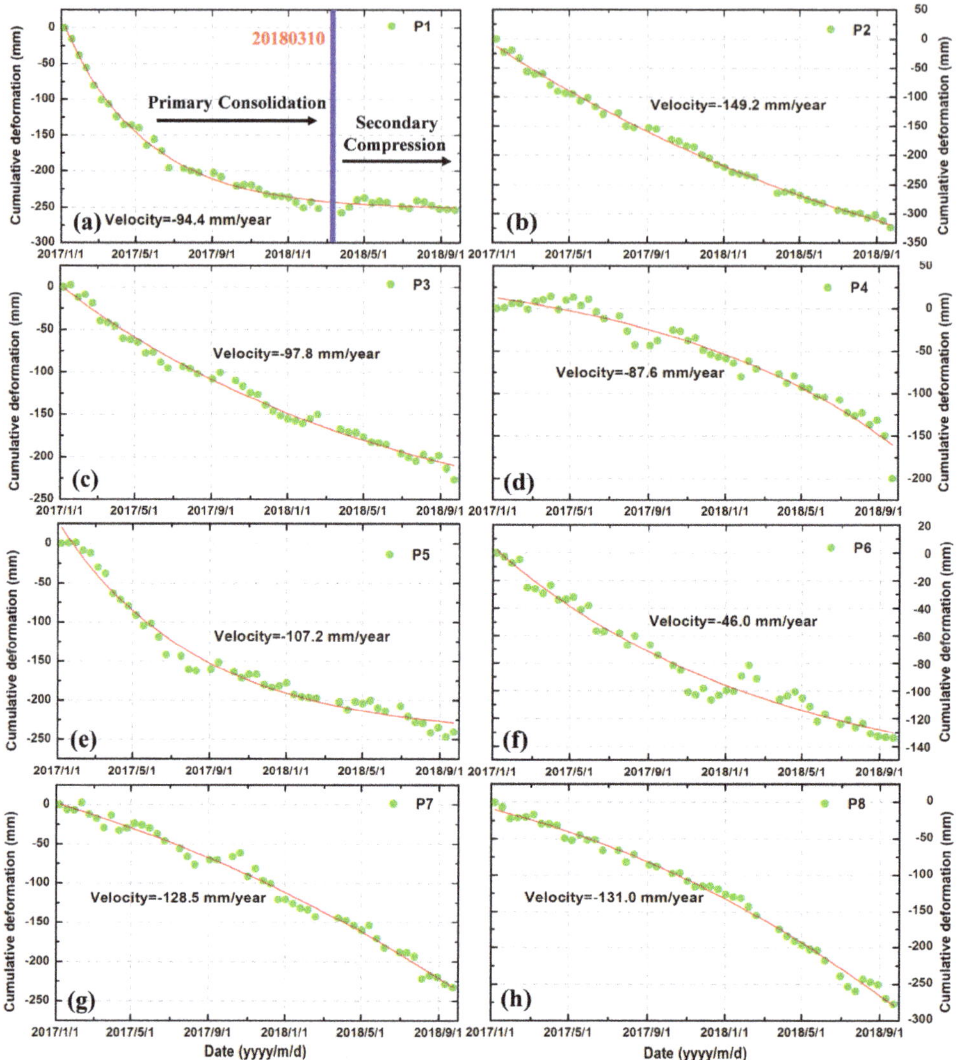

Figure 7. Time series deformation in the vertical direction of Xiamen New Airport from January 2017 to October 2018 for points P1 to P8. The locations of points P1 to P8 are shown in Figure 8. (**a**) Point P1; (**b**) point P2; (**c**) point P3; (**d**) point P4; (**e**) point P5; (**f**) point P6; (**g**) point P7; (**h**) point P8.

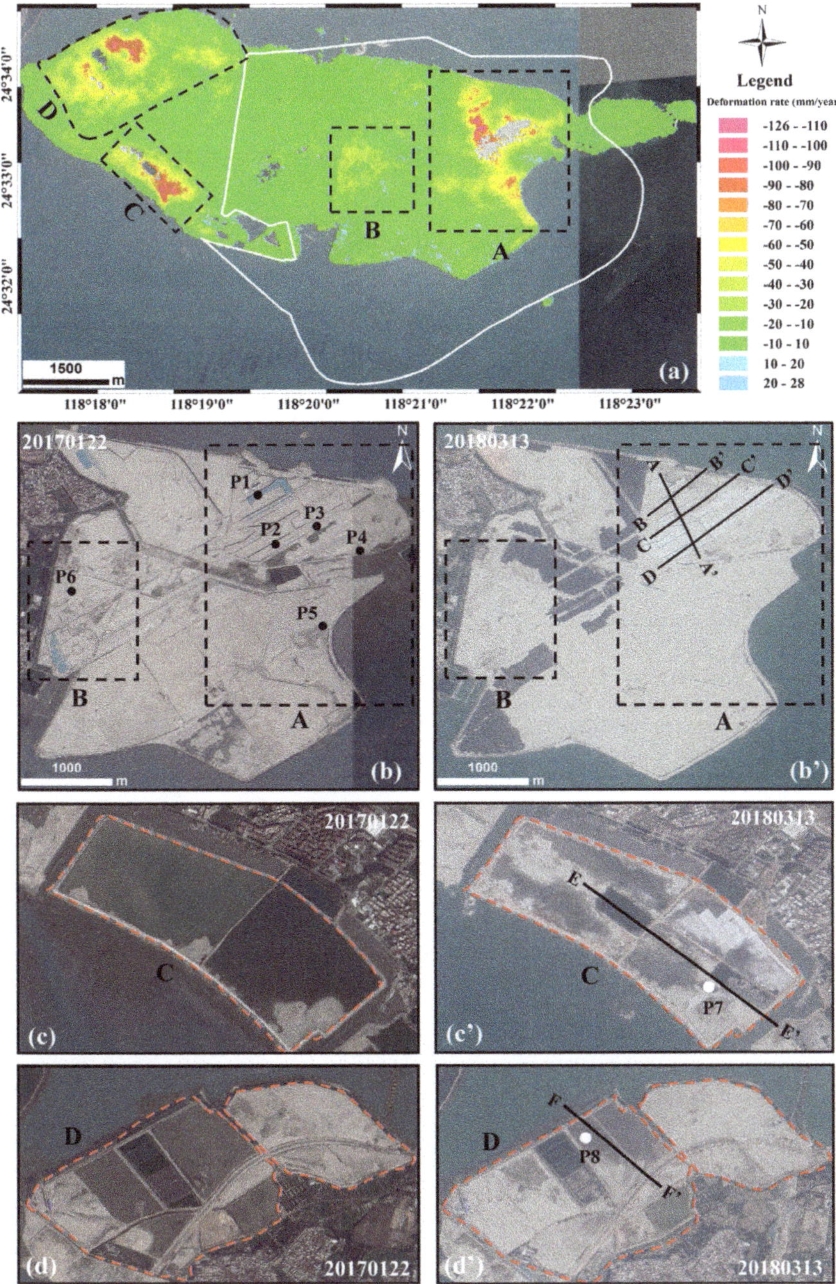

Figure 8. Ground deformation rate map (**a**) and three sections of remote sensing images acquired on 22 January 2017 in regions A, B, C, and D in (**b**), (**c**), and (**d**), respectively, and on 13 March 2018 in (**b′**), (**c′**), and (**d′**), respectively. Solid black lines from AA′ to FF′ denote profile locations, which are further shown in Figure 6. Points P1 to P8 (marked with black and white dots) are chosen to show the time series deformation in Figure 7.

To evaluate temporal deformation evolution after land reclamation at Xiamen New Airport, eight typical points, P1 to P8, located in different areas, were selected to analyze time series deformation with respect to the time table of land reclamation. The locations of the selected points are shown in Figure 8. P1 to P5 are located in Region A during the second land reclamation phase, P6 is in Region B during the first land reclamation phase, and P7 and P8 are in Regions C and D, as shown in Figure 8. From Figures 5 and 8, it can be seen that P6 was at the earliest completed reclamation region, and P2, P3, P7, and P8 were at the latest ones.

Figure 7 shows the time series vertical deformation of eight points from January 2017 to October 2018. It is evident that the eight points experienced nonlinear subsidence with various velocities. Maximum and minimum cumulative deformations were observed at P2 and P6, with a magnitude of −323 and −134 mm, respectively. In addition, large cumulative deformations were observed at P1 and P8, exceeding −250 mm in less than two years. The magnitude of the accumulated settlement also showed strong correlation with the time table of land reclamation. Furthermore, the eight points had different deformation evolutions. The subsidence trend at P1 slowed down in November 2017 and then turned into a relatively mild subsidence pattern, which suggested that the reclaimed land entered a long-term slow compression phase [7]. However, subsidence at P4, P7, and P8 still showed a large subsidence trend, which indicated that the newly reclaimed land had just been completed and would enter the phase of primary consolidation [10]. Accordingly, it is highly possible that P4, P7, and P8 could still undergo remarkable subsidence in the near future. Therefore, fieldwork should be done to further monitor the dense time series ground settlement. The detailed analysis of deformation evolution after land reclamation at Xiamen New Airport is provided in Section 5.5 based on Terzaghi theory of soil mechanisms [36].

5.3. Coastal Land Subsidence and Uplift

According to the InSAR measurements in Figure 4, there is remarkable subsidence associated with land reclamation that could be observed in most coastal areas of Xiamen, especially in Regions E and F. Detailed deformation rate maps for Regions E and F are shown in Figure 9, where Figure 9c shows the deformation rate from November 2015 to December 2016, and Figure 9d shows the deformation rate from January 2017 to October 2018. In addition, two remote sensing images acquired on 22 July 2016 and 13 March 2018, are shown in Figure 9a,b to facilitate deformation analysis. To reveal the spatial characteristics of land subsidence, three profiles located in different areas were selected to extract the deformation rates; profile locations are shown in Figure 9b, and deformation rates along the profiles are shown in Figure 10. Furthermore, the time series deformation for the four points shown in Figure 9b was obtained to further analyze the temporal evolution of land subsidence, as shown in Figure 11.

It can be seen from Figure 9a,b that Regions E and F were reclaimed on 22 July 2016 and were basically completed on 13 March 2018. However, obvious water seepage can still currently be observed in the reclamation areas, which indicates that the reclamation area is still in an unstable state. We can see from Figure 9c that the reclaimed areas for Regions E and F, from November 2015 to December 2016, suffered very large deformation. The deformation rate in the LOS direction reached −130 mm/year. From Figure 9d, we can see that the spatial deformation characteristics of Regions E and F changed after 2017. The magnitude of deformation slightly dropped, and results demonstrate that land subsidence gradually slowed down with the passing of time after land reclamation. However, the area of deformation greatly increased. The area of deformation between November 2015 and December 2016 was about 1.8 km^2, and it increased to 5.1 km^2 from January 2017 to October 2018. Obvious nonuniform land subsidence can be observed in Regions E and F in Figure 10, which is particularly evident in the location of Profile GG′. Such a nonuniform subsidence pattern could largely be attributed to the different completion times of land reclamation in different areas. Some obvious land subsidence funnels were detected at the locations of profiles HH′ and II′, and deformation rates in the vertical direction reached −105 and −61 mm/year from January 2017 to October 2018, respectively.

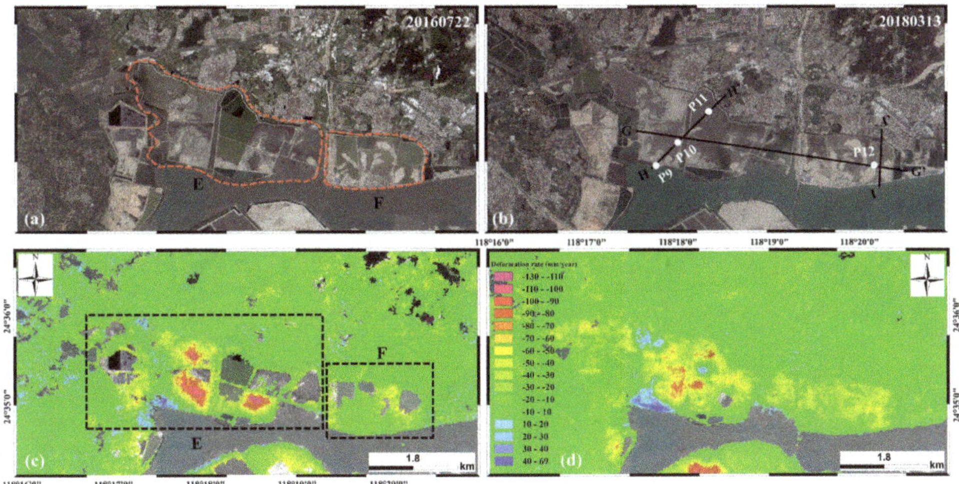

Figure 9. Ground deformation rate maps and remote sensing images of Regions E and F. (**a**) Remote sensing image acquired on 22 July 2016; (**b**) remote sensing image acquired on 13 March 2018; (**c**) average ground deformation rate map from November 2015 to December 2016; (**d**) average ground deformation rate map from January 2017 to October 2018.

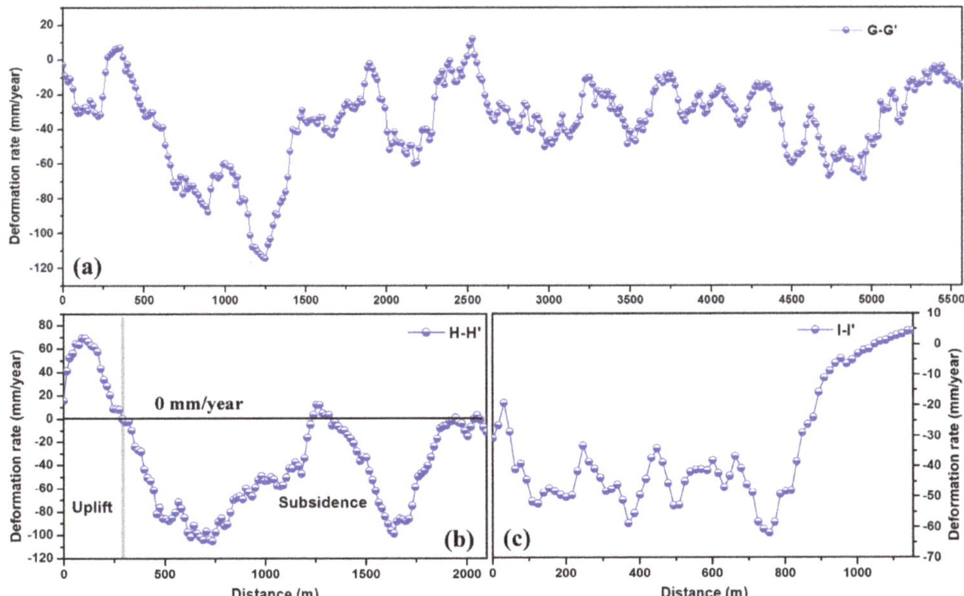

Figure 10. Average deformation rates in the vertical direction of Region E and F from January 2017 to October 2018 along three profiles (positions are indicated as black solid lines in Figure 9b). (**a**) Profile G–G′; (**b**) profile H–H′; (**c**) profile I–I′.

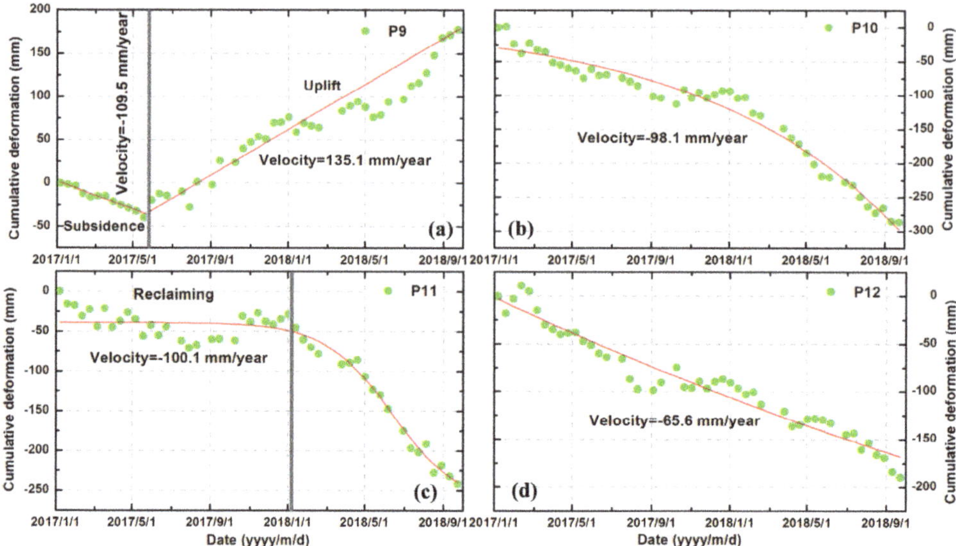

Figure 11. Time series deformation in the vertical direction of Regions E and F from January 2017 to October 2018 for P9–P12, which are indicated as white dots in Figure 9b. (**a**) Point P9; (**b**) point P10; (**c**) point P11; (**d**) point P12.

Different temporal evolution characteristics of deformation can be seen from Figure 11. It is clear that P10 and P12 suffered nonlinear subsidence from January 2017 to October 2018, while P11 had a different temporal deformation pattern. P11 deformation fluctuated before January 2018, then changed to rapid deformation until October 2018, which indicated that P11 was in the reclamation stage before January 2018 and began to deform after reclamation was completed in January 2018. These deformation patterns are in good agreement with the prediction pattern of unsaturated soil consolidation theory [37].

Besides land subsidence, several areas in Region E experienced remarkable uplift. P9 in Figure 11a was in the subsidence stage before August 2017, afterward turning to uplift. Such a phenomenon was also observed in other land reclamation cases [7], which can be explained by the well-known compression mechanisms of hydraulic fill. Generally, ground deformation associated with land reclamation is induced by primary consolidation and long-term second compression of alluvial clay deposits beneath the reclamation [10,16]. For newly reclaimed areas, such as P11, primary consolidation takes place for some time. Then, it moves on to the second stage, which is a slight rebound after long-term compression, such as in P9. Finally, land subsidence is in a stable state after long-term changes, such as in P1 [7]. Therefore, we can infer that the P9 uplift was caused by the rebound of reclamation fill materials after long-term compression.

5.4. Subsidence Along the Road

In addition to deformation caused by extensive land reclamation, typical deformation caused by road construction was also observed in the northeastern part of Xiamen, as shown in Region G in Figure 4b. Figure 12 shows the ground deformation rate of Region G from January 2017 to October 2018, and two remote sensing images acquired on 6 February 2015 and 18 May 2018. It is evident from Figure 12a that the road in the northern section was constructed before 6 February 2015, while the remaining sections were basically completed before 18 May 2018 (Figure 12c). It can be seen from Figure 12b that there was obvious deformation that occurred on the road built in the later stage, which did suffer slow-rate subsidence, with the deformation rate in the LOS direction of about −20 to

−40 mm/year from January 2017 to October 2018. No deformation could be observed for the road built in earlier years, which can be explained because the earlier built road was completely compacted, whereas the newly built one is still under the compaction process.

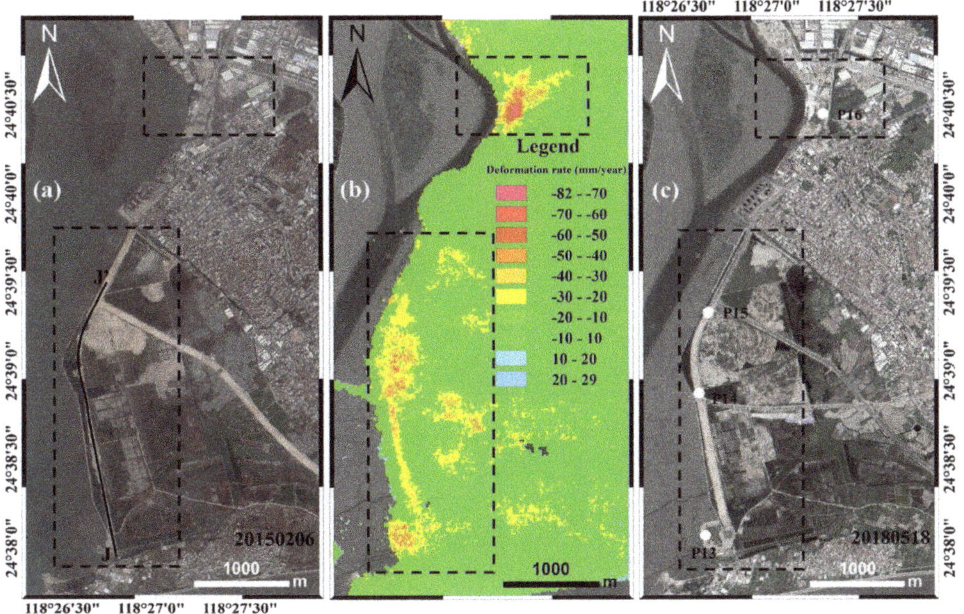

Figure 12. Ground deformation rate and remote sensing images of Region G. (**a**) Remote sensing image acquired on 6 February 2015; (**b**) average deformation rate map in the LOS direction from January 2017 to October 2018; (**c**) remote sensing image acquired on 18 May 2018.

The deformation rate along the road indicated by J–J′ (whose position is shown in Figure 12a) was extracted, and it is shown in Figure 13. It can be seen that obvious nonuniform deformation occurred along the newly built road. Deformation in the middle part of the road is greater than that at either end, and several subsidence funnels were observed in the middle section of the road. The maximum deformation rate in the vertical direction reached −57 mm/year. In order to further investigate the temporal evolution of deformation after road construction, three typical points for P13 to P15, located at the two ends and the middle section of the road, were selected to show time series deformation. The locations of the selected points are shown in Figure 12c, and time series deformation is shown in Figure 14. We can see from Figure 14 that the road experienced continuous deformation from January 2017 to October 2018. Maximum cumulative subsidence occurred in the middle section of the road, which reached −125 mm in the vertical direction in less than two years. In addition, it can be found that there were obvious fluctuations in road deformation during the InSAR monitoring period. There is a strong possibility that the results of the InSAR measurement were affected by road construction, as can be seen from the remote sensing images (see Figure 12a,c), since the road was still under construction during the InSAR monitoring period. To verify our speculation, P16 (see Figure 12c) was selected to obtain time series deformation far away from the road construction area, as shown in Figure 14d. It is evident that the time series deformation of P16 did not show obvious fluctuation.

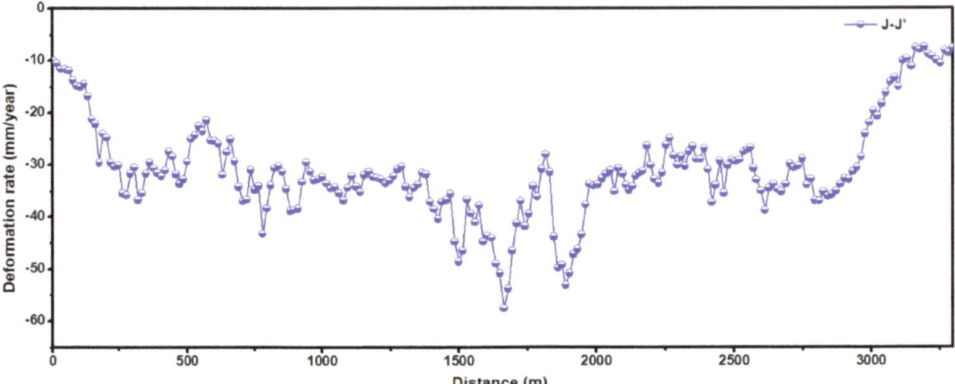

Figure 13. Average deformation rate in the vertical direction of Region G from January 2017 and October 2018 along Profile J–J', whose position are indicated as solid black lines in Figure 12.

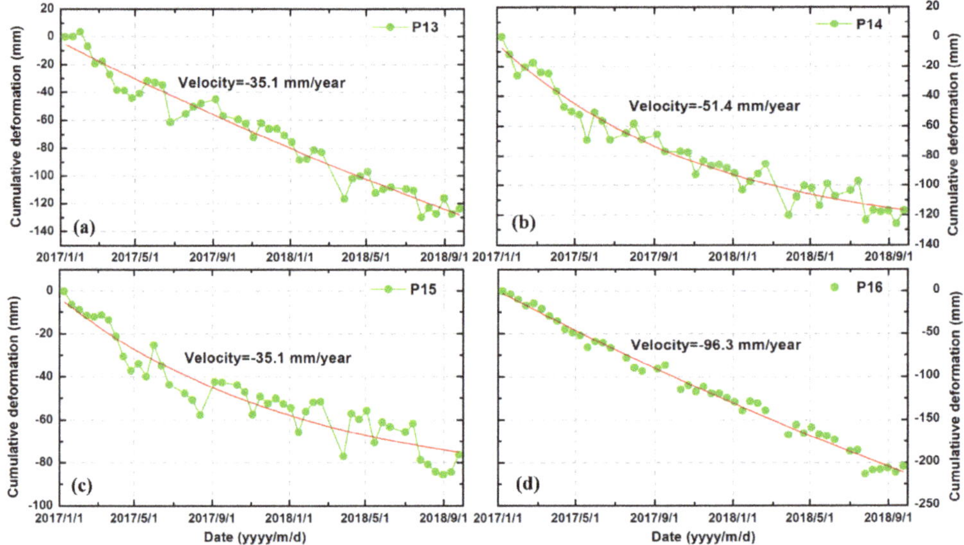

Figure 14. Time series deformation in the vertical direction of Region G from January 2017 and October 2018 for P13 to P16, which are indicated as white dots in Figure 12c. (**a**) Point P13; (**b**) point P14; (**c**) point P15; (**d**) point 16.

5.5. Detailed Analysis of Land Subsidence at Xiamen New Airport

According to previous investigations [7,10,16,36,37], land subsidence associated with land reclamation is principally induced by three mechanisms: primary consolidation, long-term second compression of alluvial clay deposits beneath the reclamation, and creep with the reclamation fill. The magnitude and velocity of land subsidence after land reclamation largely depends on the types and thickness of the reclaimed materials, the thickness of the underlying alluvial deposits, duration after completion of the reclamation, and the effect of foundation treatment [10,38]. The largest proportion of total subsidence is owed to the primary consolidation of the alluvial clays. In the case of airport land reclamation settlement, it usually amounts to 70% or more [10]. In addition, the subsidence process of primary consolidation is much faster than that of secondary compression and filling creep. Such a phenomenon can be explained by the well-known Terzaghi theory of consolidation [36]. A typical time-settlement prediction curve for both primary consolidation and second compression of alluvial clay under the reclamation is shown in Figure 15 [10]. It can be seen that it underwent sharply rapid deformation (i.e., primary consolidation) after reclamation has just been completed; then, deformation slowed down and finally gradually ceased (i.e., secondary compression). The actual deformation pattern observed by InSAR in Xiamen New Airport is similar to such a time-settlement prediction curve of alluvial clay. Therefore, we can infer that early reclaimed land, like P1 in Figure 7a, completed the primary consolidation stage, and is currently suffering long-term compression. It can also be inferred from Figure 7a that P1 was in the primary consolidation stage before 10 March 2018, and it suffered rapid deformation. Then, it changed to the long-term compression stage after 10 March 2018, and it is currently almost stable. However, newly reclaimed areas completed after August 2016 (see Figure 8), like P4, P8, P10, and P11 (see Figures 7 and 11), are currently still in the primary consolidation stage; accordingly, significant ground deformation will continue for some time.

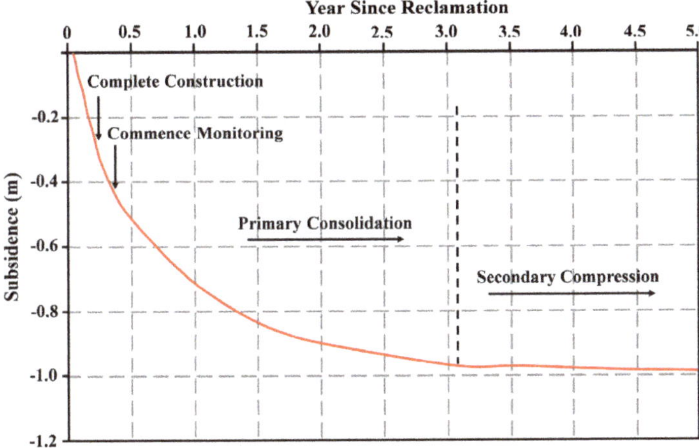

Figure 15. Typical time-settlement prediction curve of alluvial clay under the reclamation, i.e., Terzaghi theory of consolidation [10,36].

6. Conclusions

In this study, the time series InSAR technique was employed to characterize and monitor the ground deformation of Xiamen New Airport after land reclamation. In addition, some land subsidence areas associated with land reclamation were identified in Xiamen's coastal area. A total of 82 ascending Sentinel-1 images were used, which were acquired from August 2015 to October 2018. The pattern and spatiotemporal evolution characteristics of the surface deformation of Xiamen New Airport after land reclamation were fully revealed, providing important deformation information on Xiamen New Airport. The results are of great guiding significance for airport land reclamation, design, and next-stage construction. Furthermore, the airport's deformation characteristic was also successfully validated based on the compression mechanism of the hydraulic fill (i.e., Terzaghi theory of consolidation). The main conclusions that can be drawn are as follows.

A total of 20 land subsidence areas associated with land reclamation were identified in the coastal area of the city of Xiamen from November 2015 to December 2016, which increased to 26 from January 2017 to October 2018. The significant land subsidence areas are mainly concentrated in the central coastal area of Xiamen, i.e., Xiang'an district.

The land subsidence of Xiamen New Airport is mainly concentrated in the area of the second phase of land reclamation. Most reclamation areas underwent a rapid deformation process after completion; then, deformation gradually slowed down and finally became basically stable. Such a deformation characteristic is similar to the typical time-settlement prediction curve of the alluvial clay under the reclamation, i.e., the Terzaghi theory of consolidation.

Early reclaimed areas are currently in the stage of long-term compression, and deformation is currently small. However, newly reclaimed areas are still in the primary consolidation stage, where rapid deformation is currently underway. In addition, ground surface uplift caused by the rebound of reclamation fill materials after long-term compression was observed in some areas.

Author Contributions: X.L. and C.Z. performed the experiments and produced the results. X.L. drafted the manuscript and C.Z. finalized the manuscript. Q.Z., C.Y., and J.Z. contributed to the discussion of the results. All authors conceived the study, and reviewed and approved the manuscript.

Funding: This research was funded by the Natural Science Foundation of China (Grants No. 41874005, 41731066) and the Fundamental Research Funds for the Central Universities (Grants No. 300102269712).

Acknowledgments: The Sentinel-1A data used in this study were freely provided by Copernicus and ESA, and the one-arc-second SRTM DEM data were freely downloaded from the website http://e4ftl01.cr.usgs.gov/MODV6_Dal_D/SRTM/SRTMGL1.003/2000.02.11/. We thank the three anonymous reviewers as well as the editor for their helpful comments and suggestions.

Conflicts of Interest: The authors declare no conflict of interest.

References

1. Wali, M.K. *Practices and Problems of Land Reclamation in Western North America*; University of North Dakota Press: Grand Forks, ND, USA, 1975.
2. Douglas, I.; Lawson, N. Airport construction: Materials use and geomorphic change. *J. Air Trans. Manag.* **2003**, *9*, 177–185. [CrossRef]
3. De Mulder, E.; Van Bruchem, A.; Claessen, F.; Hannink, G.; Hulsbergen, J.; Satijn, H. Environmental impact assessment on land reclamation projects in the Netherlands: A case history. *Eng. Geol.* **1994**, *37*, 15–23. [CrossRef]
4. Suzuki, T. Economic and geographic backgrounds of land reclamation in Japanese ports. *Mar. Pollut. Bull.* **2003**, *47*, 226–229. [CrossRef]
5. Zhao, Q.; Pepe, A.; Gao, W.; Lu, Z.; Bonano, M.; He, M.L.; Wang, J.; Tang, X. A DInSAR Investigation of the Ground settlement Time Evolution of Ocean-Reclaimed Lands in Shanghai. *IEEE Sel. Top. Appl. Earth Obs. Remote Sens.* **2015**, *8*, 1763–1781. [CrossRef]
6. Breber, P.; Povilanskas, R.; Armaitiene, A. Recent evolution of fishery and land reclamation in curonian and lesina lagoons. *Hydrobiologia* **2008**, *611*, 105–114. [CrossRef]
7. Yang, M.S.; Yang, T.L.; Zhang, L.; Lin, J.X.; Qin, X.Q.; Liao, M.S. Spatial-Temporal Characterization of a Reclamation Settlement in the Shanghai Coastal Area with Time Series Analyses of X-, C-, and L-band SAR datasets. *Remote Sens.* **2018**, *10*, 329. [CrossRef]
8. Pepe, A.; Bonano, M.; Zhao, Q.; Yang, T.L.; Wang, H.M. The Use of C-/X-Band Time-Gapped SAR data and Geotechnical Models for the Study of Shanghai's Ocean-Reclaimed Lands through the SBAS-DInSAR Technique. *Remote Sens.* **2016**, *8*, 911. [CrossRef]
9. Baek, W.K.; Jung, H.S.; Jo, M.J.; Lee, W.J.; Zhang, L. Ground settlement observation of solid waste landfill park using multi-temporal radar interferometry. *Int. J. Urban Sci.* **2018**, 1–16. [CrossRef]
10. Jiang, L.M.; Lin, H. Integrated analysis of SAR interferometric and geological data for investigating long-term reclamation settlement of Chek lap KoK Airport, Hong Kong. *Eng. Geol.* **2010**, *110*, 77–92. [CrossRef]
11. Yu, L.; Yang, T.L.; Zhao, Q.; Liu, M.; Pepe, A. The 2016-2016 ground displacements of the Shanghai coastal area inferred a combined COSMO-SkyMed/Sentinel-1 DInSAR analysis. *Remote Sens.* **2017**, *9*, 1194. [CrossRef]
12. Liu, P.; Chen, X.F.; Li, Z.H.; Zhang, Z.G.; Xu, J.K.; Feng, W.P.; Wang, C.S.; Hu, Z.W.; Tu, W.; Li, H.Z. Resolving surface displacements in Shenzhen of China from time series InSAR. *Remote Sens.* **2018**, *10*, 1162. [CrossRef]
13. Aslan, G.; Cakır, Z.; Ergintav, S.; Lassere, G.; Renard, R. Analysis of secular ground motions in Istanbul from a long-term InSAR time series (1992–2017). *Remote Sens.* **2018**, *10*, 408. [CrossRef]
14. Zebker, H.A.; Villasensor, J. Decorrelation in interferometric radar echoes. *IEEE Trans. Geosci. Remote Sens.* **1992**, *30*, 950–959. [CrossRef]

15. Yague-Martinez, N.; Prats-Iraola, P.; Gonzalez, F.R.; Brcic, R.; Shau, R.; Geudtner, D.; Eineder, M.; Bamler, R. Interferometric Processing of Sentinel-1 TOPS Data. *IEEE Trans. Geosci. Remote Sens.* **2016**, *54*, 2220–2234. [CrossRef]
16. Xu, B.; Feng, G.C.; Li, Z.W.; Wang, Q.J.; Wang, C.C.; Xie, R.G. Coastal subsidence monitoring associated with land reclamation using the point target based on SBAS-InSAR method: A case study of Shenzhen, China. *Remote Sens.* **2016**, *8*, 652. [CrossRef]
17. Berardino, P.; Fornaro, G.; Lanari, R.; Sansosti, E. A new algorithm for surface deformation monitoring based on small baseline differential SAR interferograms. *IEEE Trans. Geosci. Remote Sens.* **2002**, *40*, 2375–2383. [CrossRef]
18. Mora, O.; Mallorquí, J.J.; Broquetas, A. Linear and nonlinear terrain deformation maps from a reduced set of interferometric SAR images. *IEEE Trans. Geosci. Remote Sens.* **2003**, *41*, 2243–2253. [CrossRef]
19. Usai, S. A least squares database approach for SAR interferometric data. *IEEE Trans. Geosci. Remote Sens.* **2003**, *41*, 753–760. [CrossRef]
20. Ferretti, A.; Prati, C.; Rocca, F. Permanent scatterers in SAR interferometry. *IEEE Trans. Geosci. Remote Sens.* **2001**, *39*, 8–20. [CrossRef]
21. Kampes, B.M. *Radar Interferometry: Persistent Scatterer Technique*; Springer: New York, NY, USA, 2006.
22. Hooper, A.; Zebker, H.; Segall, P.; Kampes, B. A new method for measuring deformation on volcanoes and other natural terrains using InSAR persistent scatterers. *Geophys. Res. Lett.* **2004**, *31*, L23611. [CrossRef]
23. Xiang'an, Xiamen City, International Airport. Available online: https://baike.baidu.com/item/%E5%8E%A6%E9%97%A8%E7%BF%94%E5%AE%89%E5%9B%BD%E9%99%85%E6%9C%BA%E5%9C%BA/3898600?fr=aladdin (accessed on 16 November 2018).
24. Werner, C.; Wegmüller, U.; Strozzi, T.; Wiesmann, A. GAMMA SAR and interferometric processing software. In Proceedings of the ERS-Envisat Symposium, Gothenburg, Sweden, 16–20 October 2000.
25. Sowter, A.; Moh, B.C.A.; Cigna, F.; Marsh, S.; Athab, A.; Alshammari, L. Mexico City land subsidence in 2014-2015 with Sentinel-1 IW TOPS: Results using the Intermittent SBAS (ISBAS) technique. *Int. J. Appl. Earth Obs. Geoinf.* **2016**, *52*, 230–242. [CrossRef]
26. Scheiber, R.; Moreira, A. Coregistration of interferometric SAR images using spectral diversity. *IEEE Trans. Geosci. Remote Sens.* **2000**, *38*, 2179–2191. [CrossRef]
27. Jiang, M.; Ding, X.; Hanssen, R.F.; Malhotra, R.; Chang, L. Fast Statistically Homogeneous Pixel Selection for Covariance Matrix Estimation for Multitemporal InSAR. *IEEE Trans. Geosci. Remote Sens.* **2015**, *53*, 1213–1224. [CrossRef]
28. Jiang, M.; Miao, Z.; Gamba, P.; Yong, B. Application of Multitemporal InSAR Covariance and Information Fusion to Robust Road Extraction. *IEEE Trans. Geosci. Remote Sens.* **2017**, *55*, 3611–3622. [CrossRef]
29. Goldstein, R.; Werner, C. Radar interferogram filtering for geophysical applications. *Geophys. Res. Lett.* **1998**, *21*, 4035–4038. [CrossRef]
30. Liu, Y.Y.; Zhao, C.Y.; Zhang, Q.; Yang, C.S. Complex surface deformation monitoring and mechanism inversion over Qingxu-Jiaocheng, China with multi-sensor SAR images. *J. Geodyn.* **2018**, *114*, 41–52. [CrossRef]
31. Costantini, M. A novel phase unwrapping method based on network programming. *IEEE Trans. Geosci. Remote Sens.* **1998**, *36*, 813–821. [CrossRef]
32. Kang, Y.; Zhao, C.Y.; Zhang, Q.; Lu, Z.; Li, B. Application of InSAR Techniques to an analysis of the Guanling Landslide. *Remote Sens.* **2017**, *9*, 1046. [CrossRef]
33. Liu, X.J.; Zhao, C.Y.; Zhang, Q.; Peng, J.B.; Zhu, W.; Lu, Z. Multi-Temporal Loess Landslide Inventory Mapping with C-, X- and L-Band SAR Datasets—A Case Study of Heifangtai Loess Landslides, China. *Remote Sens.* **2018**, *10*, 1756. [CrossRef]
34. Liao, M.S.; Wang, T. *Time Series InSAR Technique and Application*; Science Press: Beijing, China, 2014; pp. 69–71. (In Chinese)
35. Wang, H.; Wright, T.J.; Yu, Y.; Lin, H.; Jiang, L.; Li, C.; Qiu, G. InSAR reveals coastal subsidence in the Pearl River Delta. *China Geophys. J. Int.* **2012**, *191*, 1119–1128.
36. Terzaghi, K.; Peck, R.B.; Mesri, G. *Soil Mechanics in Engineering Practice*; John Wiley and Sons: Hoboken, NJ, USA, 1996; pp. 1–592.

37. Plant, G.W.; Covil, C.S.; Hughes, R.A. *Site Preparation for the New Hong Kong International Airport—The Design, Construction and Performance of the Airport Platform*; Thomas Telford: London, UK, 1998.
38. Pickles, A.R.; Tosen, R. Settlement of reclaimed land for the new Hong Kong International Airport. *Proc. ICE Geotech. Eng.* **1998**, *131*, 191–209. [CrossRef]

Article

Deriving High Spatial-Resolution Coastal Topography From Sub-meter Satellite Stereo Imagery

Luís Pedro Almeida [1,2,*], Rafael Almar [3], Erwin W. J. Bergsma [2], Etienne Berthier [4], Paulo Baptista [5], Erwan Garel [6], Olusegun A. Dada [7] and Bruna Alves [3]

1. Instituto de Oceanografia, Universidade Federal do Rio Grande (IO-FURG), 96203-000 Rio Grande, Brazil
2. Centre National d'Études Spatiales (CNES-LEGOS), 31400, Toulouse, France; erwin.bergsma@legos.obs-mip.fr
3. Institut de recherche pour le développement (IRD-LEGOS), 31400 Toulouse, France; rafael.almar@legos.obs-mip.fr (R.A.); oc.bruna@gmail.com (B.A.)
4. Centre National de la Recherche Scientifique (CNRS-LEGOS), 31400 Toulouse, France; etienne.berthier@legos.obs-mip.fr
5. Departamento de Geociências, Centro de Estudos do Ambiente e do Mar (CESAM), Universidade de Aveiro, Campus de Santiago, 3810-193 Aveiro, Portugal; renato.baganha@ua.pt
6. Centre for Marine and Environmental Research (CIMA), University of Algarve, 8005-139 Faro, Portugal; egarel@ualg.pt
7. Department of Marine Science and Technology, Federal University of Technology, 340252 Akure, Nigeria; oadada@futa.edu.ng

* Correspondence: melolp@gmail.com

Received: 15 January 2019; Accepted: 6 March 2019; Published: 12 March 2019

Abstract: High spatial resolution coastal Digital Elevation Models (DEMs) are crucial to assess coastal vulnerability and hazards such as beach erosion, sedimentation, or inundation due to storm surges and sea level rise. This paper explores the possibility to use high spatial-resolution Pleiades (pixel size = 0.7 m) stereoscopic satellite imagery to retrieve a DEM on sandy coastline. A 40-km coastal stretch in the Southwest of France was selected as a pilot-site to compare topographic measurements obtained from Pleiades satellite imagery, Real Time Kinematic GPS (RTK-GPS) and airborne Light Detection and Ranging System (LiDAR). The derived 2-m Pleiades DEM shows an overall good agreement with concurrent methods (RTK-GPS and LiDAR; correlation coefficient of 0.9), with a vertical Root Mean Squared Error (RMS error) that ranges from 0.35 to 0.48 m, after absolute coregistration to the LiDAR dataset. The largest errors (RMS error > 0.5 m) occurred in the steep dune faces, particularly at shadowed areas. This work shows that DEMs derived from sub-meter satellite imagery capture local morphological features (e.g., berm or dune shape) on a sandy beach, over a large spatial domain.

Keywords: Pleiades; photogrammetry; LiDAR; RTK-GPS; beach topography

1. Introduction

Accurate topographic data are frequently needed for the assessment of rapid morphological changes and for the implementation of models that can predict coastal evolution. High spatial resolution coastal Digital Elevation Models (DEMs—defined here as the representation of the terrain surface elevations at regularly spaced intervals) are used to support vulnerability and risk assessment of a range of coastal hazards, such as beach erosion and sedimentation, storm surges, inundation, and sea level rise [1]. For such studies, the availability of a topographic dataset is fundamental, in particular for coastal systems characterized by a complex, rapidly evolving morphology.

Among topographic survey methods of suitable quality, those based on Global Navigation Satellite Systems (GPS), such as Real Time Kinematic GPS (RTK-GPS), have been used extensively to map and monitor coastal morphology [2]. Beach topographic surveys using RTK-GPS method can be performed either by walking and carrying a GPS receiver, or driving a mobile unit (e.g., quad bike). In both cases, the vertical precision is approximately 0.05 to 0.1 m, depending on the terrain relief [2]. This method typically requires an intense human effort, which normally is optimized by reducing the number of measurements to a limited number of cross-shore sections of the beach. Nevertheless, this limited spatial coverage results in an incomplete representation of topographic spatial patterns and evolving features, especially in the case of complex topographies such as steep and unconsolidated slopes. In such cases, interpolation methods are typically required, introducing additional uncertainty into the DEM [3].

Remote sensing techniques, such as airborne LiDAR (Light Detection and Ranging) and Unmanned Aerial Vehicle (UAV), emerge in this context as a solution to overcome the limited spatial coverage of the RTK-GPS method [4–9]. The use of airborne LiDAR to measure geomorphological changes in coastal areas is relatively new. This instrumentation acquires millions of x, y, z points per hour, with a horizontal spacing of typically 1 to 3 m. This high spatial resolution, together with the capacity to survey over large areas (from 10^1 to 10^5 m), allows overcoming traditional survey limitations found with RTK-GPS [2]. The vertical accuracy of LiDAR ranges from 0.05 m to 0.15 m [5], which is in the same order as RTK-GPS and appropriate for studying beach morphology. Nonetheless, LiDAR-based DEMs are costly [5,6], which limits the frequent (e.g., monthly or-post-storm) acquisition of large-scale topographic data adequate for the evaluation of coastal changes.

Airborne optical remote sensing and 3D-mapping have been serving the needs of regional-scale low-altitude imaging and geospatial information [10]. The enhanced usability of recent UAV equipment with onboard accurate positioning, such as off-the-shelf drones, has resulted in a large change in their practical application. The RTK-GPS positioning of the camera, combined with the large number of overlapping images, makes any additional ground surveys trivial. Moreover, the high degree of automation of UAVs and the absolute vertical precision, of approximately 0.2 m, achieved by the DEMs suggests possible uses in the fields of natural hazards, disaster response, and high-resolution terrain analysis [6]. Despite these advantages, a few disadvantages still remain such as the cost of the photogrammetric software and computer power that can be relatively high [7], the difficulty in removing dense vegetation to obtain bare earth elevation estimates [11], the need for electric batteries for longer flight duration, or the usage limitations related to weather conditions [12].

Sub-meter satellite imagery can potentially provide an alternative to these field-based techniques in order to collect high spatial resolution topographic data over large areas. The first civil satellite constellation that acquired stereoscopic imagery and applied DEM reconstruction over large areas was the French SPOT mission (Satellite Pour l'Observation de la Terre) in 1986 [12]. Since then, several very high spatial resolution satellites with stereo capabilities were launched in response to an increased demand [13]. Among them, the Pleiades constellation (built by the French Space Agency (CNES), commercialized by AIRBUS Defence & Space), consists of two high spatial resolution optical spacecrafts: Pleiades –1A and –1B. Both satellites fly over the same near-polar sun-synchronous orbits at an altitude of 694 km with a 180° phase and descending node. The optical sensors of these satellites have the capability to obtain images with sub-meter image resolution (0.7 m pixel size, resampled to 0.5 m) over a maximum area of 350 km × 20 km (swath width of 20 km at nadir). An important aspect of Pleiades is the capacity to revisit any location in the world within 1 day, which is of great interest to monitor rapidly changing processes (e.g., coastal erosion due to storm events). Recent studies based on Pleiades-1A stereo-imagery include snow height mapping in mountainous areas [14], large landmass deformations due to earthquakes [15], surface reconstruction after landslides [16], and glacier topography [17,18].

The aim of the present work is to explore the use of Pleiades satellite stereo-imagery to develop a high resolution DEM of a 40-km-long sandy coastal section. The satellite-derived DEM is compared to RTK-GPS cross-shore profiles and an airborne LiDAR-derived DEM. The differences between the concurrent methods are quantified and the precision and accuracy of Pleiades-DEM analysed.

2. Study Site and Data Acquisition

A 40-km stretch of sandy coast in the South West of France was selected as the study site for the present work (Figure 1). This section of the Aquitanian coast presents a relatively low shoreface bordered by aeolian dunes with an average crest elevation of about 15 m [19–21]. The sediment consists of fine to medium quartz, with mean grain sizes ranging from 200 to 400 μm [21]. This section of the coast is characterized by a macro-tidal regime, with an average tidal range of 3.2 m that can reach 5 m during spring tides [22]. The coast is exposed to high energy North Atlantic swells travelling mainly from the W–NW sector [23].

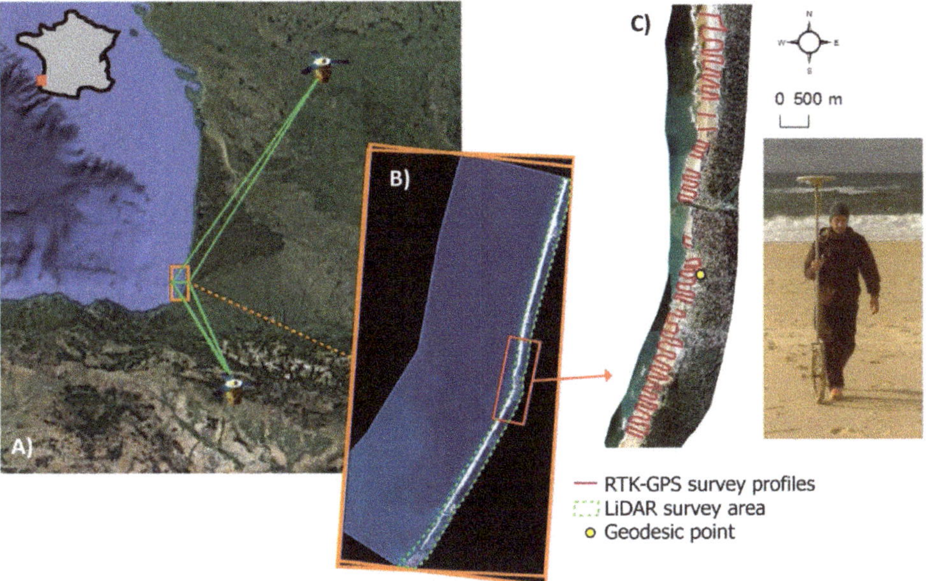

Figure 1. Satellite image of the West of France (source: Google Earth Pro 2018) showing the location where Pleiades stereo-pair was obtained (orange rectangle located in the Southwest of France) on the 14th of November 2017 (**A**). Zoom in of the Pleiades mosaic showing the area where the airborne-LiDAR topographic survey was performed (polygon with dashed green outline) and the region where RTK-GPS topographic measurements were undertaken (**B**). Panel (**C**) shows the RTK-GPS survey lines (red) and photograph of the surveyor with the GPS rover unit.

The Pleiades-HR 1A (hereinafter referred to as PL1A) stereo-pair was acquired on 14 November 2017 over a predefined area (orange box in Figure 1). The optical stereo-pair was obtained between 11:15 a.m. and 11:16 a.m. with a 40-s time-lag. The satellite orbits at 694 km altitude (base to height ratio of 0.36) and follows a descending orbit trajectory (North-South) in WGS84 decimal coordinates. A topographic DEM and ortho-image, covering the entire area of interest, was subsequently produced using NASA's AMES Stereo Pipeline [24].

In-situ RTK-GPS beach topographic measurements were used as ground-truth for inter-comparison with the Pleiades and LiDAR DEMs. The RTK-GPS survey was performed in the central section of the area of interest (coastal region of Capbreton) between the 7th and 9th of November 2017 (Figure 1). The average tidal range during the beach surveys was 3.4 m with a moderate wave climate (wave height under 3 m). Beach profiles were surveyed in continuous mode (waypoint every 1 second using a position dilution of precision—PDOD—mask of 3) from the waterline to the back of the frontal dune ridge along discrete cross-shore profiles spaced by approximately 250 m (Figure 1). Note that due to radio transmission shadowing (resulting in no real time correction) or ground obstacles (e.g., fences, walls, etc.), the survey coverage presents some spatial irregularities (Figure 1). In addition to the RTK-GPS dataset, a high resolution airborne-LiDAR topographic survey and orthophoto map (aerial imagery was orthorectified with the LiDAR observations) of the entire SW coast of France (performed by the Institut National de L'Information Geographic et Forestiere—IGN; in cooperation with the Bureau de Recherches Géologiques et Minières—BRGM) were acquired in October 2017.

3. Methods

3.1. Pleiades Stereo-imagery Acquisition and DEM Generation

In the present experiment, stereo images were acquired when the angles between the line-of-sight of the satellite camera and the horizontal plane of the ground were 72° and 76°, for the first and second images respectively. The ground projection of the Pleiades position during the stereo-pair collection was 218 and 174 km from the coast respectively. Pleiades was overlooking the coastal area of interest from the sea side. This satellite setup indicates that both images were collected close to nadir angle (90° from the ground), with an azimuth of 19° for the first and −8° for the second image, resulting in an azimuth angle difference of 27° between the two images.

The Pleiades panchromatic band of the stereo pair was processed using the Ames Stereo Pipeline, ASP [24] to generate a DEM and ortho-images at 2 and 0.5 m resolution, respectively. The ASP uses the rational polynomial coefficient (RPC) camera model format for the DEM generation. The RPC model is provided in the imagery metadata (by AIRBUS) and gives a relationship between the image coordinates and the ground coordinates. No ground control points were initially used in the DEM generation. The planimetric coordinates were referenced to the WGS84 UTM 30N coordinate system and the heights were computed above the WGS84 ellipsoid. The Pleiades DEM and ortho-images were a posteriori coregistered by applying a first order polynomial transformation defined by 37 concomitant points manually identified in the Pleiades-ortho and DEM and in the IGN/BRGM orthophoto map (used as the reference). This process was performed with QGIS software and Georeferencer GDAL plugin, with an average planimetric error of 0.5 m. To convert Pleiades altimetric data from WGS1984 to the French NGF-IGN vertical datum, the Pleiades elevations were corrected from the average difference with the LiDAR elevations (Pleiades elevations 12.2 m higher) determined at the same point locations and then used for the planimetric correction.

3.2. RTK-DGPS Topographic Survey

The planimetric coordinates of the RTK-DGPS topographic survey were referenced to the World Geodetic System (WGS84) while the vertical datum was referenced to NGF-IGN 1969 datum. A GPS base station was installed in a local geodesic point (located near Capbreton - Figure 1) and provided, in real time, the corrections to the mobile GPS unit via radio-transmission. After the survey, the planimetric coordinates were converted to the same coordinate system as the Pleiades products (WGS 84 UTM 30N), and spikes in the data (erroneous measurements) were eliminated. The processed topographic measurements were subsequently divided into individual profiles and interpolated in the cross-shore direction with 2 m spacing (to match the resolution of the Pleiades DEM). Comparisons between RTK-GPS ground-truth and remotely sensed Pleiades and LiDAR DEMs were performed by extracting values from the DEMs at each profile location (using all RTK-GPS point measurements). This task was performed using QGIS software (Lyon version) and the function "sample raster maps at point location" from GRASS-GIS toolbox. The data comparisons included the calculation of statistical parameters such as the correlation coefficient (CC), root mean squared (RMS) error, and bias (BIAS) using all RTK-GPS topographic observations.

3.3. Airborne LiDAR 3D Topographic Survey

As part of a regional coastal monitoring program, the Aquitaine coastal zone is surveyed every year with airborne LiDAR by the IGN in cooperation with the BRGM. In the present work, the LiDAR topographic survey of the Aquitaine coast performed between 4 and 7 of October 2017 was used for comparison with the RTK-GPS and Pleiades observations. Note that the LiDAR survey was executed approximately a month before the field campaign and Pleiades acquisition. The airborne-LiDAR survey was performed with the Leica ALS70-HP LiDAR, mounted on an aircraft, and acquiring topographic information with a density of 8 points per square meter. The planimetric coordinates of the LiDAR point-cloud were referenced to the Lambert 93 coordinate system, with a precision of 30 cm, and terrain topography referenced to the NGF-IGN 1969 datum, with an accuracy of 15 cm (information provided by the LiDAR survey metadata). Simultaneously with the LiDAR acquisition, high resolution (10 cm) aerial orthophotography was obtained with an 8-head IGN V2 (focal length of 135 mm) camera. The position and orientation of the images were obtained from the GPS and inertial sensors embedded in the aircraft. Two products were obtained from this flight: a DEM of the study area, with a spatial resolution of 1 m, and an orthophoto map with 10 cm resolution. For comparison with the other datasets, the coordinate systems of these LiDAR products were converted to WGS 84 UTM 30N.

4. Results

4.1. Comparison between RTK-GPS, Pleiades and LiDAR Topography

Figure 2 shows the vertical difference between the three concurrent survey methods over 4138 points (i.e., the number of data points measured during the RTK-GPS survey). The differences between remote sensing methods (Pleiades and LiDAR) and RTK-GPS are normally distributed, with mean differences (BIAS) of 0.01 m and 0.03 m, and RMS errors of 0.35 m and 0.37 m for Pleiades and LiDAR respectively (both Pleiades and LiDAR elevations are slightly higher than the RTK measurements). The observed slight mean difference between the RTK-GPS and remote sensing methods are within the accuracy of the RTK-GPS, thus indicating that the different methods have similar accuracy. It is important to note that even though the LiDAR data have a larger RMS error than the Pleiades one, the error distribution is narrower and skewed for negative values. The 1:1 scatter-comparison of the surveys shown in Figure 2 indicates that the remotely-sensed beach topography is highly correlated with the RTK-GPS observations (slope = 1.01 and CC = 0.99 for both Pleiades and LiDAR). It is also observed that the correlation with Pleiades values does not vary with elevation (extending from the back of the dune to the top of the swash zone) while for LiDAR some scattering is observed in the lower part of the beach (beach face).

A close inspection of the datasets indicates the scatter (observed in the beach face) between RTK-GPS and LiDAR topography was due to morphological changes (berm erosion—Figure 3) over the beach profile that occurred between the two surveys. During this period of the year (winter season), this coastal area is under energetic waves and significant morphological changes (> 1 m) in the beach and dune face are likely to occur [23]. Considering that the RTK-GPS survey and Pleiades image acquisition were days apart, small morphological changes likely occurred in the beach face, resulting in minor differences between the two datasets (Figure 3).

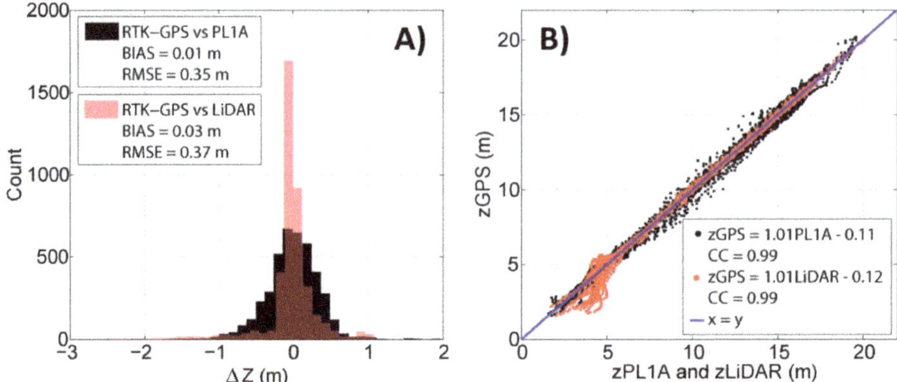

Figure 2. Pleiades (PL1A) and LiDAR survey accuracy assessed by comparison to concurrent on-ground RTK-GPS survey; histogram of the differences between elevations measured by RTK-GPS and remote sensing methods; (**A**); scatter plot of RTK-GPS elevations vs remotely sensed elevations (**B**).

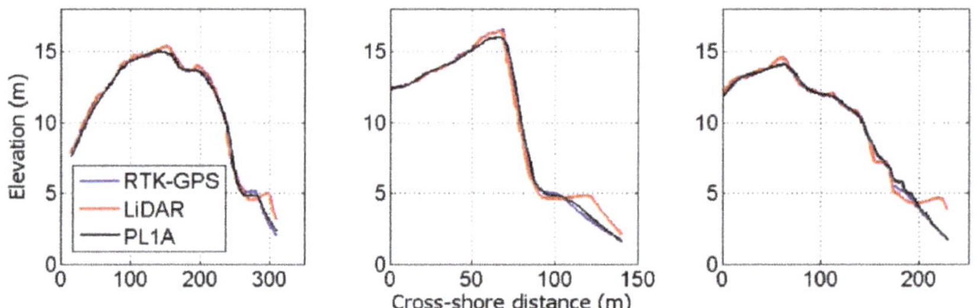

Figure 3. Three beach profiles showing the comparison between RTK-GPS, Pleiades (PL1A) and LiDAR, showing the berm erosion that occurred between the LiDAR survey and RTK-GPS and Pleiades.

4.2. 3D Beach Topography Comparisons

Figure 4 shows the two DEMs produced using Pleiades stereo imagery and LiDAR data, together with the difference between the two DEMs (Pleiades DEM minus LiDAR DEM). The northern part of the study area (North of Capbreton—Figure 4) presents higher dune fields than the South, and Pleiades DEM was able to capture this spatial variability with the same quality as the LiDAR. The difference between the two DEMs shows that over the full domain, 70% of the difference lies within $\pm$ 0.5 m. Areas with positive elevation difference (i.e., Pleiades higher than LiDAR) are located between the frontal dune face and swash zone, while a negative difference was found more often at the back of the dune. A preliminary inspection of the alongshore error distribution allowed identifying larger differences at the dune face region between Capbreton and Labenne (Figure 4) in comparison to the rest of the domain. Within the section with these particular large differences, it was possible to identify the presence of shadows (for each transect, the length of the shadow was manually digitized from the orthophotomap) in the dune face that coincided with the areas where the largest errors were observed (Figure 5).

Specific ground characteristics, such as the slope and aspect [25,26], can have an indirect impact on the remotely-sensed DEM quality in regions with high relief, such as dunes. The presence of shadows in the dune face is determined by the steepness (slope) and orientation (aspect) of the topography in relation to the Sun light (for a given Sun altitude and azimuth). Optical remote sensing images from shadowed areas have low reflectance and texture, which alters the calculation of the disparity (which is later converted into elevation) of corresponding points in the stereo-pair and leads to less accurate elevations [27,28].

In order to investigate the effect of the alongshore variability in the dune slope (front and back faces) and of the shadowed areas on the Pleiades DEM error, 266 cross-shore transects spaced 150 m apart were created. For each transect, the RMS error of the elevation difference (between Pleiades and LiDAR DEMs) over the dune face and back of the dune was computed. Figure 6 shows the RMS error variation along the 40 km of measured coastline together with the variations in the dune slope and shadow length in the dune face. The RMS error shows significant variations between transects, with values of similar magnitude on the dune face and back of the dune. The exception, as noted in Figure 4, is the coastal stretch between Capbreton and Labenne where a peak of RMS error (>0.5 m) is identified on the dune face. The alongshore location of this peak coincides with a relatively steep dune face region (large slope values) and the largest shadow regions (Figure 6). Steep dune face slopes were also present in the North section of the study area; however, the errors in this region were within the average, suggesting that the dune slope by itself cannot explain the largest errors observed. Shadowing at the dune face is more likely to represent a decisive quality factor of the produced DEM, considering the strong correlation between the RMS error and shadowing (CC = 0.77).

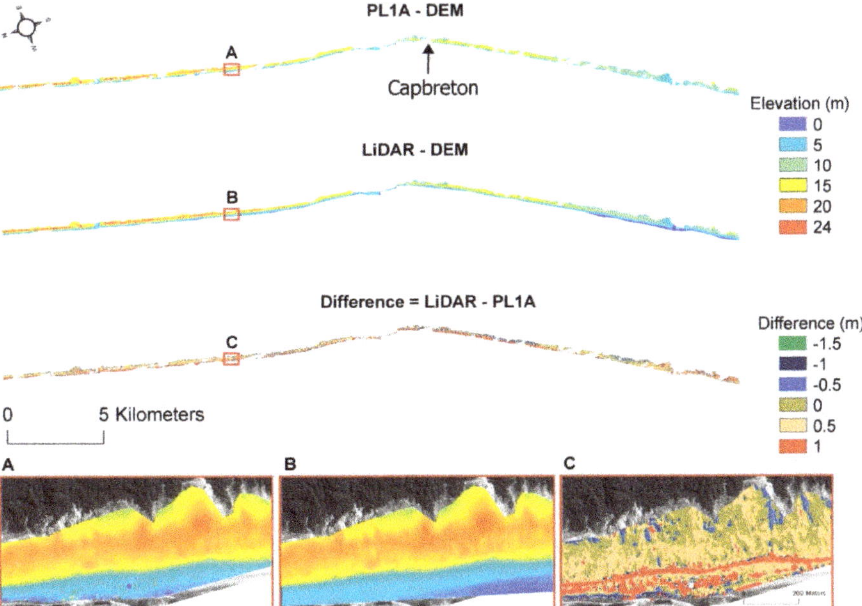

Figure 4. Results of the comparisons between the Pleiades (PL1A) and LiDAR DEMs. The lower panel presents three subsections of the study area showing the DEM produced with Pleiades (**A**), LiDAR (**B**) and the difference between the two (**C**). Note that the maps of the DEMs were rotated 110° in order to present them horizontally.

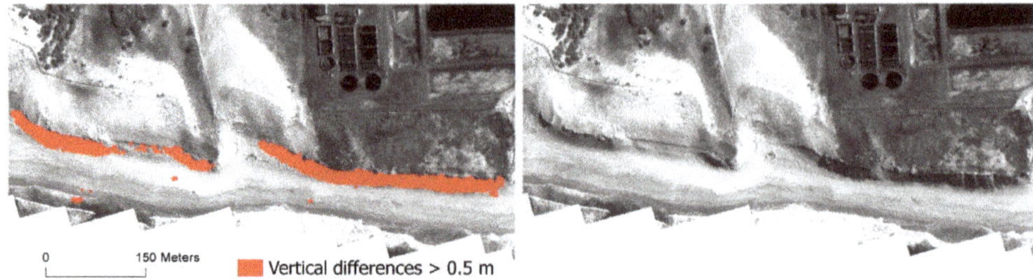

Figure 5. Example of the errors observed on the dune face, over the Pleiades orthophotomap; the presence of shadows on the dune face (**right image**) coincides with the area where the largest differences (>0.5 m in red, left image) were observed (**left image**).

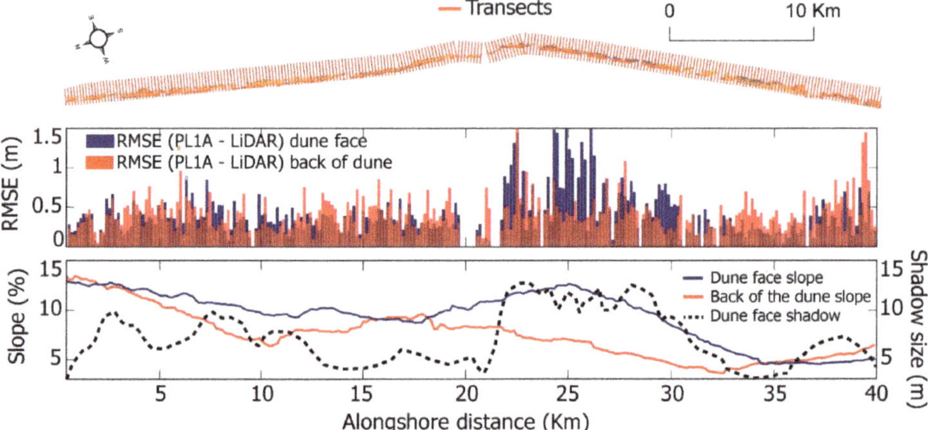

Figure 6. Results of the RMS error, slope and shadow calculations for the 266 cross-shore transects. The second panel (from the top) presents the RMS error of the difference between Pleiades (PL1A) and LiDAR topographic elevations, computed for the dune face and back of the dune; the bottom panel shows the dune face and back of the dune slope (bottom panel) and the shadow size (length of the shadow over each transect).

Figure 7 show the statistical evaluation of the comparison between the Pleiades and LiDAR elevations extracted over the 266 cross-shore transects (Figure 6). Data from the beach face, containing natural morphological changes (not related to the method), were removed from this statistical comparison. The error distribution indicates that differences between the LiDAR and Pleiades DEMs over the full domain are normally distributed with a mean difference of −0.015 m and RMS error of 0.48 m (LiDAR DEM is slightly higher). The 1:1 comparison (Figure 7) shows that Pleiades and LiDAR beach topographies are highly correlated (slope = 1.01 and CC = 0.99).

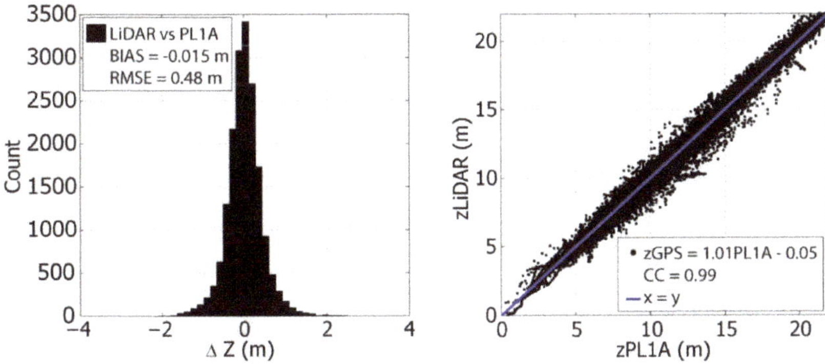

Figure 7. Pleiades (PL1A) topographic measurements accuracy assessed by comparison with LiDAR topographic measurements over points collected along 266 transects.

5. Discussion

The Use of Pleiades to Survey and Monitor Coastal Areas

The comparisons presented in Sections 4.1 and 4.2 indicate that Pleiades DEM reaches a vertical accuracy (BIAS) and precision (RMS error) similar to state-of-the-art survey methodologies used in coastal areas such as the RTK-GPS and LiDAR. Although in this study a comparison between Pleiades and UAVs stereo-topographic DEM was not presented, there are a few differences between these two methods that are important to take into consideration when deciding which of the methods to use. Both methods obtain the DEM based on photogrammetric techniques that estimate a 3D point-cloud of the ground surface using a large number of matching object and textural features automatically detected in the overlapping images. The low altitude and capacity to obtain images from many different view angles allows UAVs to obtain detailed 3D reconstruction of the ground surface with centimetric spatial resolution [7,29]. These characteristics represent an important advantage of the UAV technique when surveying highly complex coastal features (e.g., irregular rocky coastlines; [30]) or avoid the influence of sunlight exposure in the target features, such as the presence of shadows.

As it was observed in the present work, the presence of shadows can affect the quality of the DEM from stereo-satellite imagery (Figures 5 and 6). A similar problem was identified by [13] when validating the Pleiades tri-stereo digital surface model on an urban area. The fact that Pleiades, stereo and tri-stereo, only obtains images along a fixed trajectory, this limits the view angles of the ground surface (one backward looking, one forward looking, plus a third near-nadir image, in the tri-stereo configuration), making difficult an appropriate DEM estimation from features affected by shadows.

Despite this limitation, the present results show that Pleiades stereo imagery has the capability to capture local beach features such as berms or dunes crests and troughs over large domains, highlighting its incomparable advantage to any other existing methodology. These unique skills allow the Pleiades constellation to potentially overcome traditional survey challenges in coastal areas, such as the acquisition stereo-imagery measurements in large coastal segments within a short period of time (i.e., minutes). For traditional survey methods, this would mean a gigantic logistical challenge and expensive exercise. Furthermore, Pleiades' capacity to acquire imagery anywhere on the globe within 1 day is ideal for rapid response assessment of changes in the coastal zone (e.g., to assess morphological changes after storms).

A potential drawback of the Pleiades constellation is the on-demand availability, which is different from other observation missions like Landsat (NASA) or Sentinel (ESA) that acquire optical imagery on a regular basis without any previous request. In addition to this, another potential limitation of our processing flow is the dependence of ground control points to correct the vertical offset of the DEM and geometric inaccuracies of the raw data. This aspect limits the use of Pleiades DEM for applications where the absolute error in relation to a specific datum is required. Nevertheless, for cases where the offset to the local vertical datum is secondary, the inter-comparison between consecutive Pleiades DEMs is likely to result in errors within the precision of the method.

6. Conclusions

The beach topography of a 40-km-long sandy coastal stretch, located in the southwest of France, was surveyed by satellite Pleiades stereo-imagery. The computed DEM was compared with those obtained from traditional survey methodologies (RTK-GPS and LiDAR). Present findings indicate that Pleiades stereo-imagery allows the acquisition of high resolution DEM with a RMS error that ranges from 0.35 to 0.48 m. The largest errors were observed at the dune face, in regions with large shadow patches. Near-perfect agreement between Pleiades and concurrent methods (all computed CC were above 0.9) provides strong indications that this method can be used as a surveying tool to monitor detailed coastal morphological changes over large spatial domains.

Author Contributions: Conceptualization, L.P.A., R.A. and E.B.; data curation, L.P.A. and E.B.; methodology, E.B. and L.P.A.; validation, L.P.A.; writing—original draft preparation, L.P.A., B.A., E.W.J.B. and R.A.; writing—review and editing, all authors; field data acquisition, L.P.A., R.A., E.W.J.B., P.B., E.G. and O.A.D.; administration and funding acquisition, L.P.A. and R.A.

Funding: Luís Pedro Almeida was funded by a post-doctoral fellowship from the French Space Agency (CNES) to develop this work. Etienne Berthier acknowledges fundings from CNES through the TOSCA program. The work of Erwan Garel was supported by FCT research contract IF/00661/2014/CP1234. Paulo Baptista acknowledges the financial support to CESAM (UID/AMB/50017/2019), to FCT/MCTES through national funds, and the co-funding by the FEDER, within the PT2020 Partnership Agreement and Compete 2020.

Acknowledgments: The authors would like to thank the CNRS/LEGOS for funding the project the COMBI (2017) that allowed the development of the field campaign for the validation of the Pleiades products. We would like to thank the Mairie and Port de Capbreton for offering accommodation during the field campaign, and all the amazing support given during the experiment.

Conflicts of Interest: The authors declare no conflict of interest.

References

1. Hawker, L.; Bates, P.; Neal, J.; Rougier, J. Perspectives on Digital Elevation Model (DEM) Simulation for Flood Modeling in the Absence of a High-Accuracy Open Access Global DEM. *Front. Earth Sci.* **2018**, *6*, 233. [CrossRef]
2. Brasington, J.; Rumsby, B.T.; Mcvey, R.A. Monitoring and Modelling Morphological Changes in a Braided Gravel-Bed River Using HIgh Resolution GPS-Based Survey. *Earth Surf. Process. Landf.* **2000**, *25*, 973–990. [CrossRef]
3. Amante, C.J. Estimating Coastal Digital Elevation Model Uncertainty. *J. Coast. Res.* **2018**, *34*, 1382–1397. [CrossRef]
4. Sallenger, A.H., Jr.; Krabill, W.B.; Swift, R.N.; Brock, J.; List, J.; Hansen, M.; Holman, R.A.; Manizade, S.; Sontag, J.; Meredith, A.; et al. Evaluation of airborne topographic lidar for quantifying beach changes. *J. Coast. Res.* **2003**, *19*, 125–133.
5. Nelson, A.; Reuter, H.I.; Gessler, P. DEM Production Methods and Sources. In *Geomorphometry Concepts, Software, Applications*; Hengl, T., Reuter, H.I., Eds.; Elsevier: Amsterdam, The Netherland, 2009; pp. 65–85.
6. Mancini, F.; Dubbini, M.; Gatteli, M.; Stecchi, F.; Fabbri, S.; Gabbianelli, G. Using Unmanned Aerial Vehicles (UAV) for High-Resolution Reconstruction of Topography: The Structure from Motion Approach on Coastal Environments. *Remote Sens.* **2013**, *5*, 6880–6898. [CrossRef]
7. Gonçalves, J.A.; Henriques, R. UAV photogrammetry for topographic monitoring of coastal areas. *ISPRS J. Photogramm. Remote Sens.* **2015**, *104*, 101–111. [CrossRef]
8. Turner, I.L.; Harley, M.D.; Drummond, C.D. UVA's for coastal surveying. *Coast. Eng.* **2016**, *14*, 19–24. [CrossRef]
9. Chen, B.; Yang, Y.; Wen, H.; Ruan, H.; Zhou, Z.; Luo, K.; Zhong, F. High-resolution monitoring of Beach topography and its change using unmanned aerial vehicle imagery. *Ocean Coast. Manag.* **2018**, *160*, 103–116. [CrossRef]

10. Colomina, I.; Molina, P. Unmanned Aerial Systems for Photogrammetry and Remote Sensing: A Review. *ISPRS J. Photogramm. Remote Sens.* **2014**, *92*, 79–97. [CrossRef]
11. Niethammer, U.; James, M.; Rothmund, S.; Travelletti, J.; Joswig, M. UAV-based remote sensing of the Super-Sauze landslide: Evaluation and results. *Eng. Geol.* **2012**, *128*, 2–11. [CrossRef]
12. Gleyzes, J.P.; Meygret, A.; Fratter, C.; Panem, C.; Ballarin, S.; Valorge, C. SPOT5—System overview and image ground segment. In Proceedings of the IGARSS Conference, Toulouse, France, 21–25 July 2003.
13. Panagiotakis, E.; Chrysoulakis, N.; Charalampopoulou, V.; Poursanidis, D. Validation of Pleiades Tri-Stereo DSM in Urban Areas. *Int. J. Geo-Inf.* **2018**, *7*, 118. [CrossRef]
14. Marti, R.; Gascoin, S.; Berthier, E.; de Pinel, M.; Houet, T.; Laffly, D. Mapping snow depth in open alpine terrain from stereo satellite imagery. *Cryosphere* **2016**, *10*, 1361–1380. [CrossRef]
15. Zhou, Y.; Parsons, B.; Elliot, J.R.; Barisin, I.; Walker, R.T. Assessing the ability of Pleiades stereo imagery to determine height changes in earthquakes: A case study for the El Mayor-Cucapahepicentral area. *J. Geophys. Res. Solid Earth* **2015**, *120*, 8793–8808. [CrossRef]
16. Stumpf, A.; Malet, J.P.; Allemand, P.; Ulrich, P. Surface reconstruction and land-slide displacement measurements with Pleiades satellite images. *ISPRS J. Photogramm. Remote Sens.* **2014**, *95*, 1–12. [CrossRef]
17. Wagnon, P.; Vincent, C.; Arnaud, Y.; Berthier, E.; Vuillermoz, E.; Gruber, S.; Ménégoz, M.; Gilbert, A.; Dumont, M.; Shea, J.; et al. Seasonal and annual mass balances of Mera and Pokalde glaciers (Nepal Himalaya) since 2007. *Cryosphere* **2013**, *7*, 1769–1786. [CrossRef]
18. Berthier, E.; Vincent, C.; Magnusson, E.; Gunnlaugsson, A.; Pitte, P.; Le Meur, E.; Masiokas, M.; Ruiz, L.; Pálsson, F.; Belart, J.M.; et al. Glacier topography and elevation changes derived from Pleiades sub-meter stereo images. *Cryosphere* **2014**, *8*, 2275–2291. [CrossRef]
19. Abadie, S.; Butel, R.; Mauriet, S.; Morichon, D.; Dupuis, H. Wave climate and longshore drift on the South Aquitaine coast. *Cont. Shelf Res.* **2006**, *26*, 1924–1939. [CrossRef]
20. Pedreros, R.; Howa, H.; Michel, D. Application of grain size trens analysis for the determination of sediment transport pathways in intertidal areas. *Mar. Geol.* **1996**, *135*, 35–49. [CrossRef]
21. Castelle, B.; Guillot, B.; Marieu, V.; Chaumillon, E.; Hanquiez, V.; Bujan, S.; Poppeschi, C. Spatial and temporal patterns of shoreline change of a 280-km high-energy disrupted sandy coast from 1950 to 2014: SW France. *Estuar. Coast. Shelf Sci.* **2018**, *200*, 212–223. [CrossRef]
22. Castelle, B.; Bonneton, P.; Dupuis, H.; Sénéchal, N. Double bar beach dynamics on the high-energy meso-macrotidal French Aquitanian Coast: A review. *Mar. Geol.* **2007**, *245*, 141–159. [CrossRef]
23. Butel, R.; Dupuis, H.; Bonneton, P. Spatial Variability of Wave Conditions on the French Atlantic Coast using In-Situ Data. *J. Coast. Res.* **2002**, *36*, 96–108. [CrossRef]
24. Shean, D.E.; Alexandrov, O.; Moratto, Z.M.; Smith, B.E.; Joughin, I.R.; Porter, C.; Morin, P. An automated, open-source pipeline for mass production of digital elevation models (DEMs) from very-high-resolution commercial stereo satellite imagery. *ISPRS J. Photogramm. Remote Sens.* **2016**, *116*, 101–117. [CrossRef]
25. Bufton, J.L.; Garvin, J.B.; Cavanaugh, J.F.; Ramos-Izquierdo, L.; Clem, T.D.; Krabill, W.B. Airborne lidar for profiling of surface topography. *Opt. Eng.* **1991**, *30*, 72–78. [CrossRef]
26. Tsutsui, K.; Koya, K.; Kato, T. An investigation of continuous-angle laser light scattering. *Rev. Sci. Instrum.* **1998**, *69*, 3482–3486. [CrossRef]
27. Giles, P.T. Remote sensing and cast shadows in mountainous terrain. *Photogramm. Eng. Remote Sens.* **2001**, *67*, 833–839.
28. Ibarra-Delgado, S.; Cózar, J.R.; González-Linares, J.M.; Gómez-Luna, J.; Guil, N. Low-textured regions detection for improving stereoscopy algorithms. In Proceedings of the 2014 International Conference on High Performance Computing & Simulation (HPCS), Bologna, Italy, 21–25 July 2014.
29. Duffy, J.P.; Shutler, J.D.; Witt, M.J.; De Bell, L.; Anderson, K. Tracking Fine-Scale Structural Changes in Coastal Dune Morphology Using Kite aerial Photography and Uncertainty-Assessed Structure-from-Motion Photogrammetry. *Remote Sens.* **2018**, *10*, 1494. [CrossRef]
30. Cook, K.L. An evaluation of the effectiveness of low-cost UAVs and structure from motion for geomorphic change detection. *Geomorphology* **2017**, *278*, 195–208. [CrossRef]

Article

Photon-Counting Lidar: An Adaptive Signal Detection Method for Different Land Cover Types in Coastal Areas

Yue Ma [1,2,3], Wenhao Zhang [1], Jinyan Sun [4], Guoyuan Li [5], Xiao Hua Wang [2,3], Song Li [1] and Nan Xu [6,*]

1 School of Electronic Information, Wuhan University, Wuhan 430072, China; mayue19860103@163.com or yue.ma3@adfa.edu.au (Y.M.); wenhao@whu.edu.cn (W.Z.); ls@whu.edu.cn (S.L.)
2 School of Science, University of New South Wales, Canberra, BC 2610, Australia; x.h.wang@unsw.edu.au
3 Sino Australian Research Centre for Coastal Management, University of New South Wales, Canberra, BC 2610, Australia
4 Anhui Province Key Laboratory of Water Conservancy and Water Resources, Anhui & Huaihe River Institute of Hydraulic Research, Hefei 230088, China; sjy@ahwrri.org.cn
5 Satellite Surveying and Mapping Application Center of NASMG, Beijing 100830, China; ligy@sasmac.cn
6 Ministry of Education Key Laboratory for Earth System Modeling, Department of Earth System Science, Tsinghua University, Beijing 100084, China
* Correspondence: xun14@mails.tsinghua.edu.cn; Tel.: +86-1501-096-3820

Received: 15 December 2018; Accepted: 21 February 2019; Published: 25 February 2019

Abstract: Airborne or space-borne photon-counting lidar can provide successive photon clouds of the Earth's surface. The distribution and density of signal photons are very different because different land cover types have different surface profiles and reflectance, especially in coastal areas where the land cover types are various and complex. A new adaptive signal photon detection method is proposed to extract the signal photons for different land cover types from the raw photons captured by the MABEL (Multiple Altimeter Beam Experimental Lidar) photon-counting lidar in coastal areas. First, the surface types with 30 m resolution are obtained via matching the geographic coordinates of the MABEL trajectory with the NLCD (National Land Cover Database) datasets. Second, in each along-track segment with a specific land cover type, an improved DBSCAN (Density-Based Spatial Clustering of Applications with Noise) algorithm with adaptive thresholds and a JONSWAP (Joint North Sea Wave Project) wave algorithm is proposed and integrated to detect signal photons on different surface types. The result in Pamlico Sound indicates that this new method can effectively detect signal photons and successfully eliminate noise photons below the water level, whereas the MABEL result failed to extract the signal photons in vegetation segments and failed to discard the after-pulsing noise photons. In the Atlantic Ocean and Pamlico Sound, the errors of the RMS (Root Mean Square) wave height between our result and in-situ result are −0.06 m and 0.00 m, respectively. However, between the MABEL and in-situ result, the errors are −0.44 m and −0.37 m, respectively. The mean vegetation height between the East Lake and Pamlico Sound was also calculated as 15.17 m using the detecting signal photons from our method, which agrees well with the results (15.56 m) from the GFCH (Global Forest Canopy Height) dataset. Overall, for different land cover types in coastal areas, our study indicates that the proposed method can significantly improve the performance of the signal photon detection for photon-counting lidar data, and the detected signal photons can further obtain the water levels and vegetation heights. The proposed approach can also be extended for ICESat-2 (Ice, Cloud, and land Elevation Satellite-2) datasets in the future.

Keywords: Photon-counting lidar; MABEL; land cover; remote sensing; signal photons

1. Introduction

Equipped with more sensitive sensors (Gm-APDs (Geiger mode avalanche photodiodes) or PMTs (photomultipliers)), photon-counting lidar can respond to the presence of return photons rather than capturing the return waveforms of traditional lidar [1,2]. Benefitting from photon-counting sensors, the lasers of photon-counting lidar achieve lower energy (several tens of μJ), higher repetition rate (several KHz) and lower divergence (a few tens of μrad) compared to traditional lidar (several tens of mJ, a few tens of Hz, and a few mrad). After the ICESat (ice, cloud, and land elevation satellite) traditional laser altimeter, the ICESat-2 photon-counting lidar will measure the ice sheet elevation, sea ice freeboard, vegetation canopy, and ocean surface, and provide successive photon clouds of the Earth's surface with smaller laser footprints and higher spatial density (an along-track footprint interval of 0.7 m) [3,4]. Prior to the launch, an airborne photon-counting lidar, i.e., MABEL (Multiple Altimeter Beam Experimental Lidar), was used as a high-altitude prototype for the ICESat-2 lidar [5].

Due to the lower transmitted laser energy, the mean signal photons per shot varies from 0.1 to 10 photons [4,6]. The return-signal photon clouds of the surface are very noisy and suffer from background noise, backscatter noise, detector dark noise, and after-pulsing noise photons [7]. In the daytime, the solar background noise rate is approximately several MHz, which makes the number of background noise photons exceed the number of signal photons within the range gate. Compared to background noise, the detector dark noise rate is only several KHz and can be neglected [8]. The backscatter effect arising from clouds and aerosols and the after-pulsing detector effect introduce noise photons into the signal photons above and below the ground surface, respectively [9]. To use photon clouds for monitoring environmental changes, weak laser signal photons should be precisely extracted from the noisy raw datasets of photon-counting lidar.

Many methods have been proposed to successfully process the laser point clouds captured by full waveform lidars [10,11], especially for Mallet's achievements on waveform processing for different targets [12,13]. However, most methods cannot be used to effectively detect the signal photons from the much noisy raw photons captured by photon-counting lidars because the raw photons from a photon-counting lidar correspond to the energy at only a single photon level (approximately 1000 repeated measurements are needed to construct an accumulated waveform), whereas the points from a full waveform lidar correspond to the energy at thousands of photons level (these photons can directly construct a return waveform). Apart from the ice sheet surface (because ice sheet surface is very smooth and flat and can be assumed as identical in an area of hundred-meter size), 1000 repeated measurements will correspond to different Earth's surfaces (a high-altitude aircraft will pass tens of meters and a low earth orbit satellite will pass a few hundred meters during the time duration of 1000 repeated measurements). In addition, a photon-counting lidar suffers from the dead-time effect and after-pulsing effect, which significantly influence the distribution of captured photon clouds. The dead-time effect is that after a response to a received photon, a photon-counting detector needs a time duration to recover. In the recovering time duration (i.e., the dead-time), the photon-counting detector will not respond to any received photons. The after-pulsing effect is that if a photon-counting detector is triggered by incidence photons, the photon-counting detector may be self-triggered after its response to these incidence photons (normally tens of ns later). If the incidence energy is larger (with more incidence photons), the photon-counting detector will be more likely to be self-triggered [7].

Many signal detecting methods for photon-counting lidar data have been developed. First, general signal photon detection methods were proposed and used for simulated datasets or the raw datasets captured by photon-counting lidar in labs, e.g., the correlation range receiver (CRR) method [14] and adaptive ellipsoid searching (AES) method [15]. For the MABEL raw datasets, a surface-finding algorithm was proposed to detect surface profiles from the raw data captured in regions of sea ice and ice sheets [16–18]. For datasets of ice sheet surfaces, an adaptive window size with a recursive nearest-neighbor analysis was proposed for discarding noise photons [19]. For datasets in urban and forested regions, an adaptive ellipsoid searching filter [20], an adaptive density-based model [21], the contour active models [22], the spatial statistical and discrete mathematical concepts [8], and a noise removal algorithm based on localized statistical analysis [23] were derived to detect the surface profiles. For the datasets of ocean surfaces, a surface detection method was proposed based on the wave spectrum and nonlinear least-squares fitting [9].

In coastal areas, the raw photon clouds are more sophisticated because a variety of Earth's surface types (e.g., open sea, shoals, wetlands, banks, sandy beaches, and docks) correspond to different surface profiles and reflectance. There is not a specific method to detect the various types of surface profiles. Kwok proposed a method to identify and detect sea ice and open water in the Arctic [24]. However, this surface finding method cannot effectively detect the surface profiles of wetlands. To detect the surface profiles of different types from the MABEL raw datasets in coastal areas, a new method is derived in this study. First, the surface types of the MABEL trajectory with a resolution of 30 m are obtained via matching the geographic coordinates of the MABEL trajectory with the geographic coordinates in NLCD (National Land Cover Database). Second, an improved DBSCAN (Density-Based Spatial Clustering of Applications with Noise) algorithm with adaptive thresholds and a JONSWAP (Joint North Sea Wave Project) spectrum algorithm are integrated to detect surface profiles of different types. The Pamlico Sound (in North Carolina, USA), where the MABEL lidar flew over the open sea, sound water, wetlands, shoals, banks, sandy beaches, and docks, is selected as the study area and surface profiles of different types are extracted via this new method.

2. Study Area and Datasets

2.1. Study Area and the In-Situ Data

Pamlico Sound in North Carolina is the largest lagoon along the East Coast of the United States (with a length of 130 km and width of 24 to 48 km), and it is the second largest estuary in the United States (with over 7500 km^2 of open water) (from https://en.wikipedia.org/wiki/Pamlico_Sound). In the east, the Outer Banks separate the Pamlico Sound from the Atlantic Ocean. In the west, the area is covered by various vegetations (e.g., shrub/scrub, cultivated crops, woody wetlands, and herbaceous wetlands). Many sandy barrier islands are inside the Pamlico Sound. According to the NLCD 2011 products, there are seven types of land cover near the Pamlico Sound (i.e., open water, mixture of constructed materials and vegetation, barren land, shrub/scrub, cultivated crops, woody wetlands, and herbaceous wetlands). Figure 1 shows the map of the Pamlico Sound and its location in the USA.

In this area, the in-situ data of two nearby stations provided by NOAA (National Oceanic and Atmospheric Administration) are used to evaluate the calculated results (i.e., the water level and RMS (Root Mean Square) wave height). The Oregon Inlet Marina Station (ID: 8652587) is inside the sound, and Duck Station (ID: 8651370) is on a platform above the Atlantic Ocean. These stations provide historical datasets including mean sea level and wind speed every six minutes, and the data can be downloaded from the website (https://www.co-ops.nos.noaa.gov). The water level heights of these stations are on the benchmark of NAVD88 (North American Vertical Datum of 1988) [25].

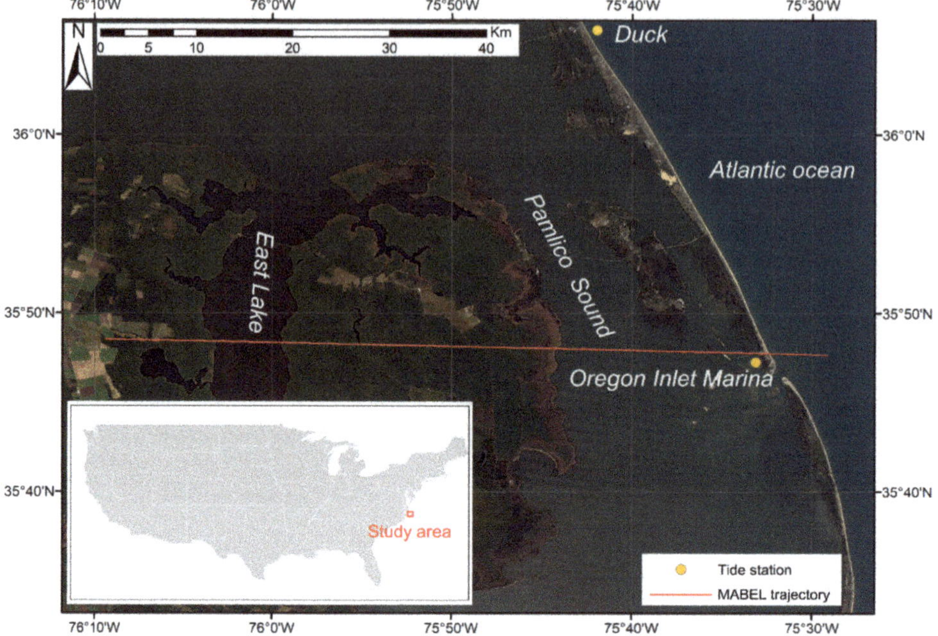

Figure 1. Pamlico Sound and its location in the USA. The MABEL trajectory (red lines) was from the ocean (east) to the land (west). The Oregon Inlet Marina Station is inside the sound, and Duck Station is on a platform above the Atlantic Ocean (yellow filled circles).

2.2. National Land Cover Database (NLCD)

The NLCD, supported by the MultiResolution Land Characteristics (MRLC) Consortium (http://www.mrlc.gov), is a widely used national scale land cover product [26]. The NLCD has 16 types of land cover and related information for 2001, 2006 and 2011. In the NLCD 2011 product, every type of land cover has a 30-m spatial resolution (i.e., the pixel size of Landsat imagery). The NLCD project was conducted to monitor the land cover and its change over time for various applications in ecology, climate change, land management and environmental planning. Landsat 5 Thematic Mapper (TM) imagery was used to generate the NLCD 2011 products. All Landsat images were acquired from the United States Geological Survey (USGS) Earth Resources Observation and Science (EROS) Center Landsat archive with radiometrical and geometrical calibration. All the production is geo-registered to the Albers Equal Area projection grid and resampled to a 30-m spatial resolution. The single-date overall accuracy of the NLCD 2011 products is 88% at level I and 82% at level II [27].

2.3. MABEL Datasets

The MABEL photon-counting lidar that was flown aboard the ER-2 and Proteus aircraft at an altitude of 20 km captured signal and noise photons (i.e., the raw datasets) and recorded the corresponding time tags [17]. The MABEL lidar has a transmitting laser pulse of 0.65 ns (1 sigma), a laser divergence of 0.1 mrad (the laser footprint is 2 m in diameter at 20 km altitude), a mean received energy of several 10^{-19} J (~1 photons @ 532 nm and 1064 nm), and a laser pulse repetition rate of 5~25 kHz. In the study area, the MABEL was at a laser repetition rate of 5 kHz and at an aircraft ground speed of approximately 215 m/s corresponding to an interval of 4 cm between contiguous laser pulses along the track. Up to 16 channels for the 532 nm green laser and eight channels for the 1064 nm near-infrared laser with different viewing angles are available for the MABEL, and each channel can independently and separately capture data photons [2]. For all available channels, every MABEL dataset records the raw data photons for a 1-min duration of the along-track flight.

The MABEL datasets used in this study were captured on 21/09/2012 near the Pamlico Sound (North Carolina, USA) and can be downloaded from the NASA ICESat-2 website (https://icesat.gsfc.nasa.gov/icesat2/data/mabel/data/browse/index.html). The weather was very clear at that time. The MABEL trajectory was from the ocean (east) to the land (west) in the study area as illustrated in Figure 1 (using red lines). In the MABEL Level 2A dataset, each photon represents the WGS84 coordinate frame and has a unique geographic coordinate of latitude, longitude and elevation. After the ranging corrections and systematic bias calibration, the MABEL signal photons are estimated to have a vertical accuracy from 13 to 24 cm (1-sigma) [28].

2.4. Global Forest Canopy Height Data

The Global Forest Canopy Height (GFCH) datasets are used for the comparison of the vegetation height calculated from the detecting signal photons. The GFCH dataset provides global vegetation canopy heights with a 1 km spatial resolution based on a fusion of the ICESat Geoscience Laser Altimeter System (GLAS) data over 20 May 2005 to 23 June 2005 and ancillary geospatial data (e.g., the annual mean precipitation, precipitation seasonality, annual mean temperature, temperature seasonality, elevation, tree cover and protection status) [29]. The GFCH data is available on the Google Earth Engine platform and the GFCH data within our study area can be downloaded from this platform (https://earthengine.google.com/).

3. Method

Our proposed method includes three parts for detecting signal photons from the raw data of photon-counting lidar. First, the surface types of the MABEL trajectory with a resolution of 30 m are obtained via matching its geographic coordinates with those in the NLCD datasets (the details are in Section 3.1). Second, for all segments, the raw data are filtered to discard the noise photons via different methods according to their corresponding land cover types. For the 'vegetation' and 'mixture' types, an improved DBSCAN surface-finding algorithm is proposed and used with different neighborhood radii, and the neighborhood radius and other parameters in this new DBSCAN algorithm are adaptively adjusted according to the spatial statistical characteristics of the inputting photon clouds (the details are described in Section 3.2). For the 'open water' type, a JONSWAP spectrum surface-finding algorithm is used (the details are described in Section 3.3). Finally, after the regular process, an extra noise filtering process is undertaken to discard the remaining after-pulsing noise photons to obtain the final signal photons (the details are in Section 3.4). The flow chart of the total process is illustrated in Figure 2. With the NLCD land cover datasets and MABEL raw data, all the steps in this method are automatically processed based on the ArcGIS (to obtain the surface types in Section 3.1) and MATLAB (to detect the signal photons under different surface types in Sections 3.2–3.4) software environment.

The verification of used methodology is conducted by three steps. First, a comparison of detecting signal photons between our method and the NASA surface finding method is conducted. Referring to various corresponding land covers from the images, the raw data photons, the result photons from the NASA surface finding method, and the result photons from our method are illustrated, and the performance of different methods can be clearly distinguished in this visual comparison. Second, the in-situ data from two nearby NOAA stations are used to evaluate the performance of different methods. The surface parameters are separately calculated using the detecting result photons of our method and the NASA surface finding method and compared with the in-situ surface parameters by comparing the errors between the calculated surface parameters from different methods and the in-situ surface parameters. Third, the average vegetation height in the study area is calculated using the detected signal photons from our method and from the NASA surface finding method. The GFCH dataset that provides forest canopy heights at a global scale is used to evaluate the calculated vegetation height.

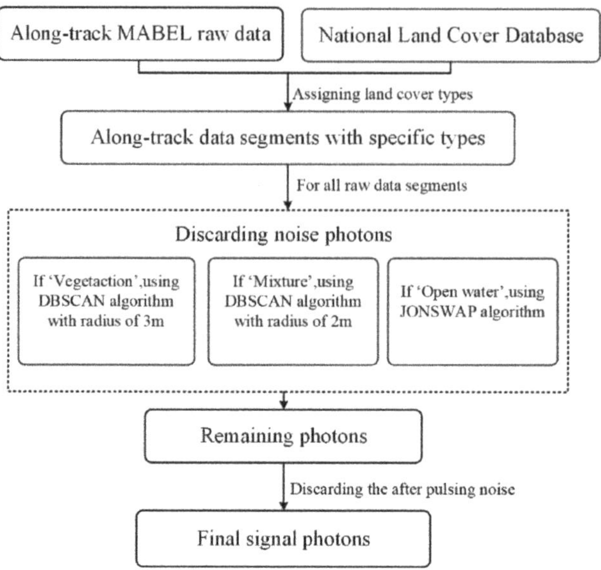

Figure 2. Flow chart of the method for detecting signal photons.

3.1. *Matching Land Cover Types for MABEL Raw Data*

Detailed information on latitude, longitude, elevation, and the time tags of raw data photons can be obtained from the MABEL datasets. All photons are transformed from three-dimensional coordinates (i.e., the geographic coordinates including the latitude, longitude, and elevation based on the WGS84 coordinate frame) to two-dimensional coordinates (i.e., the along-track coordinates including the along-track distance and elevation). The along-track distance starts from the beginning of the MABEL trajectory (in the farthest east of the trajectory and where the distance is zero). It is more convenient to discard the noise photons in a two-dimensional coordinate frame. Each raw photon has a unique index and after the noise filtering process, the remaining signal photons can be represented in three-dimensional geographic coordinates again.

The trajectory of the MABEL datasets is generated by extracting the latitude and longitude coordinates every 5 m along-track. For each trajectory location, the latitude and longitude coordinates are used to match corresponding land cover information from the NLCD 2011 products via the Function "Extract Values to Points" in ArcGIS. Owing to the similar physical properties, we merge some similar land cover types. The 'shrub/scrub', 'cultivated crops', 'woody wetlands', and 'herbaceous wetlands' types are combined and named as 'vegetation' because all these types belong to vegetation land cover and have very rough profiles. The 'barren land' type is merged into 'mixture of constructed materials and vegetation' and named 'mixture' because both types have relatively flat surfaces and higher reflectance, and the laser photon density of these land cover types is higher compared to vegetation types. The land cover types ('vegetation', 'mixture', and 'open water') are assigned to the MABEL trajectory every 5 m along-track. The MABEL raw photons can be divided into many along-track segments with specific land cover types. In each along-track segment, the raw photons correspond to an identical land cover type. We merge segments whose along-track distance is less than 50 m into its front segment.

3.2. *Improved DBSCAN Surface-Finding Algorithm for Land and Vegetation*

To extract the signal photons from the raw data corresponding to land types, an improved DBSCAN method is proposed. The DBSCAN algorithm was originally proposed to detect clusters in large noisy spatial databases [30] and was then modified to extract signal photons from the raw data of photon-counting lidar [21]. For every point in a cluster, if the point density in its neighborhood (within a specific radius) exceeds a specific threshold, this point will be classified as a "signal point" based on the criteria of the DBSCAN algorithm. The neighborhood is defined as the Euclidean distance for given points a and b, i.e., dist(a, b). In the DBSCAN algorithm, the clustering parameters R (i.e., the neighborhood radius of a point, defined by dist(a, b) $\leq R$) and $MinPts$ (i.e., the minimum number of points within the neighborhood radius) are essential [30]. The surface profiles of vegetation are much more fluctuant than artificial structures. The laser can penetrate the vegetation canopy into the undergrowth vegetation and ground, whereas the laser normally cannot penetrate the surface of artificial structures. Therefore, the signal photon density reflected by vegetation is less than by artificial structures. The neighborhood radius for the 'vegetation' type is set to R = 3 m and the neighborhood radius for the 'mixture' type is set to R = 2 m. Next, an adaptive algorithm is proposed to determine the $MinPts$. By inputting the raw data points in the current along-track segment with the 'vegetation' or 'mixture' type and its corresponding neighborhood radius R, the $MinPts$ can be automatically calculated using the following steps.

(A) The raw data photons along the vertical direction are divided into M segments (M is equal to 50 in this paper) for each MABEL dataset. For each vertical segment, the elevation length h can be expressed as $h = R_g/M$, where R_g is the range gate (approximately 1500 m for the MABEL). Then, the histogram of photon numbers for each vertical segment can be obtained. If the number of total raw photons in a dataset is N_t, the average photon number in all vertical segments is N_t/M.

(B) According to the histogram, we calculate the number of segments M_2 in which the photon number is smaller than the average photon number N_t/M and calculate the total photon number N_2 in these M_2 segments. In these M_2 vertical segments, most photons are noise photons. Similarly, one can calculate the number of segments M_1 ($M_1 = M - M_2$) in which the photon number is larger than the average number N_t/M and calculate the total photon number N_1 ($N_1 = N_t - N_2$) in these M_1 segments. Both signal and noise photons are included in these M_1 vertical segments.

(C) The total along-track distance l in each dataset can be calculated, and the elevation length in each vertical segment is $h = R_g/M$. For the M_2 segments (corresponding to noise photons), the photon density ρ_2 (count/m^2) per unit along-track distance and unit elevation length can be expressed as $\rho_2 = N_2/(h \cdot l \cdot M_2)$. Similarly, for the M_1 segments (corresponding to signal and noise photons), the photon density ρ_2 is expressed as $\rho_1 = N_1/(h \cdot l \cdot M_1)$.

(D) The area S can be calculated as $S = \pi \cdot R^2$ for a given circular zone. Next, for the M^2 segments, the expected photon number of noise SN_2 within this circular area can be expressed as in Equation (1).

$$SN_2 = \rho_2 \cdot S = \frac{\pi R^2 N_2}{hlM_2} \tag{1}$$

Similarly, the expected photon number of signal and noise SN_1 can be expressed as in Equation (2). Then, the expected number of signals can be approximately estimated using $SN_1 - SN_2$.

$$SN_1 = \rho_1 \cdot S = \frac{\pi R^2 N_1}{hlM_1} \tag{2}$$

(E) Given the expected number of noise and signal photons within a neighborhood area, the threshold of photon numbers (the minimum number of points) can be expressed as [14]

$$MinPts = \frac{2SN_1 - SN_2 + \ln(M_2)}{\ln\left(2SN_1/SN_2\right)} \tag{3}$$

The $MinPts$ can be calculated via the procedure (A) to (E) and the $MinPts$ can be adaptively adjusted by the input raw data. Then, by inputting the neighborhood radius R and the threshold of the minimum number of points in its neighborhood *MinPts* into the DBSCAN calculator, the signal photons are detected from the raw data photons [30].

3.3. JONSWAP Spectrum Surface-Finding Algorithm for Water Areas

In a previous study, we proposed a method to extract the ocean surface based on the JONSWAP wave spectrum and LM (Levenberg-Marquardt) nonlinear least-squares fitting [7]. This method first constructs an initial sea surface profile according to the wind speed above the sea surface. The wind speed in this area can be estimated from the wind speed data of the MABEL datasets. In each MABEL dataset, the meteorology data provide the wind speed with spatial resolution of approximately 15 m in the east and north at different altitudes. For each MABEL dataset, we can obtain an approximate wind speed by calculating the RSS (Root Sum Square) of the eastward and northward wind. This wind speed is sufficient to be an initial value and the actual surface profile of waters can be automatically adjusted via the LM fitting in the following procedures.

In the JONSWAP spectrum surface-finding algorithm, it is similar to the surface-finding algorithm for land and vegetation, a preliminary data processing procedure is needed to eliminate the gross error that may interfere with the fitting process. The raw data photons are uniformly divided into many segments in the vertical direction, and the photons with a lower point density will be directly discarded. Then, all the remaining photons are used to fit the water surface profile several times. In each fitting process, the 2-sigma criteria are used to discard photons corresponding to larger fitting error. Finally, the remaining data photons are divided into along-track segments with a specified distance (e.g., with an along-track distance of 500 m) and are separately filtered by the LM fitting and 2-sigma discarding. Undergoing these steps, the remaining photons are considered the laser signal photons from water surfaces.

3.4. Discarding the After-Pulsing Noise Photons

Due to the after-pulsing effect of photon-counting detectors, some noise photons will emerge after the laser pulse signal photons. The time tags of after-pulsing noise photons are normally a few tens of nanoseconds (corresponding to approximately 1.5 meters distance) later than the time tags of laser signal photons, and the number of after-pulsing noise photons are approximately one-tenth of the laser signal photons [31]. If the after-pulsing noise photons cannot be discarded, a ranging bias will be introduced (the range between the lidar and target will be overestimated, and the surface elevation of the target will be underestimated). For oceans and other water areas, the JONSWAP spectrum surface-finding algorithm has been proven to be able to eliminate the after-pulsing noise photons [7]. In a coastal area, the elevation of the ground surface is normally higher than the current water level. Therefore, for areas corresponding to the 'vegetation' and 'mixture' types, the after-pulsing noise photons are discarded as follows. First, for each segment of the 'vegetation' and 'mixture' types, the nearest water area is searched, and its local water level and RMS wave height are calculated by averaging the elevations of signal photons and calculating their standard deviations. Second a threshold is equal to the local water level subtracting the tripled RMS wave height (i.e., 3-sigma criteria). We discard the photon in the 'vegetation' and 'mixture' segments, if the elevation of the photon is lower than the threshold.

4. Results

Figure 3 illustrates the MABEL trajectory on Google Maps (using a red line), the captured photons by the MABEL lidar, and the photon results. The top figure in Figure 3 (Figure 3A) shows that on 21/09/2012 from 21:37 to 21:41 GMT (Greenwich Mean Time), when the MABEL lidar flew over Pamlico Sound, it flew over the open water of the Atlantic Ocean, across the Outer Banks of Pamlico Sound (entered into Pamlico Sound), across many small shoals and islands inside Pamlico Sound, over the open water of Pamlico Sound, entered into the land mainly covered by vegetation (flew over a very small, slim lake in the middle of this route), flew over the open water of East Lake, and finally entered the land covered by vegetation. The trajectory was from the east to the west with a path length of 60 km. The middle figure in Figure 3 (Figure 3B) illustrates the along-track raw photons captured by the MABEL lidar and the result of signal photons via the official surface-finding method. The bottom figure (Figure 3C) illustrates the along-track raw photons and the photons resulting from our method. In both Figure 3B,C, the boundaries between different land cover types of the along-track trajectory are illustrated using dashed vertical lines. These boundaries are obtained by matching the geographic coordinates of the MABEL trajectory with the land cover types in the NLCD 2011 products, and these boundaries are in accordance with the land cover image in Figure 3A, which proves that the NLCD 2011 products can provide the land cover types to help detect signal photons from the noisy raw data of a photon-counting lidar.

The raw photons were captured by channel no. 44 at 1064 nm, a near-infrared wavelength. Channel no. 44 has the best quality because it corresponds to a nearly nadir incidence with a better signal-noise ratio. Additionally, the wavelength of 1064 nm is located in the atmospheric window with lower energy loss during the transmission in the atmosphere and has much lower penetration into the water volume compared to the wavelength of 532 nm. The MABEL raw data illustrated using green filled circles are very noisy because they contain both the reflected laser photons (i.e., signal photons) and the noise photons caused by the background, backscatter, and detector noise (including the dark noise and after-pulsing effect). The signal photons of the MABEL result illustrated in Figure 3B (by blue circles) are processed by the official surface-finding algorithm and read from the MABEL dataset 'ph_class', which provides the flags for all photons that are classified as "noise", "buffer", "low", "medium", and "high". The MABEL results discard most of the noise photons; however, some noise photons remain (some noise photons construct a sublayer surface below the actual surfaces of the water and ground) and some signal photons are discarded, especially for the type of vegetation.

In Figure 3B, the MABEL result failed to detect all the vegetation signal photons because they have a lower point density. The after-pulsing noise photons (in the blue box of the left bottom in Figure 3C) were incorrectly detected because they have higher point density than the vegetation signal photons. The after-pulsing noise photons construct a sublayer surface below the actual water surface and introduce a minus elevation bias to the MSL (mean sea level) or ground surface. In Figure 3C, vegetation signal photons are extracted well and the after-pulsing noise photons below the actual water and ground surfaces are successfully discarded. The signal photons of our result look much better than those of the MABEL result.

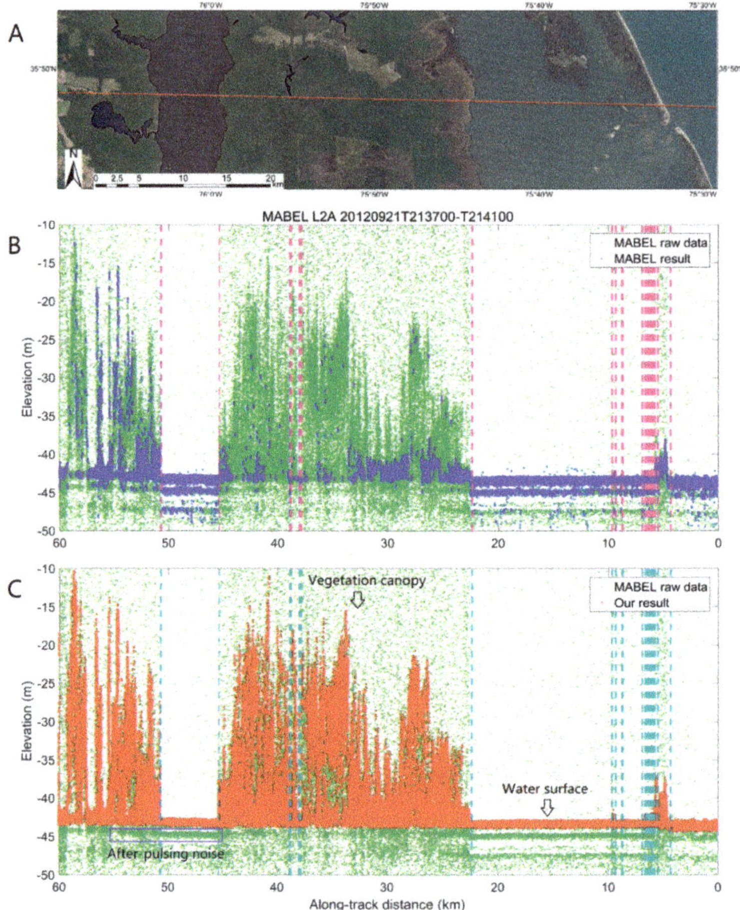

Figure 3. MABEL trajectory on Google Maps (**A**), the captured photons by the MABEL lidar (**B**), and the results of detected signal photons (**C**). The top figure shows the trajectory on 21/09/2012 when the MABEL lidar flew over Pamlico Sound (in North Carolina, USA). The MABEL raw data illustrated using green filled circles are very noisy because they contain both reflected laser photons (i.e., signal photons) and noise photons caused by the background, backscatter, and detector noise. The signal photons of the MABEL result are illustrated by blue circles in the middle figure (**B**), and the signal photons of our result are illustrated by red circles in the bottom figure (**C**). In the middle and bottom figures, the abscissa represents the along-track distance and the vertical coordinate represents elevation. All photons are transformed from three-dimensional coordinates (i.e., geographic coordinates including latitude, longitude, and elevation) to two-dimensional coordinates (i.e., along-track coordinates including distance and elevation). The elevation is based on the WGS84 ellipsoidal height. The along-track distance starts from the beginning of the MABEL trajectory (in the farthest east and where the distance is zero). It should be noted that to show the signal photons more clearly, only photons within the elevation range of −50 to 10 m are illustrated (the range gate of the MABEL lidar is approximately 1500 m). In both the middle and bottom figures, the boundaries of different land cover types of the along-track trajectory are illustrated using dashed vertical lines.

Figure 4 shows the enlarged details of the along-track distance segment from 4 to 10 km in Figure 3. This segment contains multiple land cover types (i.e., ocean, banks, sand, rocks, vegetation, artificial construction, water, and shoals). As mentioned above, all types are considered 'mixture' except for the ocean and water surfaces. The boundaries of different land cover types are much denser in this segment, and the NLCD datasets provide acceptable land cover types. The signal photon of the MABEL and our result both successfully extract the surface profile. Some noise photons remain below the water surface in the MABEL result, whereas our result eliminates these after-pulsing noise photons. In the blue box of the top (Figure 4A), middle (Figure 4B), and bottom figures (Figure 4C), a tower was captured by the MABEL lidar. Our result extracts the signal photons reflected by this tower. With our results, the height of this tower can be measured as approximately 6 m above the ground surface.

Figure 5 shows the enlarged details of the along-track distance segment from 44 to 52 km in Figure 3. This segment only contains water surface and vegetation. The boundaries between different land cover types of the NLCD datasets precisely divided this along-track segment into open water and vegetation. The signal photon of the MABEL fails to detect the vegetation photons and fails to discard the noise photons below the water surface, whereas our result successfully detects the vegetation photons and eliminated the noise photons below the water surface. From our results, we can estimate the vegetation height from the signal photons. The average height is approximately 5 m near the water boundary, and the maximum vegetation height is approximately 20 m. In Figure 5C, the after-pulsing noise photons are located in the blue box at the bottom. The elevation of after-pulsing noise photons is approximately 1.5 meters below the water and ground surface.

From the overall result figure (Figure 3), most of the land cover types of the along-track signal photons are precisely classified as 'Vegetation', 'Open water', and 'Mixture' in the exception of some very short along-track segments. The surface types are not correctly detected in these very short along-track segments because the segments whose along-track distance is less than 50 m are merged into its front segment. In both the improved DBSCAN algorithm and the JONSWAP algorithm, the raw data with at least 50 m along-track distance is essential to automatically calculate the key parameters (i.e., the minimum number of points within the neighborhood radius (*MinPts*)) in the DBSCAN algorithm or fit the parameters of the water surface profile (i.e., the amplitudes, angular frequencies, and phases) in the JONSWAP algorithm.

In the enlarged detailed Figure 4, the surface types are classified as 'Open water' and 'Mixture'. In the enlarged detailed Figure 5, the surface types are classified as 'Open water' and 'Vegetation'. Therefore, the proposed method is useful and successful to detect the main surface types in coastal area. Near the land/sea boundary, the tide has a significant effect and can change the surface types from land to sea when the tide is rising and vice versa. Inside the sound, the tide effect is much weaker. The spatial resolution of the surface types from the NLCD products is 30 m, and the spatial accuracy of the surface types is estimated better than 2 pixels (60 m in length) in the exception of the boundary between the land and sea surface.

In Figure 5, for the sea surface, the after-pulsing photons is very obvious and form a sub-layer below the sea surface because a 1064 nm laser nearly cannot penetrate the water volume and the sea surface is relatively flat (compared with the vegetation surface). However, the laser pulse can penetrate the vegetation canopy, so the after-pulsing photons are mixed with the reflected photons by the lower-layer vegetation as well as the ground, which makes the after-pulsing photons not obvious or seems weaker than the sea surface.

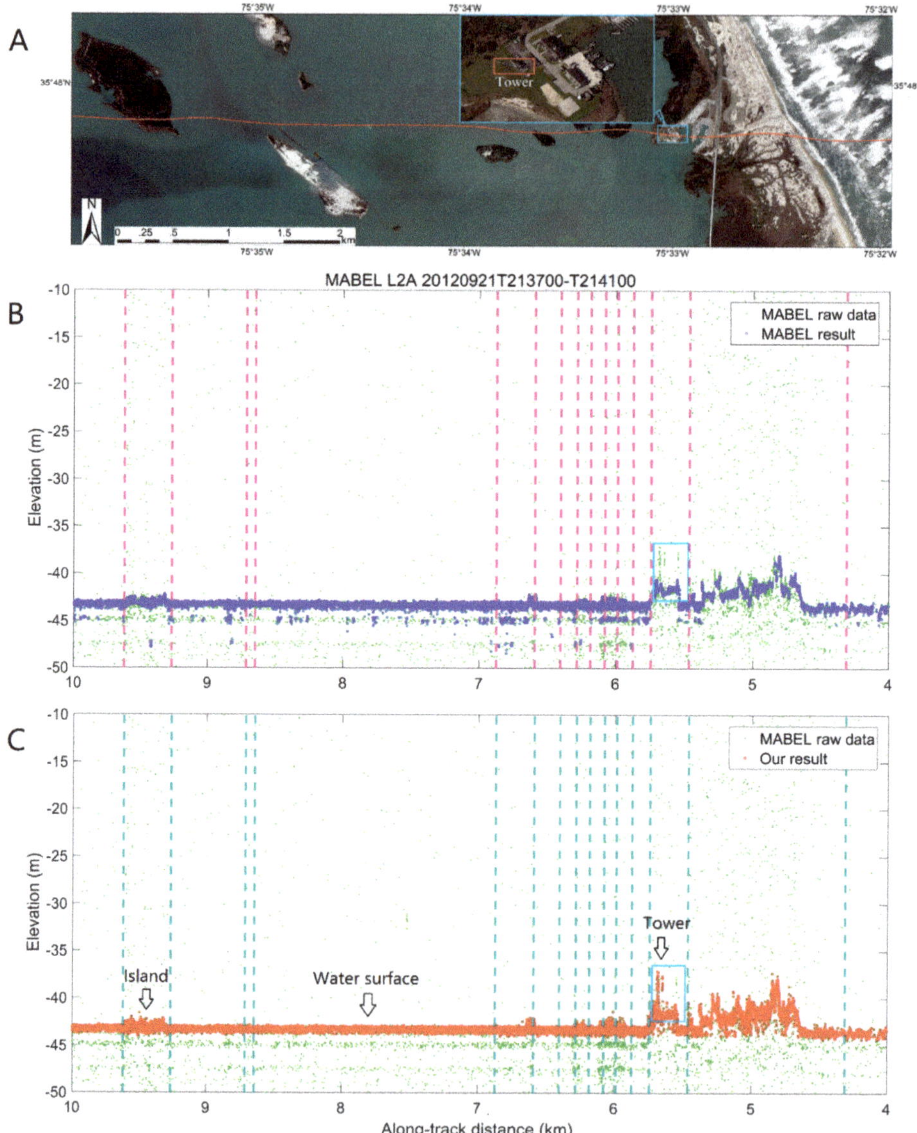

Figure 4. The enlarged details of the along-track distance segment from 4 to 10 km in Figure 3. The top figure (**A**) illustrates the corresponding high-resolution aerial image from Google Earth. The middle figure (**B**) illustrates the MABEL result of extracted signal photons, and the bottom figure (**C**) shows our result. In the blue box of the top (**A**), middle (**B**), and bottom figures (**C**), a tower was captured by the MABEL lidar.

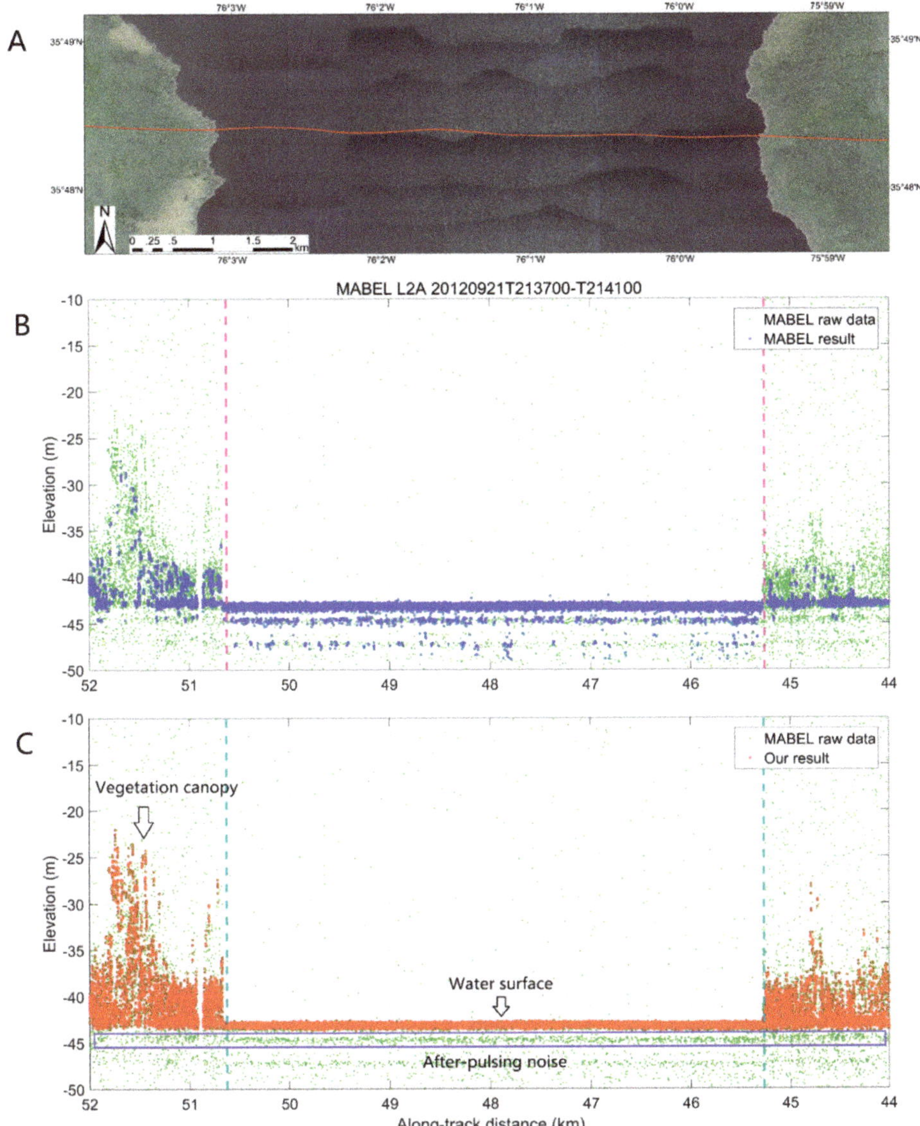

Figure 5. The enlarged details of the along-track distance segment from 44 to 52 km in Figure 2. The top figure illustrates (**A**) the corresponding high-resolution aerial image from Google Earth. The middle figure (**B**) illustrates the MABEL result of extracted signal photons, and the bottom figure (**C**) shows our result.

5. Discussions

In the coastal areas, the proposed method performed better in extracting the signal photons than the MABEL result. With the detected signal photons, water levels can be estimated by averaging their elevations within water segments. Three segments in the total trajectory corresponded to the water surface, i.e., Segment 1 from 0 to 4.3 km on the ocean surface (in the Atlantic Ocean), Segment 2 from 9.9 to 22.1 km on the water surface (in Pamlico Sound), and Segment 3 from 45.3 to 50.6 km on the water surface (in East Lake). Water levels are −43.62 m (Segment 1 in the Atlantic Ocean), −43.31 m (Segment 2 in Pamlico Sound), and −43.14 m (Segment 2 in East Lake). The water level shows a descending trend from the inner lake and sound to the ocean, in accordance with the water level trend in the study area at that time. The result is because, as shown in Figure 1, the three water regions are linked and according to the water level data of the in-situ stations, the tide was falling when the MABEL lidar flew over this area on 21/09/2012 from 21:37 to 21:41 GMT.

However, if we use the MABEL result to calculate the water level, the water levels will be −43.75 m (in the Atlantic Ocean), −43.46 m (in Pamlico Sound), and −43.37 m (in East lake). The after-pulsing noise photons introduce minus biases to the water level from 13 cm to 23 cm in three segments. At 21:36 GMT, the mean sea levels of Duck Station (corresponding to Segment 1 in the Atlantic Ocean) and the Oregon Inlet Marina Station (corresponding to Segment 2 in Pamlico Sound) were −0.25 m and 0.07 m, respectively. It should be noted that the water levels calculated by the signal photons were on the benchmark of WGS84 ellipsoidal height, whereas the water levels of the in-situ stations were on the benchmark of NAVD88 geoid height. Therefore, we can only make a comparison of the elevation difference between the water levels of the Pamlico Sound and the Atlantic Ocean. The in-situ elevation difference was 0.32 m (the Pamlico Sound had a higher water level than the Atlantic Ocean), and the elevation difference calculated by our result was 0.31 m, whereas the elevation difference calculated by the MABEL result was 0.29 m. Table 1 lists a detailed comparison between the in-situ water level and the water levels calculated by our result and MABEL result photons. The error of water level difference of the Atlantic Ocean and Pamlico Sound between our result and the in-situ result is 0.01 m, whereas the water level error between the MABEL result and in-situ data is 0.03 m.

Table 1. Water level comparison between the in-situ data and the calculated result from signal photons.

At 21:36 GMT, 21/09/2012	Atlantic Ocean	Pamlico Sound	East Lake	Difference between Pamlico Sound and Atlantic Ocean	Error Between the Calculated Water Level Difference and in-situ Difference
in-situ water level on NAVD88 geoid height (m)	−0.25 (from Duck Station)	0.07 (from Oregon Inlet Marina Station)	None	0.32	-
Water level calculated by our result on WGS84 ellipsoidal height (m)	−43.62	−43.31	−43.14	0.31	0.01
Water level calculated by MABEL result on WGS84 ellipsoidal height (m)	−43.75	−43.46	−43.37	0.29	0.03

In addition, the RMS wave heights of three segments can be calculated using the signal photons within the water segments and they are 0.32 m (Segment 1 in the Atlantic Ocean), 0.20 m (Segment 2 in Pamlico Sound), and 0.20 m (Segment 3 in East lake). The RMS wave height in the ocean (0.32 m) is larger than the inner water (0.20 m). At 21:36 GMT, the wind at Duck Station was 3.90 m/s at a height of 28.1 feet (approximately 8.58 m) above sea surface, and the wind at Oregon Inlet Marina Station was 3.30 m/s at a height of 21.8 feet (approximately 6.66 m) above the sea surface. According to Businger's theory [32], these winds can be transferred to 12.5 m above the sea surface as Equation (4).

$$U_{12.5} = U_z \frac{\ln(12.5/z_0)}{\ln(z/z_0)} \tag{4}$$

$U_{12.5}$ is the wind speed at 12.5 m above the sea surface, U_z is the wind speed at the height of z, and z_0 is 0.0023 m when the wind speed is less than 7 m/s. At the height of 12.5 m, the winds at Duck Station (corresponding to Segment 1 in the Atlantic Ocean) and the Oregon Inlet Marina Station (corresponding to Segment 2 in Pamlico Sound) were 3.56 m/s and 4.08 m/s, respectively. The RMS wave height is related to the wind speed above the sea surface and can be expressed as [33]

$$RMS_{wh} = 0.016U_{12.5}^2 \tag{5}$$

If we substitute the wind speeds of Pamlico Sound and the Atlantic Ocean at the height of 12.5 m into Equation (5), the RMS wave heights were 0.26 m and 0.20 m, respectively, in accordance with those calculated by our result photons (0.32 m and 0.20 m, respectively). However, the RMS wave heights calculated by the MABEL result are much larger (0.70 m and 0.57 m in Pamlico Sound and Atlantic Ocean) and completely different from the result calculated by the in-situ winds. Table 2 lists a detailed comparison between the RMS wave heights calculated by the in-situ winds, our result photons, and MABEL result photons. In the Atlantic Ocean and Pamlico Sound, the errors of RMS wave heights between our result and the in-situ result are −0.06 m, and 0.00 m, respectively. However, between the MABEL and in-situ results the errors are −0.44 m, and −0.37 m, respectively. With the RMS wave heights, we can further calculate the significant wave height (SWH) as SWH = 4 × RMS wave height.

Table 2. RMS wave height comparison between the in-situ data and the calculated result from signal photons.

At 21:36 GMT, 21/09/2012	Atlantic Ocean	Error Between the Calculated and in-situ Result in Atlantic Ocean	Pamlico Sound	Error Between the Calculated and in-situ Result in Pamlico Sound	East Lake
RMS wave height calculated by the in-situ winds (m)	0.26 (from Duck Station)	-	0.20 (from Oregon Inlet Marina Station)	-	None
RMS wave height calculated by our result photons (m)	0.32	−0.06	0.20	0.00	0.20
RMS wave height calculated by MABEL result photons (m)	0.70	−0.44	0.57	−0.37	0.77

Moreover, the average vegetation height is calculated using the detected signal photons from our method. The signal photons on the vegetation canopy and the ground are extracted, respectively. The maximum elevation in every 50 m along-track is searched as the canopy elevation in current along-track segment, and the minimum elevation in every 50 m along-track is searched as the ground elevation. The average values of all canopy elevations and all ground elevations are separately calculated. The average canopy height is finally obtained via subtracting the average canopy elevation by the average ground elevation. The GFCH dataset (that is derived from GLAS ICESat data and provides forest canopy heights at a global scale) is used for comparison. The profile of vegetation height is extracted from the raster format GFCH data using the ArcGIS. The mean vegetation height between the East Lake and Pamlico Sound (from 23 km to 45 km within the along-track distance of Figure 3) is also calculated as 15.17 m using the detecting signal photons from our method, which agrees well with the results (15.56 m) from the GFCH dataset. The average vegetation height derived from our results are in accordance with the average vegetation height from the GFCH dataset, whereas the signal photons from MABEL standard result (using the NASA surface finding method) failed to detect the vegetation canopy and cannot obtain the average vegetation height.

With a 30 m spatial resolution, the NLCD datasets provide acceptable land cover types to divide the MABEL along-track raw data into segments with different land cover types. The gross land cover types are verified in this paper, and it is very helpful to detect the signal photons because their clusters are very different for different land cover types due to different surface profiles and reflectance. The MABEL lidar was used as a high-altitude prototype for the ICESat-2 lidar and had similar data photons. Apart from the national scale NLCD datasets, other large-scale land cover products (e.g., GlobeLand30 [34], NLUD-C [35] can also be used to provide the land cover information for the ICESat-2 datasets in the future. With a given land cover type, the signal photons can be effectively detected from the raw data in different areas via this new proposed method. Then, the parameters of different land cover types can be calculated, e.g., the water level, the wave height, the vegetation height, and the tower height.

6. Conclusions

A novel method is proposed to detect the surface profiles of different land cover types from the MABEL raw datasets in coastal areas. First, the surface types of the MABEL trajectory with resolution of 30 m are obtained via matching the geographic coordinates of the MABEL trajectory with the NLCD datasets. The MABEL raw photons can be divided into many along-track segments with specific land cover types. For each along-track segment, the raw photons correspond to an identical land cover type. Second, an improved DBSCAN algorithm with adaptive thresholds is proposed to detect the signal photons from the raw data for the segments of 'vegetation' and 'mixture' types with different neighborhood radii, whereas an improved JONSWAP wave algorithm is integrated to detect the surface profile of water.

In Pamlico Sound, the result indicates that this new method can effectively detect the signal photons in vegetation, open water, and mixture segments and successfully eliminate the noise photons below the water surface; however, the MABEL result failed to extract the signal photons in vegetation segments and failed to discard the after-pulsing noise photons. With the detected signal photons, the water level, wave height, vegetation height, and tower height are estimated. The water level and RMS wave height calculated by our result achieved a better accordance with the in-situ data. The error of water level difference of the Atlantic Ocean and Pamlico Sound between our result and the in-situ result is 0.01 m, whereas the water level error between the MABEL result and in-situ data is 0.03 m. In the Atlantic Ocean and Pamlico Sound, the errors of RMS wave heights between our result and the in-situ result are −0.06 m, and 0.00 m, respectively. However, the errors of RMS wave heights between the MABEL result and in-situ result is −0.44 m, and −0.37 m, respectively. The mean vegetation height between the East Lake and Pamlico Sound was also calculated as 15.17 m using the detecting signal photons from our method, which agrees well with the results (15.56 m) from the GFCH dataset.

In coastal areas, the land cover types are various and complex; therefore, the distribution and density of signal photons are very different due to differences in surface profiles and reflectance. In this paper, it has been proven that our new proposed method and the NLCD datasets have the potential to provide land cover information for the improvement of the signal photon detection from the MABEL datasets, which is also useful for the ICESat-2 datasets in the future. Even though the land cover types are various and complex in coastal areas, the signal photons can be effectively detected from the raw data via this new proposed method and the detected signal photons can further to obtain the water levels and vegetation heights.

Author Contributions: Conceptualization, Y.M. and N.X.; Data curation, Y.M.; Funding acquisition, Y.M. and S.L.; Methodology, Y.M., W.Z., G.L. and N.X.; Project administration, Y.M. and N.X.; Software, W.Z. and J.S.; Supervision, X.H.W.; Writing—original draft, Y.M.; Writing—review & editing, J.S., G.L., X.H.W., S.L. and N.X. All authors read the manuscript, contributed to the discussion, and gave valuable suggestions to improve the manuscript.

Funding: This work was supported by The National Natural Science Foundation of China [Grant numbers 41506210, 41801261]; the National Science and Technology Major Project [Grant numbers 11-Y20A12-9001-17/18, 42-Y20A11-9001-17/18]; the Postdoctoral Science Foundation of China [Grant numbers 2016M600612, 20170034]; the Youth Science and Technology Innovation Fund Project of Anhui Province Key Laboratory of Water Conservancy and Water Resources [Grant number KY201703].

Acknowledgments: We thank the Goddard Space Flight Center for distributing the MABEL data, the MultiResolution Land Characteristics for distributing the National Land Cover Database, and the National Oceanic and Atmospheric Administration for distributing the in-situ measurement data. This is publication No. 65 of the Sino-Australian Research Centre for Coastal Management.

Conflicts of Interest: The authors declare no conflict of interest.

References

1. McGill, M.J.; Hlavka, D.L.; Hart, W.D.; Scott, V.S.; Spinhirne, J.D.; Schmid, B. Cloud physics lidar: Instrument description and initial measurement results. *Appl. Opt.* **2002**, *41*, 3725–3734. [CrossRef] [PubMed]
2. Kwok, R.; Markus, T.; Morison, J.; Palm, S.P.; Neumann, T.A.; Brunt, K.M.; Cook, W.B.; Hancock, D.W.; Cunningham, G.F. Profiling sea ice with a multiple altimeter beam experimental lidar (MABEL). *J. Atmos. Ocean. Technol.* **2014**, *31*, 1151–1168. [CrossRef]
3. Schutz, B.E. Overview of the ICESat mission. *Geophys. Res. Lett.* **2005**, *32*, S01. [CrossRef]
4. Markus, T.; Neumann, T.; Martino, A.; Abdalati, W.; Brunt, K.; Csatho, B.; Farrell, S.; Fricker, H.; Gardner, A.; Harding, D.; et al. The ice, cloud, and land elevation satellite-2 (ICESat-2): Science requirements, concept, and implementation. *Remote Sens. Environ.* **2017**, *190*, 260–273. [CrossRef]
5. McGill, M.; Markus, T.; Scott, V.S.; Neumann, T. The Multiple Altimeter Beam Experimental Lidar (MABEL): An airborne simulator for the ICESat-2 mission. *J. Atmos. Ocean. Technol.* **2013**, *30*, 345–352. [CrossRef]
6. Jasinski, M.F.; Stoll, J.D.; Cook, W.B.; Ondrusek, M.; Stengel, E.; Brunt, K. Inland and near-shore water profiles derived from the high-altitude multiple altimeter beam experimental lidar (MABEL). *J. Coast. Res.* **2016**, *76*, 44–55. [CrossRef]
7. Ma, Y.; Liu, R.; Li, S.; Zhang, W.; Yang, F.; Su, D. Detecting the ocean surface from the raw data of the MABEL photon-counting lidar. *Opt. Express* **2018**, *26*, 24752–24762. [CrossRef] [PubMed]
8. Herzfeld, U.C.; McDonald, B.W.; Wallins, B.F.; Neumann, T.A.; Markus, T.; Brenner, A.; Field, C. Algorithm for detection of ground and canopy cover in micropulse photon-counting lidar altimeter data in preparation for the ICESat-2 mission. *IEEE Trans. Geosci. Remote Sens.* **2014**, *52*, 2109–2125. [CrossRef]
9. Ma, Y.; Li, S.; Zhang, W.; Zhang, Z.; Liu, R.; Wang, X.H. Theoretical ranging performance model and range walk error correction for photon-counting lidars with multiple detectors. *Opt. Express* **2018**, *26*, 15924–15934. [CrossRef] [PubMed]
10. Wagner, W.; Ullrich, A.; Ducic, V.; Melzer, T.; Studnicka, N. Gaussian decomposition and calibration of a novel small-footprint full-waveform digitizing airborne laser scanner. *ISPRS J. Photogramm. Remote Sens.* **2006**, *60*, 100–112. [CrossRef]
11. Wallace, A.; Nichol, C.; Woodhouse, I. Recovery of forest canopy parameters by inversion of multispectral LiDAR data. *Remote Sens.* **2012**, *4*, 509–531. [CrossRef]
12. Mallet, C.; Bretar, F. Full-waveform topographic lidar: State-of-the-art. *ISPRS J. Photogramm. Remote Sens.* **2009**, *64*, 1–16. [CrossRef]
13. Mallet, C.; Lafarge, F.; Roux, M.; Soergel, U.; Bretar, F.; Heipke, C. A marked point process for modeling lidar waveforms. *IEEE Trans. Image Process.* **2010**, *19*, 3204–3221. [CrossRef] [PubMed]
14. Degnan, J.J. Photon-counting multikilohertz microlaser altimeters for airborne and spaceborne topographic measurements. *J. Geodyn.* **2002**, *34*, 503–549. [CrossRef]
15. Milstein, A.B.; Jiang, L.A.; Luu, J.X.; Schultz, K.I. Acquisition algorithm for direct detection ladars with Geiger-mode avalanche photodiodes. *Appl. Opt.* **2008**, *47*, 296–311. [CrossRef] [PubMed]
16. Farrell, S.L.; Brunt, K.M.; Ruth, J.M.; Kuhn, J.M.; Connor, L.N.; Walsh, K.M. Sea-ice freeboard retrieval using digital photon-counting laser altimetry. *Ann. Glaciol.* **2015**, *56*, 167–174. [CrossRef]
17. Brunt, K.M.; Neumann, T.A.; Walsh, K.M.; Markus, T. Determination of local slope on the Greenland ice sheet using a multi beam photon-counting lidar in preparation for the ICESat-2 mission. *IEEE Geosci. Remote Sens. Lett.* **2013**, *11*, 935–939. [CrossRef]
18. Brunt, K.M.; Neumann, T.A.; Amundson, J.M.; Kavanaugh, J.L.; Moussavi, M.S.; Walsh, K.M.; Cook, W.B.; Markus, T. MABEL photon-counting laser altimetry data in Alaska for ICESat-2 simulations and development. *Cryosphere* **2016**, *10*, 1707–1719. [CrossRef]
19. Magruder, L.A.; Neuenschwander, A.L.; Pederson, D.; Leigh, H.W.; Greenbaum, J.; de Gorordo, A.G.; Blankenship, D.D.; Kempf, S.D.; Young, D.A. Noise filtering and surface detection techniques for IceBridge photon counting lidar data over Antarctica. In Proceedings of the AGU Fall Meeting (AGU 2012), San Francisco, CA, USA, 3–7 December 2012; p. C21B-0584.
20. Wang, X.; Glennie, C.; Pan, Z. An adaptive ellipsoid searching filter for airborne single-photon lidar. *IEEE Geosci. Remote Sens. Lett.* **2017**, *14*, 1258–1262. [CrossRef]
21. Zhang, J.; Kerekes, J. An adaptive density-based model for extracting surface returns from photon-counting laser altimeter data. *IEEE Geosci. Remote Sens. Lett.* **2014**, *12*, 726–730. [CrossRef]

22. Awadallah, M.; Abbott, L.; Ghannam, S. Segmentation of sparse noisy photon clouds using active contour models. In Proceedings of the IEEE International Conference on Image Processing, Paris, France, 27–30 October 2014; pp. 6061–6065.
23. Nie, S.; Wang, C.; Xi, X.; Luo, S.; Li, G.; Tian, J.; Wang, H. Estimating the vegetation canopy height using micro-pulse photon-counting LiDAR data. *Opt. Express* **2018**, *26*, A520–A540. [CrossRef] [PubMed]
24. Kwok, R.; Cunningham, G.F.; Hoffmann, J.; Markus, T. Testing the ice-water discrimination and freeboard retrieval algorithms for the ICESat-2 mission. *Remote Sens. Environ.* **2016**, *183*, 13–25. [CrossRef]
25. Vanicek, P. Vertical datum and NAVD88. *Surv. Land Inf. Syst.* **1991**, *51*, 83–86.
26. Wickham, J.D.; Homer, C.G.; Vogelmann, J.E.; McKerrow, A.; Mueller, R.; Herold, N.; Coluston, J. The multi-resolution land characteristics (MRLC) consortium 20 years of development and integration of USA national land cover data. *Remote Sens.* **2014**, *6*, 7424–7441. [CrossRef]
27. Wickham, J.; Stehman, S.V.; Gass, L.; Dewitz, J.; Sorenson, D.G.; Granneman, B.J.; Poss, R.V.; Baer, L.A. Thematic accuracy assessment of the 2011 national land cover database (NLCD). *Remote Sens. Environ.* **2017**, *191*, 328–341. [CrossRef]
28. Magruder, L.A.; Brunt, K.M. Performance analysis of airborne photon- counting lidar data in preparation for the ICESat-2 mission. *IEEE Trans. Geosci. Remote Sens.* **2018**, *56*, 2911–2918. [CrossRef]
29. Simard, M.; Pinto, N.; Fisher, J.B.; Baccini, A. Mapping forest canopy height globally with spaceborne lidar. *J. Geophys. Res. Biogeosci.* **2011**, *116*, G4. [CrossRef]
30. Ester, M.; Kriegel, H.P.; Sander, J.; Xu, X. A density-based algorithm for discovering clusters in large spatial databases with noise. In Proceedings of the 2nd International Conference on Knowledge Discovery and Data Mining, Portland, OR, USA, 2–4 August 1996; pp. 226–231.
31. Butcher, A.; Doria, L.; Monroe, J.; Retiere, F.; Smith, B.; Walding, J. A method for characterizing after-pulsing and dark noise of PMTs and SiPMs. *Nucl. Instrum. Methods Phys. Res.* **2017**, *875*, 87–91. [CrossRef]
32. Businger, J.A. Transfer of momentum and heat in the planetary boundary layer. In Proceedings of the Arctic Heat Budget and Atmospheric Circulation, Lake Arrowhead, CA, USA, 31 January–4 February 1966; pp. 305–332.
33. Tsai, B.M.; Gardner, C.S. Remote sensing of sea state using laser altimeters. *Appl. Opt.* **1982**, *21*, 3932–3940. [CrossRef] [PubMed]
34. Chen, J.; Chen, J.; Liao, A.; Cao, X.; Chen, L.; Chen, X.; He, C.; Han, G.; Peng, S.; Lu, M.; et al. Global land cover mapping at 30 m resolution: A POK-based operational approach. *ISPRS J. Photogramm. Remote Sens.* **2015**, *103*, 7–27. [CrossRef]
35. Zhang, Z.; Wang, X.; Zhao, X.; Liu, B.; Yi, L.; Zuo, L.; Wen, Q.; Liu, F.; Xu, J.; Hu, S. A 2010 update of National Land Use/Cover Database of China at 1: 100000 scale using medium spatial resolution satellite images. *Remote Sens. Environ.* **2014**, *149*, 142–154. [CrossRef]

Article

Automatic Semi-Global Artificial Shoreline Subpixel Localization Algorithm for Landsat Imagery

Yan Song [1,2,*], Fan Liu [1], Feng Ling [3] and Linwei Yue [1]

1 School of Geography and Information Engineering, China University of Geosciences (Wuhan), Wuhan 430074, China
2 Key Laboratory of Geological Survey and Evaluation of Ministry of Education, China University of Geosciences (Wuhan), Wuhan 430074, China
3 Key Laboratory for Environment and Disaster Monitoring and Evaluation, Institute of Geodesy and Geophysics, Chinese Academy of Sciences, Wuhan 430077, China
* Correspondence: songyan@cug.edu.cn

Received: 4 June 2019; Accepted: 25 July 2019; Published: 29 July 2019

Abstract: Shoreline mapping using satellite remote sensing images has the advantages of large-scale surveys and high efficiency. However, low spatial resolution, various geometric morphologies and complex offshore environments prevent accurate positioning of the shoreline. This article proposes a semi-global subpixel shoreline localization method that considers utilizing morphological control points to divide the initial artificial shoreline into segments of relatively simple morphology and analyzing the local intensity homogeneity to calculate the intensity integral error. Combined with the segmentation-merge-fitting method, the algorithm determines the subpixel location accurately. In experiments, we select five artificial shorelines with various geometric morphologies from Landsat 8 Operational Land Imager (OLI) data. The five subpixel artificial shoreline RMSE results lie in the range of 3.02 m to 4.77 m, with line matching results varying from 2.51 m to 3.72 m. Thus, it can be concluded that the proposed subpixel localization algorithm is effective and applicable to artificial shoreline in various geometric morphologies and is robust to complex offshore environments, to some extent.

Keywords: shoreline mapping; semi-global subpixel localization; intensity integral error

---

1. Introduction

The coastline, the boundary of land and sea, is one of the 27 most important land surface features, and is vulnerable to natural processes such as coastal erosion/accretion, sea level changes and human activities [1]. Coastline mapping is, therefore, becoming a fundamental work for coastal erosion monitoring, coastal resource management, coastal environmental protection and coastal sustainable development [2–6]. In reality, the shoreline accurate position is difficult to be localized, as the position changes continually through time, because of cross-shore and alongshore sediment movement in the littoral zone and especially because of the dynamic nature of water levels at the coastal boundary (e.g., waves, tides, groundwater, storm surges, setups, runups, etc.) [7].

With the advantages of cost-effectiveness and large spatial and temporal scales, satellite remote sensing data have been used widely for coastline mapping [1,7–9]. When shoreline changes are sufficiently large (several tens of meters), satellite remote sensing can enable semi-automated comparison of large-scale areas by providing a common protocol for all sites [10], thus making comparisons consistent [11]. However, most observed shoreline changes are presently much smaller [12], so that the coarse spatial resolution of pixels prevent the accurate determination of shoreline positions when monitoring shoreline changes [13]. In this case, shoreline change observations can only be obtained by means of repeated in situ surveys, analysis of aerial or satellite high-resolution photographs at several time intervals, or a combination of both approaches [11,14]. However, the expensive price and shortage of historical data cannot meet large-scale shoreline supervision demands and the requirements of shoreline change analysis. Thus, it is important to conduct research on how to accurately determine the shoreline's position from long-term sequences of medium spatial resolution satellite images.

In recent years, many articles have appeared on how to use super resolution mapping (SRM) or subpixel edge localization (SEL) algorithms to extract the shoreline accurately. In these articles, positioning accuracy is quantitatively evaluated by four indicators: mean absolute error (MAE), standard deviation (SD), root mean squared error (RMSE), and line matching (LM) [15].

SRM has been applied to low- or medium-resolution satellite remote sensing images to overcome the limitation of the image spatial resolution of the original image. Li et al. [16] proposed that SRM can be categorized into two groups. The first group [17–20] is directly applied to satellite images instead of the intermediate spectral unmixing result, whereas the second group [21–29] is expected to improve the result's accuracy when highly-accurate fraction images are available from a spectral unmixing model [30]. According to shoreline SRM, Foody et al. [31] presented a soft fuzzy classification utilizing a geostatistical approach to obtain accurate waterline locations. Muslim et al. [32] proposed a localized soft classification approach to predict the shoreline location by a two-point histogram and pixel-swapping algorithms. Muslim et al. [33] proposed a contouring and geostatistical method to geographically position the coastline within image pixels. Zhang et al. [34] integrated a geostatistical approach and the high-resolution spatial structure prior model to undertake super-resolution mapping, which can properly illustrate the spatial distribution of the coastline at a fine scale. Comparisons [35] have been made using three soft classification methods and three subpixel mapping methods for coastal area classification.

SEL algorithms are often designed as follows: first, the initial position is obtained by edge detection; second, a local edge model is adopted to refine the initial edge position to the subpixel level. Subpixel detection techniques can be grouped into three categories [36]: moment-based; least-squares-error-based; and interpolation-based. Concerned with subpixel shoreline localization, Pardo-Pascual et al. [37] extracted subpixel shorelines utilizing local spatial structures from Landsat TM and ETM+, where the RMSE obtained ranged from 4.69 to 5.47 m. Almonacid-Caballer et al. [38] determined the annual mean shoreline subpixel position from Landsat images, and the extracted shorelines were biased from the seaward direction by approximately 4–5 m. Qingxiang Liu [39] presented a subpixel vector-based shoreline method to monitor shoreline changes at Narrabeen–Collaroy Beach, Australia, over 29 years. The experimental results show that after the correction of tidal effects, the RMSEs of annual mean shorelines are within 5.7 m. Pardo-Pascual et al. [40] evaluated the accuracy of shoreline positions obtained from the infrared (IR) bands of Landsat 7, Landsat 8, and Sentinel-2 imagery on natural beaches, where the mean error reached 3.06 m (± 5.79 m) from Landsat 8 and Sentinel-2 images.

In fact, there are different types and various geometric morphologies of shorelines in complex offshore environments. From the point of view of different shoreline types, there are artificial shorelines and natural shorelines. A shoreline may also be considered over a slightly longer timescale, such as a tidal cycle, where the horizontal/vertical position of the shoreline could vary anywhere between centimeters and tens of meters (or more), depending on the beach slope, tidal range, and prevailing wave/weather conditions [7]. It is, therefore, more difficult and challenging to evaluate the shoreline location accurately, especially natural shorelines.

From a geometric morphology point of view, shorelines include simple straight, quasi-straight, and high curvature shorelines, or combinations of these. It is difficult for traditional subpixel shoreline algorithms to determine various geometric morphological shoreline subpixel positions. Most algorithms mentioned above obtain the most accurate shoreline position for simple straight shorelines, but fail for high curvature shorelines in which the positional error increases [37].

Owing to the complex offshore environment, which includes suspended sediment, foam, different land-cover types etc., there are many mixed pixels and noise along the shoreline. Thus, it is difficult to find pure pixels along the natural or artificial shoreline for spectral unmixing, and the pure pixels obtained by global spectral analysis may not represent shoreline local spectra finely. Meanwhile, these conditions may also lead to difficulties in modeling the local edge for SEL algorithms.

Compared with a natural shoreline, an artificial shoreline is stable and its position is not affected by tidal effects or other factors, thus the reference shoreline can be extracted from high-resolution satellite images from different imaging times. Thus, artificial shorelines have the advantage of being more easily validated than natural shorelines. In this study, we focus on how to determine the artificial shoreline position accurately. In addition, we propose a method called the semi-global shoreline subpixel localization (SGSSL) algorithm. The main thoughts underlying SGSSL are simplifying the shoreline subpixel localization problem to a segmented shoreline subpixel fitting problem, expressing a shoreline segment geometric morphology perfectly, and minimizing the intensity integral error in local windows. To express various geometric morphological shorelines, we utilize multi-scale corner points to divide the initial shoreline into relatively simpler shoreline segments. To prevent offshore environment interference on subpixel localization, we analyze the water index intensity homogeneity for designing local windows. In designed local windows, intensity integral errors are minimized to obtain the subpixel shoreline positions. The entire method is dependent not only on shoreline geometric morphology and global spectral features but also local window intensity analysis and segmented shoreline geometric morphology. Thus, the proposed method is named semi-global subpixel shoreline localization (SGSSL).

2. Study Areas & Datasets

2.1. Study Areas

With urbanized development, there are increasingly more artificial shorelines located along Chinese coastal areas. Caofeidian Port and the Xiamen coastal area were selected as the study areas. As shown in Figure 1a, the Caofeidian Port located at 118.5°E, 39°N, is adjacent to China's Beijing Tianjin Hebei urban agglomeration and is one of China's important ore transportation ports. As shown in Figure 1b, the Xiamen coastal area, located between 118°E–118.5°E and 24.35°N–24.6°N, is next to the Taiwan Strait, and Xiamen is an important port for international economic and cultural exchange. Five artificial shorelines of various geometric morphologies as experimental areas were chosen to evaluate SGSSL, and their key characteristic parameters are listed in Table 1. The Gaofen-2 (GF-2) satellite data is selected as our reference data. GF-2 is the first civil optical remote sensing satellite independently developed by China with a spatial resolution better than 1 meter [41].

Our experiments verify the proposed algorithm from the following aspects. First, the subpixel shoreline localization results are superimposed on the original data to evaluate the visual effect of the proposed algorithm. Subsequently, compared with reference shoreline from GF-2, the four error indicators of the subpixel shoreline are calculated to verify the correctness and adaptability of the algorithm to different geometric morphology shorelines. Finally, the differences between the subpixel shoreline length and the reference shoreline length are calculated, which could illustrate ability of the proposed method to preserve details.

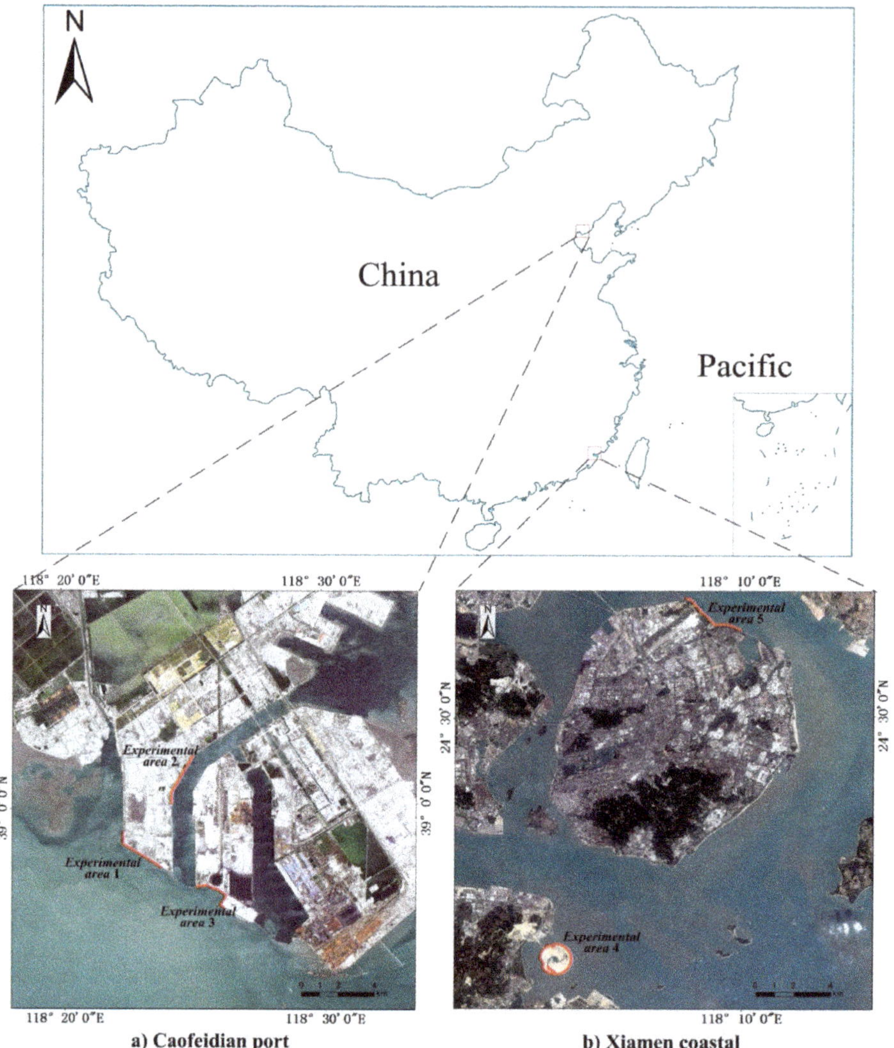

Figure 1. Five experimental areas.

Table 1. Key characteristics of the five experimental areas.

		Experimental Areas 1/2	Experimental Area 3	Experimental Areas 4/5
Location		Caofeidian Port	Caofeidian Port	Xiamen coastal area
Shoreline type		artificial	artificial	artificial
Geometric morphology		simple straight	combination of quasi-straight and curved shape	high curvature/combination of quasi-straight and curved shape
Experimental image	Data	Landsat-8 OLI images (Path 122, Row 033)	Landsat-8 OLI images (Path 122, Row 033)	Landsat-8 OLI images (Path 119, Row 043)
	Date	04/25/2015	04/25/2015	10/13/2015
	Resolution	15 m fusion image	15 m fusion image	15 m fusion image
Reference image	Data	GF-2 image	GF-2 image	GF-2 image
	Date	05/31/2015	05/31/2015	02/06/2015
	Resolution	1 m fusion image	1 m fusion image	1 m fusion image

2.2. Data Pre-Processing

First, using the rational polynomial coefficient (RPC) of the GF-2 image, ortho-rectification of the GF-2 multi-spectral (MS) data and panchromatic (PAN) data were performed separately.

Then, the Gram Schmidt [42] pan sharpening algorithm was used to fuse the MS data with the PAN data; then, the spatial resolution of the Landsat 8 OLI fusion images was 15 m and that of the GF-2 fusion images was 1 m.

The registration parameters were estimated by correspondence feature points that were selected manually and by the polynomial model. The maximum registration error between the two fusion images (15 m/pixel fused Landsat8 OLI image, 1 m/pixel fused GF-2 image) is less than 3 m. The registration error will bring uncertainty to the accuracy assessment and is discussed in Section 5.1.

3. Materials and Methods

According to the main thoughts underlying SGSSL, the overall process is shown in Figure 2. First, global spectral and geometric morphology analysis are conducted: the initial shoreline is extracted using the Otsu [43] automatic threshold method from water index images; and geometric morphology control points, abbreviated as morphology control points (MCPs), are extracted using the multi-scale Harris algorithm [44]. The primary MCPs are utilized to divide the initial shoreline. Then, semi-global analysis is performed by the segmentation-merge-fitting (SMF) method. In the SMF process, the segmented shoreline subpixel location is determined by minimizing the intensity integral error, finally obtaining a continuous subpixel shoreline vector.

In Section 3.1, an ideal image is taken as an example to illustrate the basic principle of SGSSL. In Section 3.2, the challenges faced when the basic principle is applied to real satellite images are explained. In Section 3.3, the method of conducting the global analysis for SGSSL is introduced, including obtaining the initial shoreline position and extracting the MCPs. In Section 3.4, the process of performing a semi-global shoreline analysis for SGSSL is proposed, including designing local windows and details of the SMF processes.

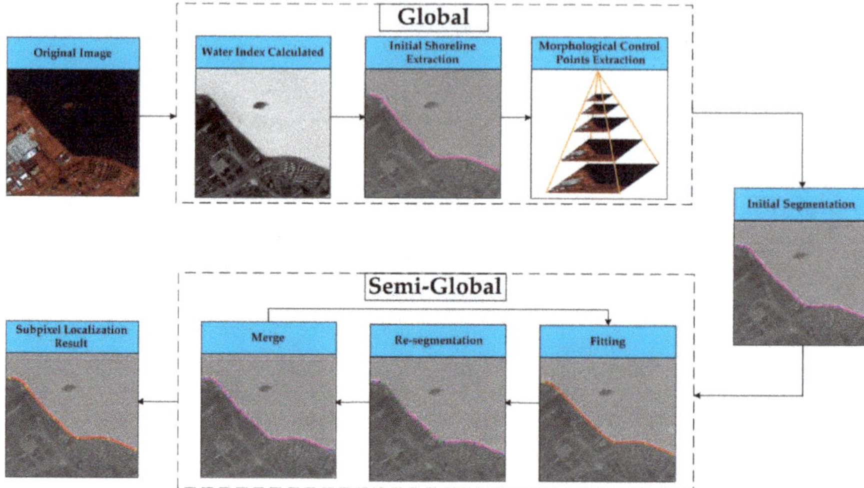

Figure 2. Overall flow chart.

3.1. Basic Principles of Subpixel Shoreline Localization

The entire shoreline can be regarded as consisting of many shoreline segments. The proposed subpixel shoreline localization algorithm is based on the following two assumptions:

Assumption 1: Any shoreline segment can be approximated by the polynomial function $y = f(x)$.

Assumption 2: The shoreline segment divides the image into two homogeneous regions with intensities A and B ($A < B$).

The ideal binary image is built in the image coordinate system $O\text{-}xy$, as shown in Figure 3, in which there is one shoreline segment.

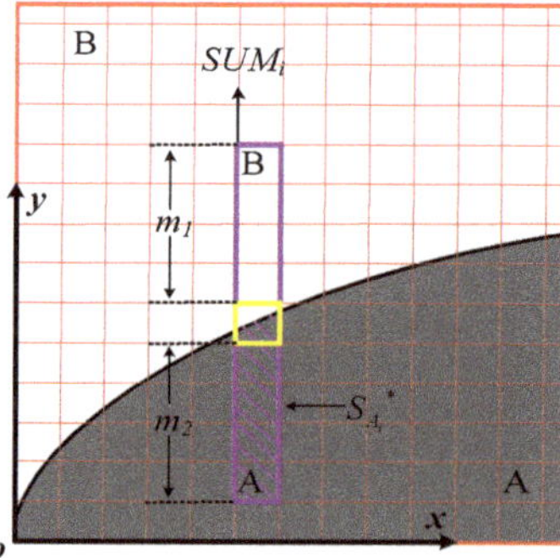

Figure 3. Ideal binary image. A shoreline segment separates the image into two homogeneous regions with intensities A and B. The ith shoreline point locates in the yellow box, and m_1, m_2 are the pixels' number above or under the shoreline segment in the local window; $S_{A_i}{}^*$ and $S_{B_i}{}^*$ are areas covered by A or B in the local window.

A local window (the purple box in Figure 3) centered on the ith shoreline pixel (x_i,y_i) is set, so the sum of intensity in the ith window is:

$$SUM_i = \sum_{j=y_i-m_2}^{y_i+m_1} G_{x_i,j} \quad {}_{(i=1,2,\ldots n)}, \tag{1}$$

where (x_i, y_i) is the current shoreline point's pixel coordinate, G the pixel's intensity, n the number of pixels in the shoreline segment, and m_1, m_2 are the pixels' number above or under the shoreline csegment in the local window, respectively. According to Assumption 2, the integral of the intensity in the ith window is:

$$SUM_i^* = A \times S_{A_i}{}^* + B \times S_{B_i}{}^*{}_{(i=1,2,\ldots n)}, \tag{2}$$

where $S_{A_i}{}^*$ and $S_{B_i}{}^*$ are areas covered by A or B in the local window, respectively:

$$S_{A_i}{}^* + S_{B_i}{}^* = m_1 + m_2 + 1. \tag{3}$$

According to Assumption 1, as the cubic function can express more geometric details and has superior morphological adaptability, the shoreline segment's polynomial function is:

$$f(x) = a + bx + cx^2 + dx^3. \tag{4}$$

So, the area under the shoreline segment in the ith local window ($S_{A_i}{}^*$) can be calculated as:

$$\begin{aligned} S_{A_i}{}^* &= \int_{x_i-1/2}^{x_i+1/2} \left(a + bx + cx^2 + dx^3 + 0.5 + m_2 - y_i\right) dx \\ &= 0.5 + m_2 - y_i + a + x_i b + \left(x_i^2 + \tfrac{1}{12}\right)c + \left(x_i^3 + \tfrac{1}{4}x_i\right)d \quad {}_{(i=1,2,\ldots n)} \end{aligned}, \tag{5}$$

With the above derivations, the intensity integral of the ith local window's (SUM_i^*) can be described as:

$$\begin{aligned} SUM_i^* &= A \times S_{A_i}{}^* + B \times S_{B_i}{}^* \\ &= (1/2 + m_2 - y_i)A + (1/2 + m_1 + y_i)B + (A - B)a + (A - B)x_i b \\ &+ (A - B)\left(x_i^2 + 1/12\right)c + (A - B)\left(x_i^3 + x_i/4\right)d \end{aligned}, \tag{6}$$

In ideal conditions,

$$\begin{cases} SUM_i = SUM_i^* \\ \sum_{j=y_i-m_2}^{y_i+m_1} G_{x_i,j} = (1/2 + m_2 - y_i)A + (1/2 + m_1 + y_i)B + (A-B)a + (A-B)x_i b + (A-B)\left(x_i^2 + 1/12\right)c + (A-B)\left(x_i^3 + x_i/4\right)d \end{cases}, \tag{7}$$

A similar idea has been researched by Trujillo-Pino et al. [36] for medical and indoor images, in which a subpixel edge location algorithm based on the partial area effect (PAE) was proposed. However, in that work, the algorithm neither considered the various geometric morphologies of the real shoreline/contour nor proved its application to actual satellite remote sensing images.

To solve Equation (7), the equation can be represented simply as:

$$p_i \beta = q_i \qquad (i = 1, 2, \ldots n) \tag{8}$$

where:

$$\begin{aligned} p_i &= (A - B)\begin{bmatrix} 1 & x_i & x_i^2 + 1/12 & x_i^3 + x_i/4 \end{bmatrix} \\ \beta &= \begin{bmatrix} a & b & c & d \end{bmatrix}^T \\ R_i &= (1/2 + m_2 - y_i)A + (1/2 + m_1 + y_i)B \\ q_i &= SUM_i - R_i. \end{aligned}$$

The following non-homogeneous equation can be obtained:

$$\begin{bmatrix} p_{11} & p_{12} & p_{13} & p_{14} \\ p_{21} & p_{22} & p_{23} & p_{24} \\ \vdots & \vdots & \vdots & \vdots \\ p_{n1} & p_{n2} & p_{n3} & p_{n4} \end{bmatrix} \beta = \begin{bmatrix} q_1 \\ q_2 \\ \vdots \\ q_n \end{bmatrix}_{(i=1,2,\ldots n)} \qquad (9)$$

$$\boldsymbol{\beta} = \left(\mathbf{P}^{\mathrm{T}}\mathbf{P}\right)^{-1}\mathbf{P}^{\mathrm{T}}\mathbf{Q}$$

The intensity integral error in the ith local window (e_i) is defined as:

$$e_i = \left|SUM_i - SUM_i^*\right| = \left|q_i - p_i\beta\right|_{(i=1,2,\ldots n)} \qquad (10)$$

By the least squares methodology, the cubic polynomial coefficients vector $\boldsymbol{\beta} = (a,b,c,d)^{\mathrm{T}}$ can be solved with intensity integral error minimization, and then the shoreline subpixel localization is determined.

3.2. Challenges for Subpixel Shoreline Localization in Remote Sensing Images

In Section 3.1, the derivation was based on the ideal image in Figure 3. When dealing with real shorelines, as Figure 4a shows, the initial shoreline (colored in purple) with various geometric morphologies does not satisfy Assumption 1. Therefore, the initial shoreline should be divided into segments of relatively simple morphologies, which can be approximated by the cubic polynomial function. For a shoreline to be divided appropriately, the multi-scale Harris corner algorithm [44] should be utilized to extract the MCPs.

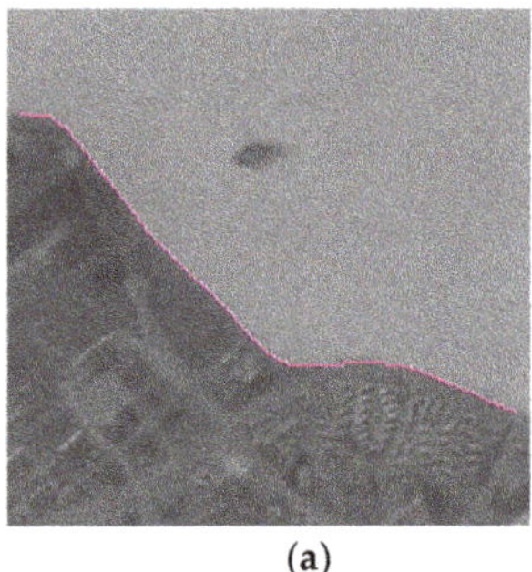

(a)

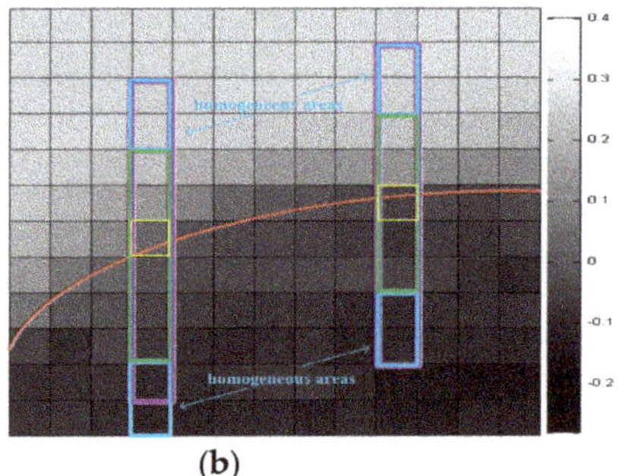

(b)

Figure 4. Actual remote sensing image. (**a**) Various shoreline morphology analysis; (**b**) homogeneous conditions in the water index image.

When processing actual satellite images, the water index result is affected by sensor imaging noise and the interaction between adjacent classes. Meanwhile, heterogeneous pixels exist in the local window with intensity changes (Figure 4b, green box). Then, Assumption 2 is not satisfied, which leads to the intensity integral error expressed in Equation (10) not equaling zero.

It is therefore necessary to design the local window to guarantee the intensity integral error approaches zero to ensure the performance of SGSSL correctly.

3.3. *Shoreline Global Analysis*

3.3.1. Determination of Initial Shoreline Position

To make full use of satellite image global information, we will conduct a shoreline global analysis, which includes a spectral feature analysis and geometric morphology analysis to determine the initial shoreline position at the pixel level and extract appropriate MCPs.

3.3.2. Determination Initial Shoreline Position

Before extracting MCPs, we should determine the initial pixel level shoreline from the original fused satellite images. This procedure includes the following steps. First, the water index image is calculated. The modified normalized difference water index (MNDWI) [45] is preferred, and the reason why MNDWI is preferred is discussed in Section 5.1. Subsequently, the Otsu method [43] is applied to the water index image, in which an optimal threshold T^* is selected automatically by maximizing the inter-class discrepancy. Third, using the optimal threshold value T^*, the water index image is divided into a binary image, which includes non-water and water classes. A series of points representing the pixel level shoreline is obtained.

3.3.3. MCP Extraction

To satisfy Assumption 1, the initial pixel level shoreline with various geometric morphologies should be divided into segments with relatively simple morphology by MCPs. As multi-scale Harris detection [44] is sensitive to corners, MCPs are extracted by multi-scale Harris detection [44] from a binarized water index image:

$$M = \mu(x, \sigma_I, \sigma_D) = \sigma_D^2 g(\sigma_I) \otimes \begin{bmatrix} L_x^2(x, \sigma_D) & L_x L_y(x, \sigma_D) \\ L_x L_y(x, \sigma_D) & L_y^2(x, \sigma_D) \end{bmatrix}, \tag{11}$$

where σ_I is the integral scale, $g(\sigma_I)$ the Gaussian convolution kernel with integral scale σ_I, and L_a the derivative computed in the a direction. The multi-scale Harris *cornerness* measure combines the trace and the determinant of the scale-adapted second moment matrix:

$$cornerness = \det(\mu(X, \sigma_I, \sigma_D)) - \alpha trace^2(\mu(X, \sigma_I, \sigma_D)) \tag{12}$$

where α is an empirical coefficient, $\alpha \in (0.04, 0.06)$. It should be noted that decreasing the value of α will increase the *cornerness*. Since our aim is to extract MCPs from the water/non-water binary image, the value of α can be set to 0.06 (the maximum empirical value). The local maximum of *cornerness* at each scale determines the scales' corner positions.

Second, each corner is verified depending on whether the Laplacian of Guassian (LOG) attains the maximum at the scale, and the LoG values are calculated by:

$$\left|LoG\,(X, \sigma_n)\right| = \sigma_n^2 \left|L_{xx}(X, \sigma_n) + L_{yy}(X, \sigma_n)\right|, \tag{13}$$

Comparing the LoG values with the adjacent two scale space images at the same position, if:

$$F(x, \sigma_n) > F(x, \sigma_l), l \in \{n-1, n+1\}, \tag{14}$$

these corners would be reserved as multi-scale Harris corners.

Considering shoreline geometric morphological changes can be classified as dramatic variations and minor variations, the MCPs should include primary MCPs and supplementary MCPs. As the primary MCPs locate the positions at which the shoreline's morphology changes drastically, theoretically, the multi-scale Harris corner points can be directly viewed as primary MCPs. However, during the process of building multi-scale image pyramids, images would be blurred and image structure details could be missed. Therefore, the initial scale's corner subset is also preserved from multi-scale Harris detection [44] as supplementary MCPs. The partial area effect (PAE) subpixel algorithm [36] should be conducted for MCP positions to obtain their accurate subpixel position. These two types of MCPs and subpixel positions will be used in the SMF process.

3.4. Shoreline Semi-Global Analysis

To utilize the semi-global information of a shoreline segment correctly, we conduct shoreline semi-global analysis, which includes a designed local window and the SMF method to determine the local homogeneous intensity and obtain proper shoreline segments for subpixel localization.

3.4.1. Designing Local Window

In this subsection, we introduce how to design the local window [36] and estimate homogeneous intensities A and B [36] to ensure that the intensity integral error e_i in Equation (10) is close to zero, so as to satisfy Assumption 2.

First, the maximum gradient direction of each shoreline point is calculated, and for every shoreline point, Sobel edge detection is utilized to calculate the gradient (G_x, G_y). Then, the larger gradient is preserved as the points' maximum gradient direction.

It should be noted that the shoreline segment consists of numerous points. Therefore, if shoreline points of a certain maximum gradient direction (G_x or G_y) have a larger proportion in the segment, then that direction will be used as the main direction for the segment.

If the main direction for the segment is G_y, the (m_1+m_2+1) ×1 local window is designed, and the cubic polynomial function of a segment is:

$$y = a + bx + cx^2 + dx^3. \tag{15}$$

If the main direction for the segment is G_x, the 1×(m_1+m_2+1) window is designed and the cubic polynomial function of segment is:

$$x = a + by + cy^2 + dy^3. \tag{16}$$

Second, since the homogeneous pixels' intensities are stable, they have a minimum gradient in the local window's direction. The algorithm adjusts m_1, m_2 to find the minimum gradient pixels (pixels in the blue box in Figure 4b). Once we find the minimum gradient pixels, their water index intensities are used to estimate the intensity A, B and their coordinates are used to set the window size [36].

In the ith window, the intensities A_i and B_i are the farthest pixels from the shoreline. To ensure the correlation between pixels in the local window, we limit $m_1 \leq 4, m_2 \leq 4$.

$$A_i = G_{i,j-m_2}, B_i = G_{i,j+m_1}, \tag{17}$$

Since there is a correlation between the intensities of adjacent points in the shoreline segment, to ensure that the homogeneity of A or B further prevents isolated noises, according to the adjacent shoreline points' relative positions, we determine slope k of this shoreline point and calculate the more homogeneous intensity estimation values A_i^* and B_i^*:

$$\begin{cases} A_i^* = \frac{A_i + A_{i+1}}{2}, B_i^* = \frac{B_i + B_{i-1}}{2}, & if k \geq 0 \\ or \quad A_i^* = \frac{A_i + A_{i-1}}{2}, B_i^* = \frac{B_i + B_{i+1}}{2}, & if k < 0 \end{cases}, \tag{18}$$

where $k = \frac{y_{i+1} - y_i}{x_{i+1} - x_i}$, ($x_i$,$y_i$) and ($x_{i+1}$,$y_{i+1}$) are the adjacent shoreline points' coordinates.

3.4.2. Segmentation-Merge-Fitting Method

The MCPs determined in Section 3.3.2 include the primary MCPs and the supplementary MCPs, which have been derived from multi-scale Harris corners to represent the major and minor shoreline geometric morphologies. However, utilizing them all without any selection it will lead to over-segmented shorelines and a high computational burden for the subpixel localization algorithm. The segmentation-merge-fitting (SMF) method is proposed to obtain appropriate shoreline segments adaptively, according to the residuals of the least-squares process and the main directions of the adjacent shoreline segments. In the SMF, the least-squares fitting process and the subpixel positions of the MCPs are added as constraints (Equation (19)) to connect adjacent shoreline segments and obtain the continuous subpixel shoreline.

$$\begin{cases} y_{mcp_i} - f(x_{mcp_i}) = 0, \\ y_{mcp_j} - f(x_{mcp_j}) = 0, \end{cases} \tag{19}$$

The detailed steps are shown in Figure 5.

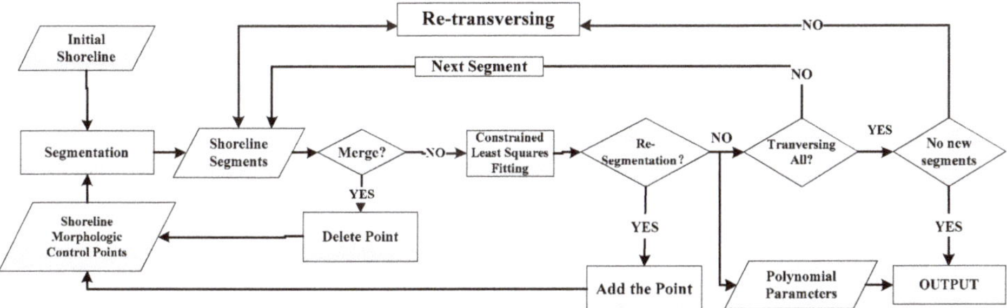

Figure 5. Flow chart of SMF method.

The detailed explanation for all of the above steps follows:

0. Preparatory work: Build the shoreline morphological control point set (SMCPS) and add all primary MCPs to the SMCPS; set the threshold t of the least squares residual to 0.08.

1. Segmentation: Utilize the SMCPS to divide shoreline to obtain shoreline segments and calculate the main direction of each shoreline segment.

2. Merge: Judge whether the main direction of the current shoreline segment is the same as the main direction of the adjacent segment:

If TRUE, merge these two adjacent segments and go to the Step 3 'Delete Point'; else, go to Step 4.

3. Delete point: Remove the current morphological control point connecting the two adjacent segments from the SMCPS.

4. Fitting: Polynomial coefficients are calculated by the constrained least squares methodology, and the least squares residuals are computed for each point. If there are four shoreline points adjacent to the shoreline MCP, the residuals of which are larger than the threshold t, this shoreline MCP should be removed from the least square constraints. Then, the current shoreline segment must be recalculated. Else, go to Step 5.

5. Judge the 're-segmentation' condition: If, in a shoreline segment, shoreline points with least-squares residuals larger than the threshold t continuously appear and these points' number is larger than 4, the shoreline segment must be re-segmented. Go to Step 6; else go to Step 7.

6. Add Point:

① If the segment is a merged shoreline in Step 2, the MCPs removed in Step 2 should be restored in the SMCPS.

② If the segment is not a merged shoreline in Step 2, the supplementary MCPs located in the segment are added into the SMCPS.

③ If there are no supplementary MCPs in the segment, the point with the largest least squares residual should be selected and added into the SMCPS.

Update the SMCPS, and go to Step 1.

7. Traversing: If all shoreline segments have been traversed, keep the subpixel results and go to Step 8; else, select the next shoreline segment and go to Step 2.

8. End: If the SMCPS is constant during this traversal, end the loop and go to Step 9; else, go to the Step 2 and re-traverse.

9. Output: The subpixel shoreline results are output.

3.5. Verification Method

The reference shorelines are extracted manually from GF-2 fusion images, which satisfy the standards of the "Technical Regulations for Satellite Remote Sensing Survey on Island & Coastal Zones" [14]. Using the SGSSL algorithm, a subpixel shoreline can be obtained. The reference shoreline and the shoreline determined by SGSSL can be compared.

We choose four error indicators to assess the SGSSL performance: the MAE, RMSE, SD and LM. The MAE (Equation (20)) is obtained by averaging all distance errors, because all the errors are obtained by calculating the absolute value of the distance from the SGSSL shoreline to the GF-2 reference shoreline. The MAE and RMSE describe the SGSSL result bias towards the reference shoreline. The SD indicates the variability around the MAE (Equation (21)):

$$MAE = \frac{\sum_{i=1}^{N} |d_i|}{N} \tag{20}$$

$$SD = \sqrt{\frac{1}{N}\sum_{i=1}^{N} (d_i - MAE)^2} \tag{21}$$

$$RMSE = \sqrt{\frac{\sum_{i=1}^{N} d_i^2}{N}} \tag{22}$$

where $|d_i|$ is the distance from the subpixel shoreline point to the reference shoreline.

As Figure 6 shows, for the calculation of the LM [15], S_Δ is the sum of the area enclosed by the SGSSL shoreline and the reference shoreline, and L_{real} is the length of the reference shoreline. In Figure 6, the black dotted line represents the reference shoreline and the solid line represents the shoreline determined by SGSSL.

$$LM = \frac{S_\Delta}{L_{real}} \tag{23}$$

where $S_\Delta = S_1 + S_2 + S_3 + S_4$.

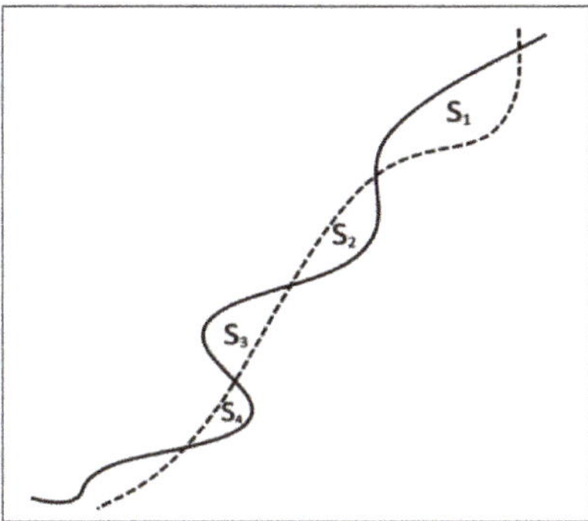

Figure 6. Line matching schematic [15].

4. Results

Here, we evaluate the results of the proposed SGSSL in terms of visual comparison, quantitative assessment, and shoreline detail preservation ability.

4.1. Visual Comparison

There are various geometric morphological shorelines in the selected coastal experimental areas. In the first two experimental areas (Figure 7a,e), the initial shoreline colored in purple has been extracted by Otsu [43] in Figure 7b,f. Owing to the fact that shoreline morphology is simple, MCPs labeled by yellow crosses can divide the initial shoreline into relatively simpler segments (Figure 7b,f). The proposed SGSSL algorithm determines the subpixel shoreline, which is represented by the red line in Figure 7c,g and which coincides with the real shoreline well. Although in the local zoomed image (Figure 7d,h) the initial pixel level shoreline points in yellow are located slightly landward, the final subpixel shoreline results (red line) still locate accurately.

In the latter three experimental areas (Figure 7i,m,q), the shoreline morphology is relatively complex, and the initial shoreline in purple has also been extracted by Otsu [43] in Figure 7 j,n,r. With the SMF method, we keep the selected MCPs labeled with yellow crosses (Figure 7j,n,r) and using them, the shoreline can be divided into relatively simpler segments to be perfectly expressed by red line in Figure 7k,o,s. Although in the local zoomed image (Figure 7l,p,t) the initial pixel level shoreline points in yellow are located slightly landward, the final subpixel shoreline results also coincide with the actual position well.

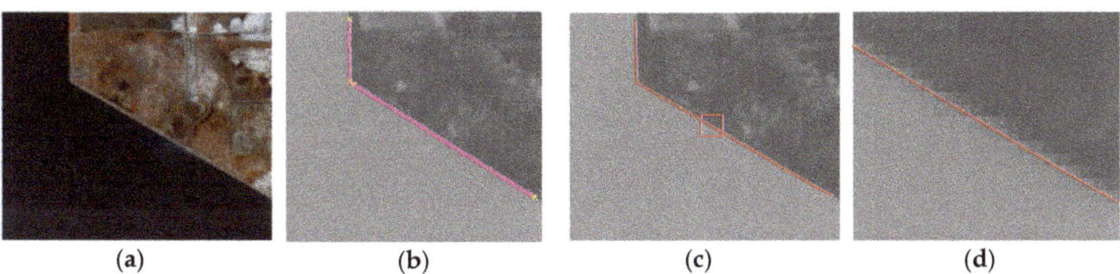

Figure 7. *Cont.*

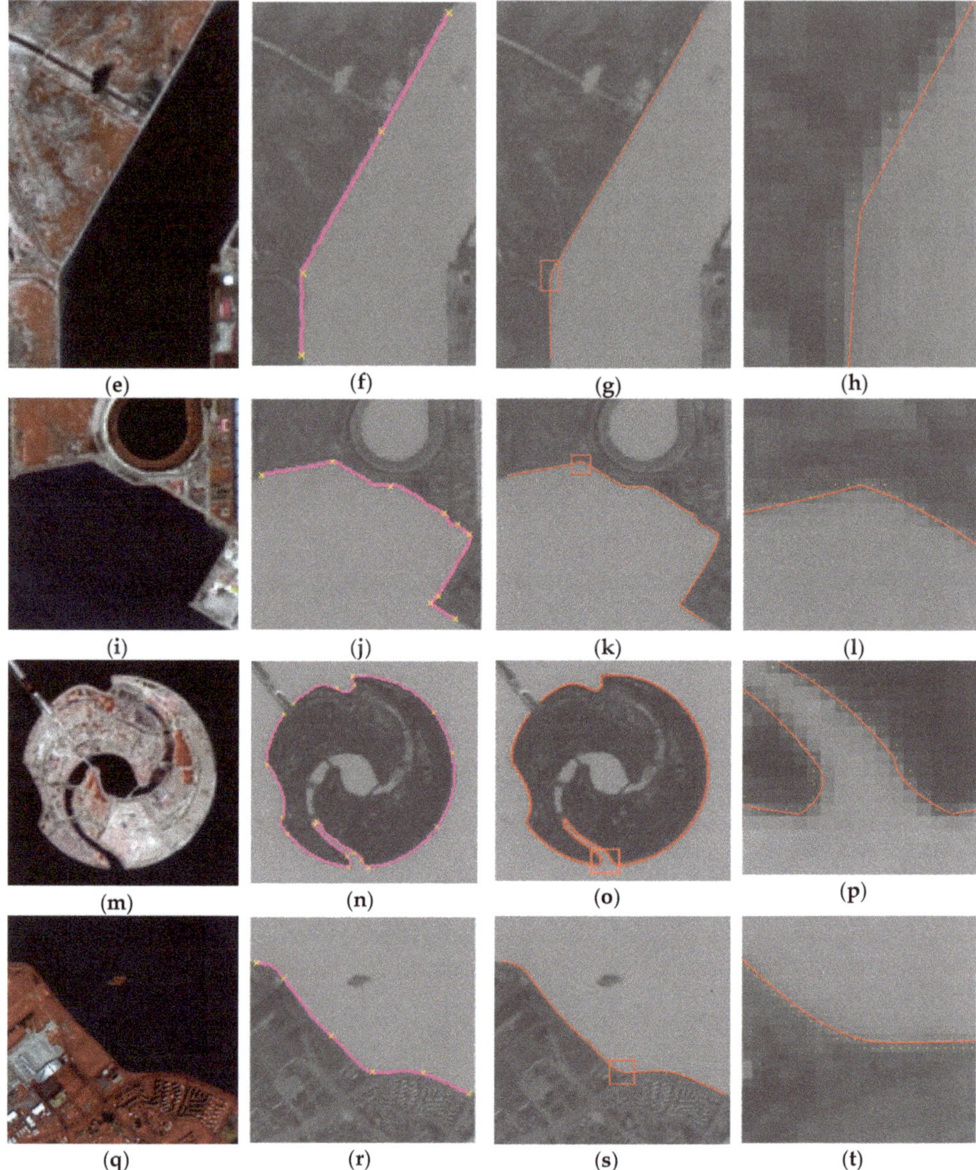

Figure 7. Visual comparison in experimental areas 1–5. Original images (R, band5; G, band4;B, band3) appear in the first column [(**a**,**e**,**i**,**m**,**q**)]; final shoreline morphological control point set (SMCPS) and segmented shorelines are in the second column [(**b**,**f**,**j**,**n**,**r**)]; semi-global subpixel shoreline localization (SGSSL) results in the third column [(**c**,**g**,**k**,**o**,**s**)]; and magnified images of the third column in the fourth column [(**d**,**h**,**l**,**p**,**t**)].

4.2. Quantitative Assessments

Table 2 summarizes the quantitative assessment results. In all experimental areas, the MAE at the subpixel level lies in the range of 2.48–3.34 m with an average of 3.03 m; the RMSE varies from

3.02–4.77 m with an average of 3.80 m; while the LM lies in the range of 2.51–3.72 m, with an average of 3.03 m. All quantitative assessments prove that the proposed SGSSL is reliable.

Table 2. The quantitative assessments in experimental areas.

Experimental Area	*MAE* (m)	*SD* (m)	*RMSE* (m)	*LM* (m)
1	2.94	1.93	3.51	2.87
2	3.34	2.16	3.97	3.30
3	3.67	3.06	4.77	3.72
4	2.72	2.61	3.77	2.77
5	2.48	1.72	3.02	2.51

It should be noted that these assessment results may be affected by registration errors. How the registration errors influence the quantitative assessments is discussed in Section 5.1.

4.3. Shoreline Detail Preservation Ability

Some shoreline details observed in the high-resolution images would be blurred or missed in low-resolution images. Hence, it is necessary to verify the detail preservation ability of SGSSL for shoreline details by comparing the lengths between subpixel results and high-resolution images.

In Table 3, the length difference ratios between subpixel results and reference shorelines are calculated, and the maximum value is less than 2.5%, which indicates that the proposed SGSSL can effectively preserve shoreline details.

Table 3. Subpixel shoreline length and reference shoreline length.

Experimental Area	1	2	3	4	5
Subpixel Shoreline Length (m)	3206.22	3000.19	2739.38	6324.41	3572.84
Reference Shoreline Length (m)	3205.25	2994.95	2675.57	6236.77	3571.32
Length Difference Ratio	0.03%	0.17%	2.33%	1.39%	0.04%

5. Discussion

5.1. Registration Error Influence on Quantitative Assessment

There are unavoidable registration errors when the accuracy assessment is conducted between the reference shorelines and SGSSL results. The registration errors will bring uncertainty to the quantitative assessment of SGSSL.

For objective analysis, three typical shoreline geometric morphologies are chosen for the registration error effect analysis.

First, the upper part of experimental area 1 is chosen. Because the upper part of experimental area 1 is a nearly vertical shoreline, the displacement of the horizontal direction will bring an apparent influence to the final MAE. Using 1 m as the displacement interval (0.067 pixels for Landsat8 OLI data) and with the maximum displacement limited to 5 m, the registration error influence in the horizontal direction across the range of [–5 m, +5 m] can be calculated and observed in Figure 8a,b.

Second, the middle part of experimental area 5 is chosen. Because the middle part of experimental area 5 is a nearly horizontal shoreline, displacement in the vertical direction will bring an apparent effect on the final MAE. Using 1 m as the displacement interval (0.067 pixels for Landsat8 OLI data) and with the maximum displacement limited to 5 m, the registration error effect at the vertical direction in the range of [–5 m, +5 m] can be calculated and observed in Figure 8c,d.

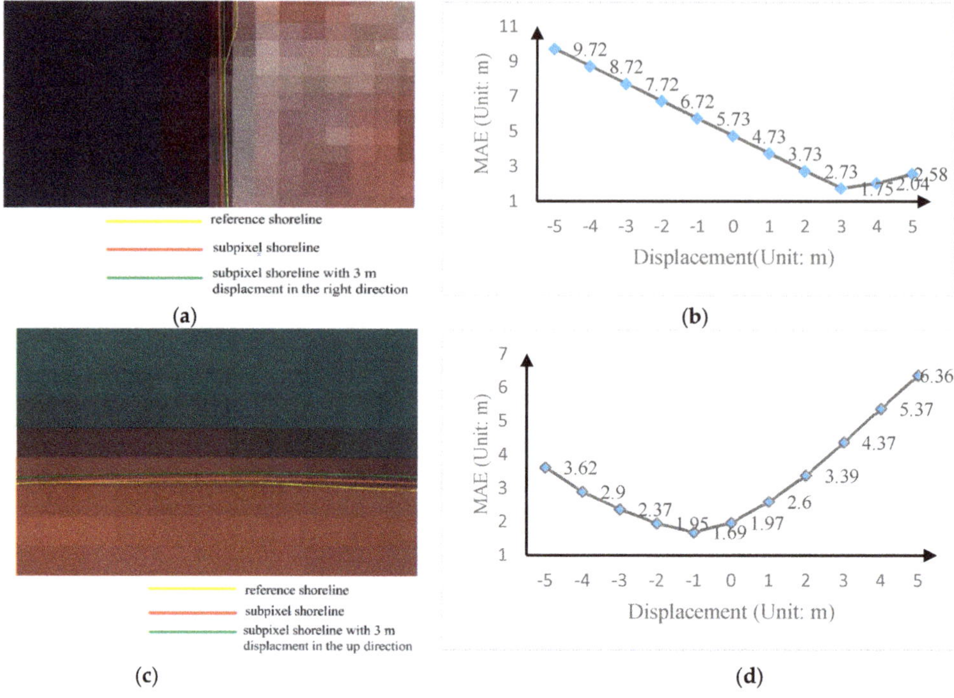

Figure 8. Mean absolute error (MAE) with registration errors compensated illustration (**a**) vertical shoreline, (**b**) registration error compensated results for the vertical shoreline, (**c**) horizontal shoreline and (**d**) registration error compensated results for the horizontal shoreline.

As Figure 8a,b show, displacements in the right direction will compensate for the registration error influence on the MAE. The MAE without registration error compensation is 5.73 m, which is 2.59 m larger than the entire experimental area 1 MAE result. With registration error compensation in the right direction, the MAE decreases. At a displacement of 3 m in the right direction, the MAE is 2.73 m; at the same direction of displacement of 4 m, the MAE reaches its most accurate value, 1.75 m. After that, MAE results cannot be compensated.

As Figure 8c,d show, for the middle part of experimental area 5, displacement will also have an effect on the final MAE. The MAE is 1.69 m without any registration error compensation, which is smaller by 0.59 m than the entire MAE of experimental area 5. With displacements in the up or down directions, the worse MAE will be obtained. Additionally, at a displacement of 3 m in the up direction, the MAE is 3.39 m; at the displacement 3 m in the down direction, the MAE is 2.9 m.

Third, experimental area 4 is chosen. Because the artificial shoreline in experimental area 4 is almost circular, the registration error will not influence the SGSSL result in certain directions. So the reference shoreline is displaced in four different directions, namely up, down, left and right, with 1 m of the displacement interval (0.067 pixels for Landsat8 OLI data), and the maximum displacement is limited to 5 m. In Figure 9, the MAE results of SGSSL with different displacements are shown.

From Figure 9, it can be observed that the MAE becomes more accurate in the 270° direction, namely in the down direction. The MAE first becomes most accurate at 2.60 m, which is the minimum MAE. The MAE then increases to 2.63 m in the 270° direction with a displacement of 2 m, and increases to 2.77 m in the same direction with a displacement of 3 m. For other directions, the MAE results are not better. The worst subpixel accuracy appears in the direction of 90°, the up direction, and the worst MAE is 3.74 m.

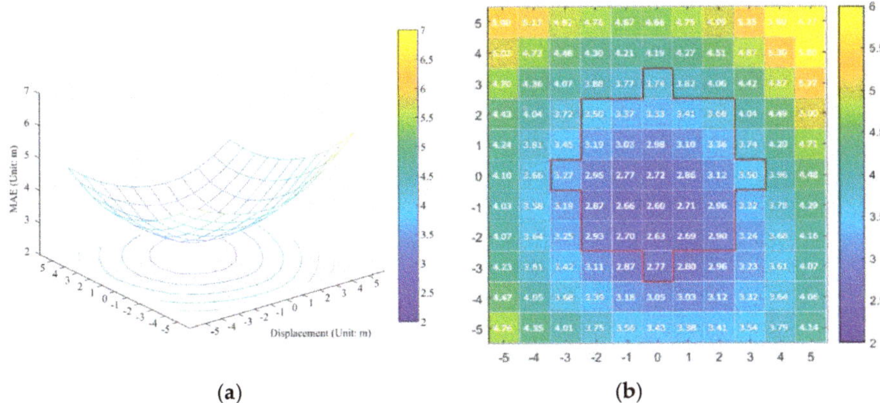

(a) (b)

Figure 9. MAE with registration errors compensated illustration (**a**) 3D surface results (**b**) 2D computation results.

From the above analysis, it can be concluded that the registration error brings uncertainty to the SGSSL quantitative assessment results. With the same registration method, different geometric morphological shorelines are influenced by registration errors to different extents. The lowest accuracy of the MAE appears at the vertical shoreline, and the MAE reaches at 5.73 m, which is still better than 0.5 pixels (15 m of the fused Landsat8 OLI image). Hence, the registration error with a maximum value of less than 3 m is acceptable for SGSSL.

5.2. Water Index

To select the water index with optimal positioning accuracy, the positioning errors of the SGSSL algorithm under three different water indices—the normalized difference water index (NDWI) [46], MNDWI, and automated water extraction index (AWEI) [47]—are calculated and compared.

As Figure 10 shows, the accuracies of shoreline positioning under three different water indices have all reached the subpixel level, indicating that the proposed algorithm is applicable to all water indices. It is obvious that the MNDWI is best in the selected experimental areas. One of the reasons is that the short-wave infrared 1 ($SWIR_1$,1566.50 – 1651.22 nm) band is used in the calculation of the MNDWI, and the most accurate and robust sub-pixel shoreline positioning results are often obtained using the $SWIR_1$ band [40]. Therefore, this paper prefers to use the MNDWI to enhance the differences between land and water, but, considering complicated offshore environments and data sources, in other coastal areas utilizing other water indices is acceptable.

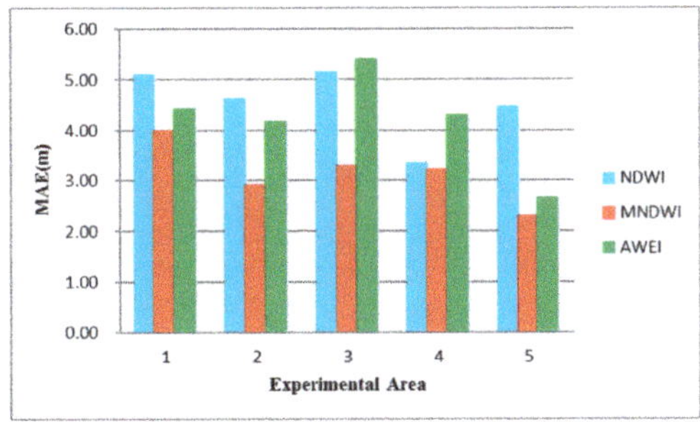

Figure 10. MAE of three water indices in different experimental areas.

5.3. Intensity Integral Error Analysis

Owing to sensor imaging noise and the interaction between adjacent classes, the intensity integral error in the ith local window is probably not equal to zero.

As Figure 11 shows, the initial shoreline pixel coordinate is (x_i,y_i), and the red line is the real shoreline that crosses the pixel (x_i,y_i). The subpixel level coordinates of the point in the real shoreline are (x_0,y_0), $(x_0 \in [x_i-1/2,\ x_i+1/2],\ y_0 \in [y_i-1/2,\ y_i+1/2])$. Once the local window size is determined by finding the minimum gradient pixels along the window direction, the window sizes m_1, m_2 and the homogeneous intensity A_i, B_i are all obtained. At shoreline point (x_0,y_0), the intensity profile is drawn along the window direction, presuming the direction lies in the y axes in Figure 11.

If the shoreline segment can be expressed by a cubic polynomial. S_1,S_2,S_3 are areas enclosed by the intensity profile and the y axis (window direction),

$$\begin{array}{c} S_1(x_0) = A_i \times (f(x_0) - (y_i - 0.5 - m_2)) \\ S_3(x_0) + S_4(x_0) = B_i \times ((y_i + 0.5 + m_1) - f(x_0)) \end{array}, \tag{24}$$

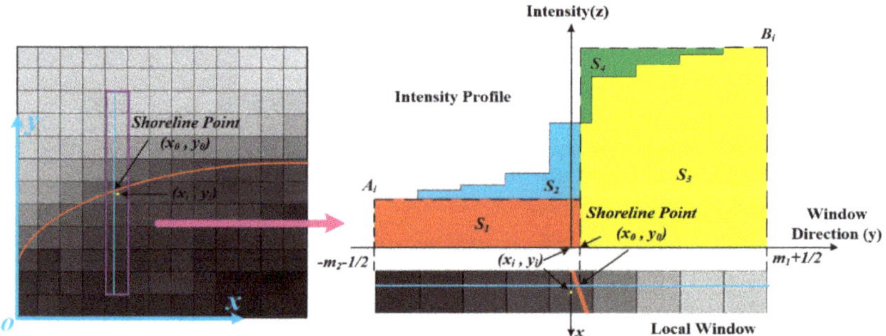

Figure 11. Intensity profile in the local window.

Therefore, the sum of the ith local window intensity is:

$$SUM_i = \sum_{j=y_i-m_2}^{y_i+m_1} G_{x_i,j} = \int_{x_i-1/2}^{x_i+1/2} (S_1(x_0) + S_2(x_0) + S_3(x_0))dx_0, \tag{25}$$

where, in the local window, G is the pixel intensity. The approximation of the sum of the ith local window intensity is:

$$\begin{aligned} SUM_i^* &= A_i \times S_{A_i}{}^* + B_i \times S_{B_i}{}^* \\ &= A_i \times \int_{x_i-1/2}^{x_i+1/2} (f(x_0) - (y_i - 0.5 - m_2))dx_0 \\ &+ B_i \times \int_{x_i-1/2}^{x_i+1/2} ((y_i + 0.5 + m_1) - f(x_0))dx_0 \\ &= \int_{x_i-1/2}^{x_i+1/2} (S_1(x_0) + S_3(x_0) + S_4(x_0))dx_0 \end{aligned}, \tag{26}$$

Thus, in the local window the intensity integral error e_i can be described as:

$$e_i = |SUM_i - SUM_i^*| = \left| \int_{x_i-1/2}^{x_i+1/2} (S_2(x_0) - S_4(x_0))dx_0 \right| \tag{27}$$

where the intensity integral error e_i is related to three factors: the window size (m_1+m_2+1), the homogeneous intensity difference (B-A), and the intensity slope at the shoreline point (x_0,y_0). Regarding the three factors, the intensity slope is determined by image information. The other two factors are determined by appropriate homogeneous intensity estimation *A*, *B* and the window's size (m_1+m_2+1), which can ensure that the e_i approaches zero.

Furthermore, to verify the correctness of the local window design method and to verify whether or not the intensity integral error e_i approaches zero, the relative error δ is calculated within the local window in Landsat OLI8 MNDWI images.

$$\delta = \frac{SUM_i^* - SUM_i}{SUM_i}. \tag{28}$$

The shorelines extracted manually from GF-2 images are viewed as the reference shorelines. In Equation (28), SUM_i^* is the integral of intensity in the *i*th window and calculated according to Equation (6). The reference shoreline coordinates are used to calculate S^*_{Ai} and S^*_{Bi}. SUM_i is the sum of intensity in the *i*th window and calculated according to Equation (3).

As Figure 12a shows, 6743 local windows covering different offshore environments over different periods of time in the study areas were sampled. The relative error distribution is shown in Figure 12b, the mean of δ is 0.0174 and the variance is 0.0278. The probability of δ less than 5% is 87.57%, and that of δ being less than 10% is 99.85%. Thus, it can be concluded that in most local windows, the relative error δ can be seen as a small number, whose absolute value approaches zero.

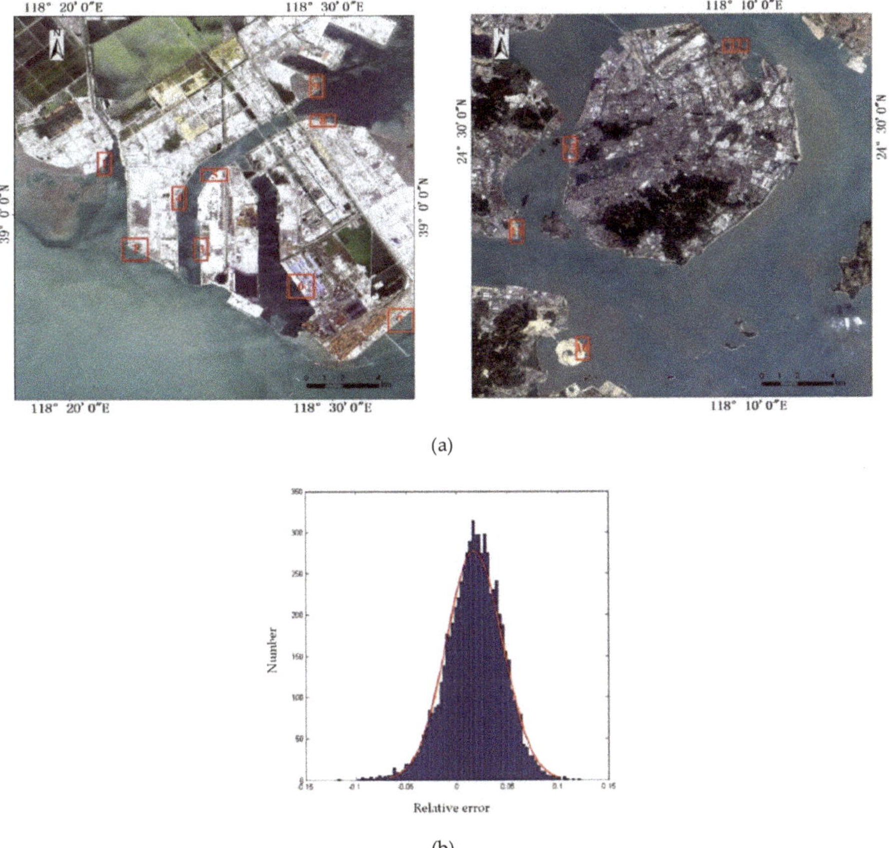

Figure 12. Local window samples and their relative error distribution. (**a**) Samples of local windows in the experimental areas; (**b**) distribution of relative error.

5.4. Segmentation-Merge-Fitting Process

As mentioned in Section 3.4.2, the initial shoreline may not be perfectly expressed by the cubic polynomial function. In Figure 13a, it can be observed that the initial shoreline could be divided into relatively simpler segments by primary MCPs (colored blue). As shown in Figure 13b,c (a magnified version of Figure 13b), after using primary MCPs and constrained least squares solving, one shoreline segment containing many shoreline points of larger fitting residuals still exists (marked by the yellow box in Figure 13b), which must be segmented further. In Figure 13d, all supplementary MCPs in green would be used to re-segment this problematic segment. Then, as the Figure 13e shows, some segments would be merged. Finally, appropriate shoreline segments that perfectly agree with polynomial functions are obtained using the SMF method, and are named seg1–seg3 in Figure 13f, the final SMCPS selected from MCPs are labeled by yellow crosses. As shown in Figure 13g,h (a magnified version of Figure 13g), the final subpixel positioning results coincide well with the real shoreline position.

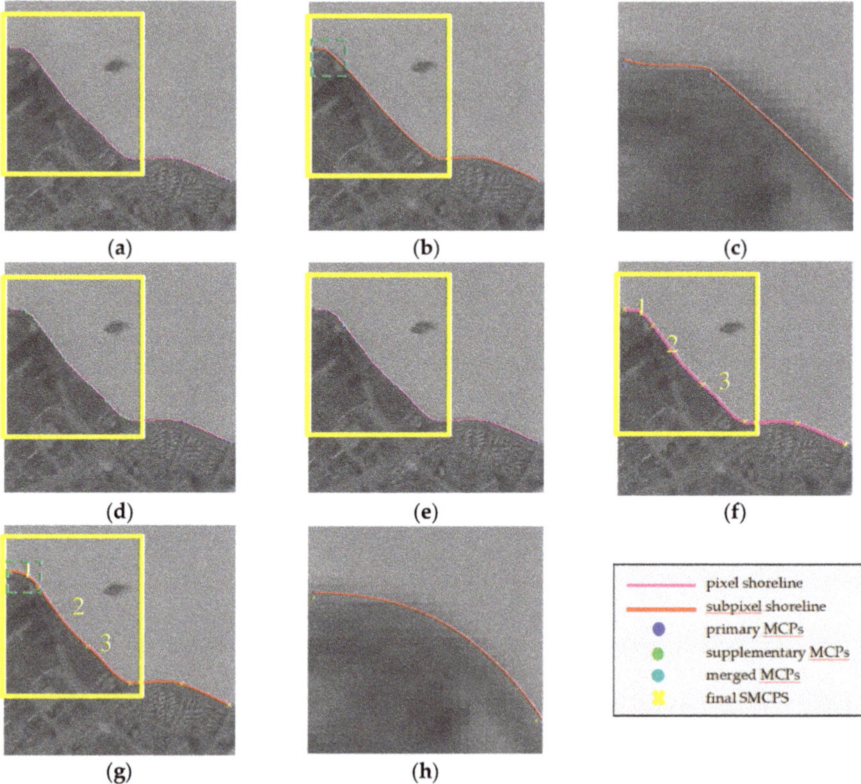

Figure 13. SMF process. (**a**) Initial shoreline and primary MCPs; (**b**) one problematic shoreline segment in yellow box; (**c**) the magnified version of green box in (**b**); (**d**) supplementary MCPs in the problematic shoreline segment; (**e**) the merged MCPs in the problematic shoreline segment; (**f**) the final shoreline morphological control point set; (**g**) the fitted subpixel shoreline segments; (**h**) the magnified version of green box in (**g**).

Table 4 lists the subpixel assessment indicators (MAE and SD) of shoreline segments marked by yellow boxes in Figure 13 during the SMF process. For the initial longer shoreline segment divided by primary MCP, the MAE and SD of subpixel localization results are 10.53 m and 12.11 m, respectively, which indicate that the initial subpixel localization result is problematic. With the SMF process, the final selected MCPs distribute appropriately along the shoreline and three correct segments are preserved. In addition, the subpixel localization MAE results lie in the range of 2.12 m to 3.22 m and the SD results from 1.84 m to 2.20 m.

Table 4. Quantitative assessments of subpixel shoreline segments.

Error Indicator	Primary Shoreline	Final Shoreline Segments Result			
		Total	Seg 1	Seg 2	Seg 3
MAE (m)	10.53	2.58	3.22	2.12	2.66
SD (m)	12.11	1.84	2.20	1.26	1.97

5.5. Robustness to Complex Offshore Environment and Salt-And-Pepper Noises

In Figure 14, due to the complex offshore environment, for example, the existence of suspended sediment (Figure 14a), the local window is difficult to obtain, which will directly affect the subpixel localization accuracy.

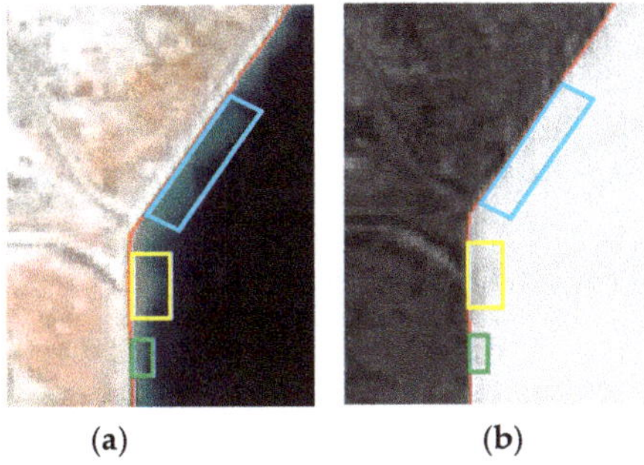

Figure 14. Semi-global subpixel results in complex offshore environment. (**a**) three categories of suspended sediments in the original image (R, band5; G, band4; B, band3); (**b**) three categories of suspended sediments in the MNDWI image.

In Figure 14a,b, suspended sediment situations can be grouped into three different categories depending on the concentration extent and accumulated area.

When the suspended sediment concentration is low (the regions in the blue boxes in Figure 14), the influence will be suppressed or even eliminated in MNDWI images, regardless of the size of the accumulated area.

When the suspended sediment concentration is high and the accumulated area is small with a width less than four pixels, this region (green boxes in Figure 14) will be regarded as the intensity variation region in the local window. In this situation, with the local window design method, the minimum gradient pixels would be found in water, whose intensity is homogeneous.

When the suspended sediment concentration is high and the accumulated area is large (in yellow boxes in Figure 14), with the local window design method, the homogeneous pixels will be selected directly in the suspended sediment region.

In conclusion, regarding the above three forms of suspended sediments, the homogeneous intensity estimations are dealt with effectively and will not reduce the subpixel localization accuracy.

Table 5 summarizes the MAE results of the selected local suspended sediment region, which lie in the range of 0.96–3.55 m, proving that our proposed SGSSL is robust to suspended sediments to some degree.

Table 5. MAE of local regions in suspended sediment environments.

	low Concentration	High Concentration	
		Large Area	Small Area
MAE (m)	0.96	2.28	3.55

Furthermore, we evaluate the proposed SGSSL performance under the influence of salt-and-pepper noises. In Figure 15, the yellow points are determined by the PAE subpixel algorithm proposed by Trujillo-Pino [36], and the red lines are determined from SGSSL, where the white and black points are salt-and-pepper noise. With increasing noise percentage, the PAE subpixel results become increasingly problematic. However, the results determined by SGSSL are always accurate and are not affected by increasing salt-and-pepper noise.

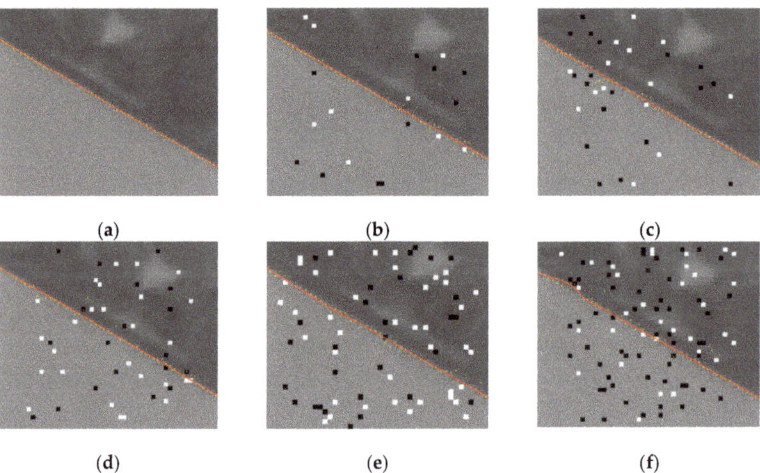

Figure 15. PAE and SGSSL results in salt-and-pepper noises. Percentage of area occupied by various levels of noise: (**a**) 0%; (**b**) 1%; (**c**) 2%; (**d**) 3%; (**e**) 4%; and (**f**) 5%.

As shown in Figure 16, the positioning accuracy results of the two methods under different percentages of salt-and-pepper noise are quantitatively evaluated. With increasing noise percentage, it is obvious that the SGSSL algorithm exhibits better accuracy, proving that the proposed SGSSL is robust to salt-and-pepper noise to some extent.

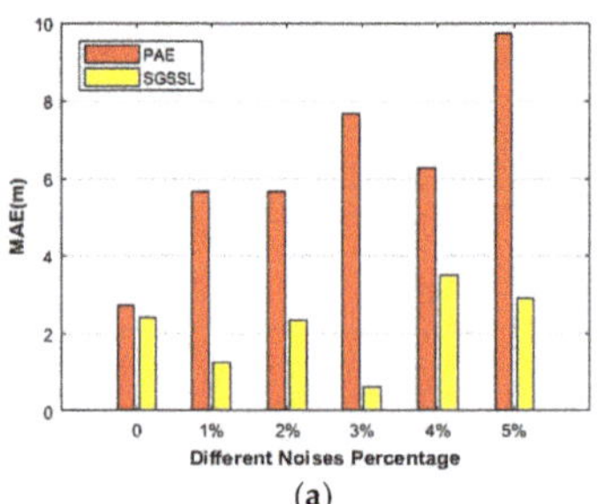

(a)

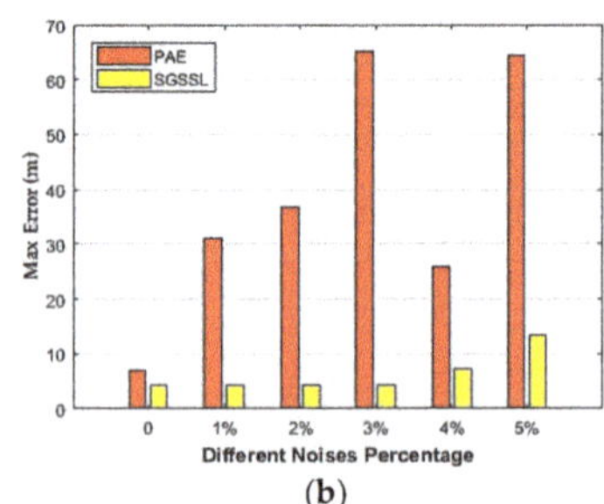

(b)

Figure 16. Positioning errors comparison of PAE and SGSSL in different noise percentages. (**a**) MAE difference between PAE and SGSSL in different noise percentages; (**b**) Max positioning errors comparison between PAE and SGSSL in different noise percentage.

6. Conclusions

With the merits of efficient, large-scale investigational capability, satellite remote sensing shoreline mapping plays an important role in the monitoring of coastal resource management. However, low spatial resolution, various shoreline geometric morphologies, and complex offshore environments prevent the accurate positioning of shorelines. In this study, therefore, we proposed a semi-global subpixel shoreline localization (SGSSL) algorithm for accurately determining artificial shorelines.

The proposed method utilizes not only global spectral information and shoreline morphological features, but also local water index homogeneity features and simplifies the entire shoreline subpixel positioning problem with a segmented shoreline fitting solution. The method considers the following factors: (1) MCPs are utilized to divide the initial shoreline into segments of relatively simple geometric morphology; (2) minimum gradient pixels are found to design a local window; (3) the intensity integral error is minimized in every local window within a segment to initially determine the subpixel location; and (4) the SMF process is presented to obtain the shoreline segments that can be perfectly expressed by a cubic polynomial function and to determine the final subpixel results.

In experiments, five artificial shorelines of various geometric morphologies from Landsat 8 OLI images were selected. The accuracy of the proposed method was evaluated using four error indicators: the MAE, SD, RMSE, and LM. The subpixel RMSE results are all less than 5 m, ranging from 3.02–4.77 m; and the LM results are all less than 4 m, ranging from 2.51–3.72 m, proving that subpixel shoreline accuracy obtained by the proposed method is stable over different experimental areas with various morphologies.

It can be concluded that the proposed algorithm is applicable to the various geometric morphologies of artificial shorelines and is robust to complex offshore environments and salt-and-pepper noise, to some extent.

Limitations of the proposed algorithm include the fact that its performance heavily depends on MCP distribution. Although the SMF process helps in obtaining optimum segments, in some experimental images a lack of MCPs will lead to irreparable subpixel accuracy loss. Another issue worth mentioning is that the proposed algorithm relies on the initial pixel level shoreline, which is the local window determination basis. More adaptation thresholding methodology should be applied to guarantee the initial pixel level shoreline's correct position. Finally, the proposed SGSSL has only been verified on a selected artificial shoreline, other types of shoreline, for example, sandy shorelines and mangrove shorelines, have not been evaluated.

In future research, the performance of the method should be improved by a more flexible MCP extraction algorithm and a more reliable initial shoreline determination method. In terms of application prospects, the method will be combined with multi-source satellite images or ground truth data in the continuous monitoring of shoreline dynamics, coastal terrain mapping and other related research topics.

Author Contributions: Y.S. conceived the idea of the study and designed it. L.F. conducted the experiments. F.L proposed the study on morphological control point distribution and method of how to verify the algorithms effectiveness. L.W.Y help to design the study and verify the results. All authors contributed equally, analyzed the data.

Funding: This work was supported by the Opening Fund of Key Laboratory of Geological Survey and Evaluation of Ministry of Education (Grant No. CUG2019ZR09) and the Fundamental Research Funds for the Central Universities. And this work was also supported by Natural Science Fund of Hubei Province (No.2016CFB690) and the Fund of Key Laboratory of Technology for Safeguarding of Maritime Rights and Interests and Application (No. SCS1610).

Acknowledgments: The authors would like to thank Haitian Zhu (National Satellite Ocean Application Service) for his useful comments from a coastal engineering point of view.

Conflicts of Interest: The authors declare no conflict of interest. The funders had no role in the design of the study; in the collection, analyses, or interpretation of data; in the writing of the manuscript, nor in our decision to publish the results.

References

1. Liu, Y.; Wang, X.; Ling, F.; Xu, S.; Wang, C. Analysis of Coastline Extraction from Landsat-8 OLI Imagery. *Water* **2017**, *9*, 816. [CrossRef]
2. Aedla, R.; Dwarakish, G.S.; Reddy, D.V. Automatic shoreline detection and change detection analysis of Netravati-Gurpur River mouth using histogram equalization and adaptive thresholding techniques. *Aquat. Procedia* **2015**, *4*, 563–570. [CrossRef]
3. Moore, L.J. Shoreline mapping techniques. *J. Coast. Res.* **2000**, *16*, 111–124.
4. Lira, C.P.; Silva, A.N.; Taborda, R.; de Andrade, C.F. Coastline evolution of Portuguese low-lying sandy coast in the last 50 years: An integrated approach. *Earth Syst. Sci. Data* **2016**, *8*, 265–278. [CrossRef]
5. Caixia, Y.U.; Wang, J.; Jun, X.U.; Peng, R.; Cheng, Y.; Wang, M.; Academy, D.N. Advance of Coastline Extraction Technology. *J. Geomater. Sci. Technol.* **2014**, *31*, 305–309. (In Chinese)
6. Dewi, R.; Bijker, W.; Stein, A.; Marfai, M. Flzzy Classification for Shoreline Change Monitoring in a Part of the Northern Coastal Area of Java, Indonesia. *Remote Sens.* **2016**, *8*, 190. [CrossRef]
7. Boak, E.H.; Turner, I.L. Shoreline Definition and Detection: A Review. *J. Coast. Res.* **2005**, *214*, 688–703. [CrossRef]
8. Alesheikh, A.A.; Ghorbanali, A.; Nouri, N. Coastline change detection using remote sensing. *Int. J. Environ. Sci. Technol.* **2007**, *4*, 61–66. [CrossRef]
9. Li, W.; Gong, P. Continuous monitoring of coastline dynamics in western Florida with a 30-year time series of Landsat imagery. *Remote Sens. Environ.* **2016**, *179*, 196–209. [CrossRef]
10. Shearman, P.; Bryan, J.; Walsh, J.P. Trends in Deltaic Change over Three Decades in the Asia—Pacific Region. *J. Coast. Res.* **2013**, *290*, 1169–1183. [CrossRef]
11. Le Cozannet, G.; Garcin, M.; Yates, M.; Idier, D.; Meyssignac, B. Approaches to evaluate the recent impacts of sea—Level rise on shoreline changes. *Earth-Sci. Rev.* **2014**, *138*, 47–60. [CrossRef]
12. Bird, E.C.F. The modern prevalence of beach erosion. *Mar. Pollut. Bull.* **1987**, *18*, 151–157. [CrossRef]
13. Gens, R. Remote sensing of coastlines: Detection, extraction and monitoring. *Int. J. Remote Sens.* **2010**, *31*, 1819–1836. [CrossRef]
14. State Oceanic Administration 908 Special Office. *Technical Regulations for Satellite Remote Sensing Survey on Island Coastal Zone*; Ocean Press: Beijing, China, 2005; ISBN 9787502764852. (In Chinese)
15. Wang, C.; Zhang, J.; Ma, Y. Coastline interpretation from multispectral remote sensing images using an association rule algorithm. *Int. J. Remote Sens.* **2010**, *31*, 6409–6423. [CrossRef]
16. Li, X.; Ling, F.; Du, Y.; Feng, Q.; Zhang, Y. A spatial–temporal Hopfield neural network approach for super—Resolution land cover mapping with multi—Temporal different resolution remotely sensed images. *ISPRS J. Photogramm.* **2014**, *93*, 76–87. [CrossRef]
17. Kasetkasem, T.; Arora, M.; Varshney, P. Super-resolution land cover mapping using a Markov random field based approach. *Remote Sens. Environ.* **2005**, *96*, 302–314. [CrossRef]
18. Tolpekin, V.A.; Stein, A. Quantification of the effects of land-cover-class spectral separability on the accuracy of Markov-random-field-based super resolution mapping. *IEEE Trans. Geosci. Remote Sens.* **2009**, *47*, 3283–3297. [CrossRef]
19. Ling, F.; Du, Y.; Xiao, F.; Li, X. Subpixel Land Cover Mapping by Integrating Spectral and Spatial Information of Remotely Sensed Imagery. *IEEE Geosci. Remote Sens. Lett.* **2012**, *9*, 408–412. [CrossRef]
20. Zhang, Y.; Du, Y.; Li, X.; Fang, S.; Ling, F. Unsupervised Subpixel Mapping of Remotely Sensed Imagery Based on Fuzzy C-Means Clustering Approach. *IEEE Geosci. Remote Sens. Lett.* **2014**, *11*, 1024–1028. [CrossRef]
21. Atkinson, P.M. Sub-pixel Target Mapping from Soft-classified, Remotely Sensed Imagery. *Photogramm. Eng. Remote Sens.* **2015**, *71*, 839–846. [CrossRef]
22. Ge, Y.; Li, S.; Lakhan, V.C. Development and Testing of a Subpixel Mapping Algorithm. *IEEE Trans. Geosci. Remote Sens.* **2009**, *47*, 2155–2164.
23. Su, Y.; Foody, G.M.; Muad, A.M.; Cheng, K. Combining Hopfield Neural Network and Contouring Methods to Enhance Super-Resolution Mapping. *IEEE J.-STARS* **2012**, *5*, 1403–1417.
24. Ling, F.; Du, Y.; Li, X.; Li, W.; Xiao, F.; Zhang, Y. Interpolation-based super-resolution land cover mapping. *Remote Sens. Lett.* **2013**, *4*, 629–638. [CrossRef]
25. Li, X.; Ling, F.; Du, Y.; Zhang, Y. Spatially Adaptive Superresolution Land Cover Mapping with Multispectral and Panchromatic Images. *IEEE Trans. Geosci. Remote Sens.* **2014**, *52*, 2810–2823. [CrossRef]

26. Li, X.; Du, Y.; Ling, F. Super-Resolution Mapping of Forests With Bitemporal Different Spatial Resolution Images Based on the Spatial-Temporal Markov Random Field. *IEEE J.-STARS* **2014**, *7*, 29–39.
27. Chen, Y.; Ge, Y.; Heuvelink, G.B.M.; Hu, J.; Jiang, Y. Hybrid Constraints of Pure and Mixed Pixels for Soft-Then-Hard Super-Resolution Mapping with Multiple Shifted Images. *IEEE J.-STARS* **2015**, *8*, 2040–2052. [CrossRef]
28. Shi, Z.; Li, P.; Jin, H.; Tian, Y.; Chen, Y.; Zhang, X. Improving Super-Resolution Mapping by Combining Multiple Realizations Obtained Using the Indicator—Geostatistics Based Method. *Remote Sens.* **2017**, *9*, 773. [CrossRef]
29. Chen, Y.; Ge, Y.; An, R.; Chen, Y. Super-Resolution Mapping of Impervious Surfaces from Remotely Sensed Imagery with Points-of-Interest. *Remote Sens.* **2018**, *10*, 242. [CrossRef]
30. Ling, F.; Li, X.; Xiao, F.; Du, Y. Super resolution Land Cover Mapping Using Spatial Regularization. *IEEE Trans. Geosci. Remote Sens.* **2014**, *52*, 4424–4439. [CrossRef]
31. Foody, G.M.; Muslim, A.M.; Atkinson, P.M. Super-resolution mapping of the waterline from remotely sensed data. *Int. J. Remote Sens.* **2007**, *26*, 5381–5392. [CrossRef]
32. Muslim, A.M.; Foody, G.M.; Atkinson, P.M. Localized soft classification for super-resolution mapping of the shoreline. *Int. J. Remote Sens.* **2007**, *27*, 2271–2285. [CrossRef]
33. Muslim, A.M.; Foody, G.M.; Atkinson, P.M. Shoreline Mapping from Coarse–Spatial Resolution Remote Sensing Imagery of Seberang Takir, Malaysia. *J. Coast. Res.* **2007**, *236*, 1399–1408. [CrossRef]
34. Zhang, Y. Super-resolution mapping of coastline with remotely sensed data and geostatistics. *J. Remote Sens.* **2010**, *14*, 157–172.
35. Liu, Q.; Trinder, J.; Turner, I. A Comparison of Sub-Pixel Mapping Methods for Coastal Areas. *ISPRS Ann. Photogramm. Remote Sens. Spat. Inf. Sci.* **2016**, *III-7*, 67–74. [CrossRef]
36. Trujillo-Pino, A.; Krissian, K.; Alemán-Flores, M.; Santana-Cedrés, D. Accurate subpixel edge location based on partial area effect. *Image Vis. Comput.* **2013**, *31*, 72–90. [CrossRef]
37. Pardo-Pascual, J.E.; Almonacid-Caballer, J.; Ruiz, L.A.; Palomar-Vázquez, J. Automatic extraction of shorelines from Landsat TM and ETM+ multi-temporal images with subpixel precision. *Remote Sens. Environ.* **2012**, *123*, 1–11. [CrossRef]
38. Almonacid-Caballer, J.; Sánchez-García, E.; Pardo-Pascual, J.E.; Balaguer-Beser, A.A.; Palomar-Vázquez, J. Evaluation of annual mean shoreline position deduced from Landsat imagery as a mid-term coastal evolution indicator. *Mar. Geol.* **2016**, *372*, 79–88. [CrossRef]
39. Liu, Q.; Trinder, J.; Turner, I.L. Automatic super-resolution shoreline change monitoring using Landsat archival data: A case study at Narrabeen-Collaroy Beach, Australia. *J. Appl. Remote Sens.* **2017**, *11*, 16036. [CrossRef]
40. Pardo-Pascual, J.; Sánchez-García, E.; Almonacid-Caballer, J.; Palomar-Vázquez, J.; Priego De Los Santos, E.; Fernández-Sarría, A.; Balaguer-Beser, Á. Assessing the Accuracy of Automatically Extracted Shorelines on Microtidal Beaches from Landsat 7, Landsat 8 and Sentinel-2 Imagery. *Remote Sens.* **2018**, *10*, 326. [CrossRef]
41. Pan, T. Technical characteristics of the Gaofen-2 satellite. *Aerospace China* **2015**, *1*, 3–9.
42. Laben, C.A.; Brower, B.V. Process for Enhancing the Spatial Resolution of Multispectral Imagery Using Pan-Sharpening. U.S. Patent 6,011,875, 4 January 2000.
43. Otsu, N.A. Threshold Selection Method from Gray-Level Histograms. *IEEE Trans. Syst. Man Cybern.* **1979**, *9*, 62–66. [CrossRef]
44. Mikolajczyk, K.; Schmid, C. Scale Affine Invariant Interest Point Detectors. *Int. J. Comput. Vis.* **2004**, *60*, 63–86. [CrossRef]
45. Xu, H. Modification of normalised difference water index (NDWI) to enhance open water features in remotely sensed imagery. *Int. J. Remote Sens.* **2006**, *27*, 3025–3033. [CrossRef]
46. Gao, B.C. A Normalized Difference Water Index for Remote Sensing of Vegetation Liquid Water from Space. *Remote Sens. Environ.* **1996**, *58*, 257–266. [CrossRef]
47. Feyisa, G.L.; Meilby, H.; Fensholt, R.; Proud, S. Automated Water Extraction Index: A New Technique for Surface Water Mapping Using Landsat Imagery. *Remote Sens. Environ.* **2014**, *140*, 23–35. [CrossRef]

Article

Analysis of Ship Detection Performance with Full-, Compact- and Dual-Polarimetric SAR

Chenghui Cao [1,2,3,4], Jie Zhang [3,4], Junmin Meng [3,4], Xi Zhang [3,4] and Xingpeng Mao [1,2,*]

1 School of Electronic and Information Engineering, Harbin Institute of Technology, Harbin 15001, China; 17b905035@stu.hit.edu.cn
2 Key Laboratory of Marine Environmental Monitoring and Information Processing, Ministry of Industry and Information Technology, Harbin 150001, China
3 First Institute of Oceanography, Ministry of Natural Resources, Qingdao 266061, China; zhangjie@fio.org.cn (J.Z.); mengjm@fio.org.cn (J.M.); xi.zhang@fio.org.cn (X.Z.)
4 Ocean Telemetry Technology Innovation Center, Ministry of Natural Resources, Qingdao 266061, China
* Correspondence: mxp@hit.edu.cn

Received: 20 August 2019; Accepted: 11 September 2019; Published: 17 September 2019

Abstract: Polarimetric synthetic aperture radar (SAR) is currently drawing more attention due to its advantage in Earth observations, especially in ship detection. In order to establish a reliable feature selection method for marine vessel monitoring purposes, forty features are extracted via polarimetric decomposition in the full-polarimetric (FP), compact-polarimetric (CP), and dual-polarimetric (DP) modes. These features were comprehensively quantified and evaluated using the Euclidean distance and mutual information, and the result indicated that the features in CP SAR are better than those of FP or DP SAR in general. The CP SAR features are thus further studied, and a new feature, named phase factor, in CP SAR mode is presented that can distinguish ships and the sea surface by the constant 0 without complex calculation. Furthermore, the phase factor is independent of the sea surface roughness, and hence it performs stably for ship detection even in high sea states. Experiments demonstrated that the ship detection performance of the phase factor detector is better than that of roundness, delta, *HESA* and CFAR detectors in low, medium and high sea states.

Keywords: polarimetric SAR; polarimetric decomposition; ship detection; Euclidean distance; mutual information; new feature

---

1. Introduction

Ship detection is of great significance in maritime traffic, immigration control, and fishing activity monitoring. Synthetic aperture radar (SAR) can work day and night with high resolution, even under cloudy conditions, and has been widely used in ship detection.

Constant false alarm rate (CFAR) detection is a classic method and has been used extensively and effectively in SAR images for ship target detection. The key to the CFAR method is the selection of a threshold, and the threshold depends on the probability density function (PDF) of the sea clutter (the backscatter of the sea surface). Many different probability density models have been proposed to simulate the sea clutter distribution, including the Log-normal, Weibull, Rayleigh, G^0, K, gamma, generalized Gamma, and generalized Gaussian Rayleigh distributions. Ni and Anfinsen [1] discussed the advantages and disadvantages of using a statistical model to describe the sea clutter in the CFAR algorithm. Although CFAR detection has a better performance in a uniform background region, the results will be greatly affected in multitarget and clutter-edge environments. Ai et al. [2] presented a new algorithm that utilizes the strong gray intensity correlation in the ship target and the 2-D joint Log-normal distribution in the clutter. Experiments demonstrated that the detection performance is much better. Qin et al. [3] proposed a novel CFAR detection algorithm for high-resolution SAR images using the generalized Gamma distribution (GΓD), and the performance of the proposed algorithm is better than the Weibull distribution. However, with the higher resolution of the SAR image, the sea clutter becomes complex in the time and spatial domains, and then the existing models are not suitable, resulting in the severe degradation of the CFAR detection performance and many false alarms [4]. Additionally, the parameter estimation is complex, and the threshold cannot be acquired easily [5].

To overcome the drawbacks of the CFAR method, ship detection methods based on new features have been studied by researchers, and many results has been achieved [6–8]. For example, based on Cloude decomposition, Wang et al. [9] used the local uniformity of the third eigenvalue of a polarization coherence matrix (T) to detect ships. Sugimoto et al. [10] combined Yamaguchi decomposition theory and the CFAR method to detect ships by analyzing the differences between the scattering mechanisms of the sea surface and ships. Shirvany et al. [11] indicated the effectiveness of the degree of polarization (DoP) in ship detection. Then, this work was further studied by Touzi et al. [12], who defined an extension of the DoP to enhance significant ship-sea contrasts. In contrast to using a single feature, Yin et al. [13] investigated the capability of m-α and m-χ decompositions in coastal ship detection. Then, three features extracted from compact-polarimetric (CP) SAR were proven to have a good performance in ship detection in [14]. Furthermore, Paes et al. [15] provided a more detailed analysis of the detection capability and sensitivity of δ together with m, μ_c,$|\mu_{xy}|$, and the entropy H_ω. Gui et al. [16] extracted a new feature from the proposed power-entropy decomposition, called the high-entropy scattering amplitude (*HESA*), to detect ships, and experiments verified that *HESA* achieves good detection performance.

The polarization features used in ship detection are extracted from different SAR polarimetric modes. With the development of radar systems, SAR data acquisition modes have been extended from single-polarimetric, dual-polarimetric (DP) and full-polarimetric (FP) SAR to CP SAR [16]. FP SAR can provide more target scattering information than single-polarimetric and DP SAR [17]. Compared with FP SAR, CP SAR is a new type of sensor with a wider swath of coverage and smaller energy budget [18]. According to the polarization state, three CP SAR modes exist, including $\pi/4$, dual circular polarization, and circular transmission and linear reception (CTLR) polarization [19–21]. The CTLR mode is simpler, more stable and less sensitive to noise than the other two modes. Furthermore, the CTLR mode achieves a better performance in self-calibration and engineering [22]. At present, RISA-1 in India, ALOS-2 in Japan, and even the future Canadian RADARSAT Constellation Mission (RCM) all support CP SAR. It can be predicted that there will be more polarization features for ship detection in the future.

Although much research has been done, there are still some drawbacks. (1) At present, there are dozens of polarization features, but most of the studies are based on just one or several features. The problem is how to choose suitable features from these features for marine vessel monitoring purposes. (2) Considering the difficulty of ship detection under complex sea states, how to develop new features to improve the ship detection rate, especially for the detection of weak and small ship targets in a high sea state, is another problem.

In this paper, we perform a comprehensive quantification and evaluation of the polarization features extracted from FP, CP and DP modes in C-band SAR data. Our motivation is to establish a reliable feature selection method for marine vessel monitoring purposes. CP SAR features [23] are further studied owing to their advantages for ship detection. In order to develop new CP SAR features that are simple and suitable for complex sea states, we analyzed the scattering difference between the ships and the sea surface by introducing the sea surface roughness. On the basis, a new feature is proposed that is stable and simple for ship detection, especially in a high sea state. Finally, experiments are carried out to verify the better ship detection performance based on the new feature compared with the roundness, delta, *HESA* and CFAR methods in low, medium and high sea states. The main parts are shown in Figure 1.

Section 2 introduces DP, FP and CP SAR data and polarization features. In Section 3, the feature selection method is analyzed by the Euclidean distance and mutual information. Three features are analyzed with the introduced sea surface roughness, and a feature is presented for ship detection in Section 4. In Section 5, the performances of different detectors are compared. Finally, conclusions are drawn in Section 6.

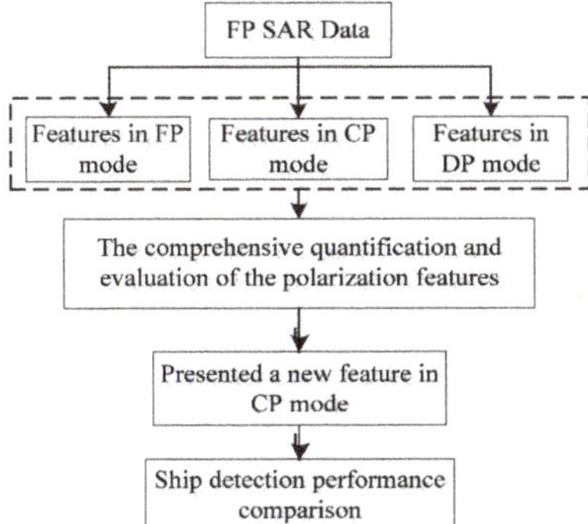

Figure 1. The main parts of this paper.

2. Data and Polarization Features

2.1. Data

In this paper, five RADARSAT-2 images are used, and information on the five images is shown in Table 1.

Table 1. Information of the five SAR images.

Scene ID	Imaging Time	Polarization Mode	Incident Angle	Resolution
01	16 December, 2008	HH/HV/VH/VV	28°	8/m
02	22 January, 2014	HH/HV/VH/VV	44°	8/m
03	25 September, 2014	HH/HV/VH/VV	40°	8/m
04	29 March, 2015	HH/HV/VH/VV	27°	8/m
05	21 November, 2015	HH/HV/VH/VV	20°	8/m

Figure 2a–e show the five RADARSAT-2 images with longitude and latitude information after geometric correction, among which, R = HH, G = HV and B = VH. The locations are the sea areas of the West Lamma Channel in Hong Kong, the Yangtze Estuary, the Yellow River Estuary, Lianyungang and Singapore, respectively. In these images, the bright dots with strong scattering echoes are ship targets, while the dark areas are the sea surface. The scattering echo intensity of the ship target is significantly greater than that of the sea surface. On the whole, many ships can be observed in Figure 2 except Figure 2d. In Figure 2a, the ships are located in the West Lamma Channel. In Figure 2c,e, the ships are mainly located near the port and shore, while in Figure 2b, there is no land area, the ships are mainly concentrated in the middle of the image, and some of the ships have strong crosswise side lobes.

The sea surface wind speeds are calculated by CMOD5 [24], which is a C-band geophysical model function for the inversion of the sea surface wind speed [25]. Combined with the Beaufort wind scale [26], the sea state in scene 04 reaches level 6, which belongs to the high sea state, and scene 03 belongs to the medium sea state; scenes 01–02 and 05 belong to the low sea state. The average wind speeds of the five images are listed in Table 2.

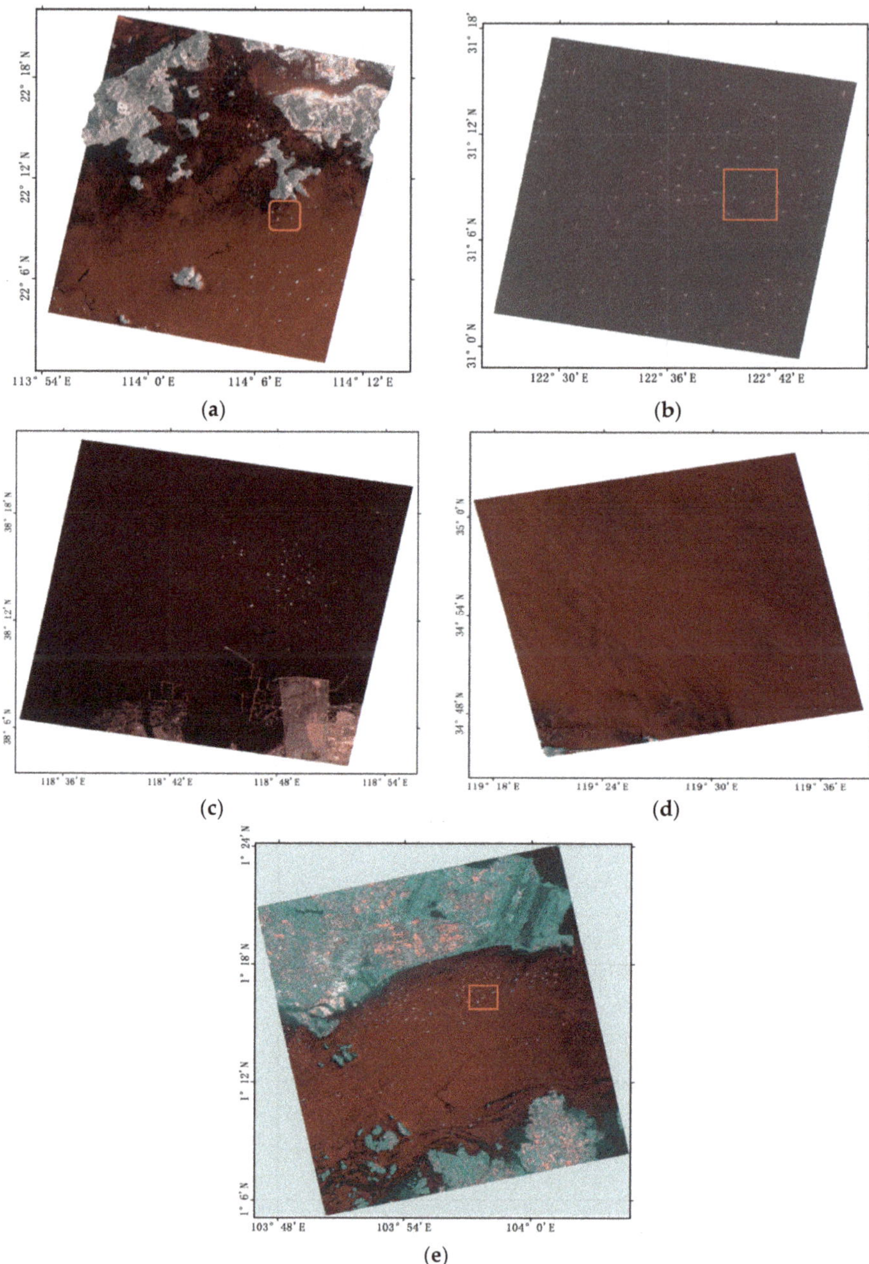

Figure 2. The RGB images of Scenes 01–05 after geometric correction. R = HH, G = HV, B = VH. (**a**) Scene 01: areas of the West Lamma Channel in Hong Kong; (**b**) scene 02: areas of the Yangtze Estuary; (**c**) scene 03: areas of the Yellow River Estuary; (**d**) scene 04: areas of Lianyungang; (**e**) scene 05: areas of Singapore.

Table 2. The average wind speed of five SAR images.

Scene ID	01	02	03	04	05
Average wind speed (m/s)	3.8	2.6	9.1	10.9	4.7
Sea state	3	2	5	6	3

The ships in these five images are all matched by the Automatic Identification System (AIS) [27,28]. The AIS was developed primarily as a tool for maritime safety. The AIS equipment aboard vessels continuously and autonomously transmits information about the vessel including its identity, position, course and speed. Figure 3 is part of scene 02, and it shows the matching result, in which the SAR image contains 33 ships.

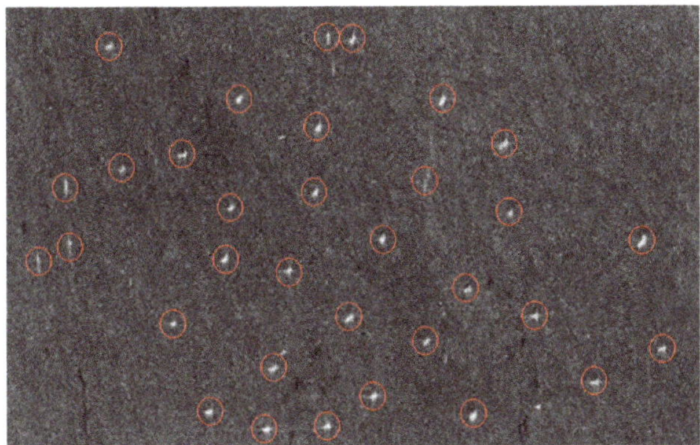

Figure 3. Matching results of a SAR image and the AIS.

2.2. Extraction of Features from FP, CP and DP Data

2.2.1. Features from FP Data

First, we extracted several features by polarimetric decomposition, and the features are shown in Table 3. The first column of Table 3 shows the features extracted from the FP data ($f1$–$f16$). The methods used in this paper are described below.

In FP mode, assuming that $S_{HV} = S_{VH}$, each pixel of an image can be represented by a linear scattering vector as follows:

$$k_L = \begin{bmatrix} S_{HH} & \sqrt{2}S_{HV} & S_{VV} \end{bmatrix}^T \tag{1}$$

where S_{HH}, S_{HV}, and S_{VV} are elements of the scattering matrix. The Pauli scattering vector enhances the scattering mechanism and is given by:

$$k_P = \frac{1}{\sqrt{2}}\begin{bmatrix} S_{HH}+S_{VV} & S_{HH}-S_{VV} & 2S_{HV} \end{bmatrix}^T \tag{2}$$

Features $f1$–$f5$ are defined as the amplitudes of the five polarization components introduced in Equations (1) and (2).

Features $f6$ and $f7$ are the polarimetric coherences of (HH, HV) and (HH, VV), respectively, and $f8$ and $f9$ are the phase differences of (HH, HV) and (HH, VV), respectively. The polarimetric coherence γ and phase difference $\Delta\phi$ between HH and HV are described by

$$\gamma_{HH/HV} = \frac{\left|\left\langle S_{HH}S^*_{HV}\right\rangle\right|}{\sqrt{\left\langle |S_{HH}|^2\right\rangle\left\langle |S_{HV}|^2\right\rangle}} \tag{3}$$

$$\Delta\phi_{HH/HV} = \arg(\langle S_{HH}S^*_{HV}\rangle) \quad (4)$$

where $\langle\rangle$ indicates averaging in an 11 × 11 window. Features $f10, f11$, and $f12$ are the entropy, alpha, and anisotropy, respectively. These features are derived from eigenvalues analysis of the averaged coherency matrix T [29,30], $\langle T\rangle = \langle k_P k_P^{*T}\rangle$.

Table 3. Polarization features extracted under FP, CP and DP SAR modes.

FP SAR	CP SAR	DP SAR
Amplitudes:		
*f*1: |HH|		
*f*2: |HV|	*c*1: |RH|	*d*1: |HH|
*f*3: |VV|	*c*2: |RV|	*d*2: |HV|
*f*4: |HH-VV|		
*f*5: |HH+VV|		
Polarimetric coherences:		
*f*6: HH/HV	*c*3: RH/RV	*d*3: HH/HV
*f*7: HH/VV		
Polarimetric phase differences:		
*f*8: HH/HV	*c*4: RH/RV	*d*4: HH/HV
*f*9: HH/VV		
Eigenvalue parameters:		
*f*10: Entropy	*c*5: Entropy	*d*5: Entropy
*f*11: Alpha	*c*6: Alpha	*d*6: Alpha
*f*12: Anisotropy		
	Cloude decomposition:	
Freeman decomposition:	*c*7: Surface	
*f*13: Surface	*c*8: Double	
*f*14: Double	*c*9: Random	
*f*15: Volume	Raney decomposition:	
Yamaguchi decomposition:	*c*10: DoP	
*f*16: Surface	*c*11: Roundness	
*f*17: Double	*c*12: Delta	
*f*18: Volume	*c*13: Surface	
*f*19: Helix	*c*14: Double	
	*c*15: Random	

Features $f13$–$f19$ are components from a model-based decomposition. Among which, $f13$–$f15$ are amplitudes of surface scattering, double scattering and random scattering from Freeman decomposition, respectively [31]. Similarly, $f16$–$f19$ are amplitudes of surface scattering, double bounce, volume scattering, and helix scattering from Yamaguchi decomposition, respectively [32]. Based on the corresponding scattering mechanisms, the surface scattering, double scattering and random scattering from the Freeman decomposition are

$$\begin{cases} P_S = f_S(1+|\beta|^2) \\ P_D = f_D(1+|\alpha|^2) \\ P_V = f_V \end{cases} \quad (5)$$

where $P_S + P_D + P_V = |S_{HH}|^2 + 2|S_{HV}|^2 + |S_{VV}|^2$ and α, β depend on the sign of $Re(\langle S_{HH}S^*_{VV}\rangle)$. If $Re(\langle S_{HH}S^*_{VV}\rangle) \geq 0$, surface scattering is dominant (α = −1); otherwise, double scattering is dominant (β = 1).

On the basis of three-component decomposition, Yamaguchi presented four-component decomposition [33]. In this decomposition method, $\langle S_{HH}S_{HV}^{*}\rangle \neq 0$ and $\langle S_{VH}S_{VV}^{*}\rangle \neq 0$ are introduced to show that the symmetry hypothesis is not true. The features $f16, f17, f18$, and $f19$ are defined as follows

$$\begin{array}{ll} P_S = \mathrm{f}_S(1+|\beta|^2), & P_D = \mathrm{f}_D(1+|\alpha|^2) \\ P_C = f_C, & P_V = f_V \end{array} \tag{6}$$

where $P_S + P_D + P_C + P_V = |S_{HH}|^2 + 2|S_{HV}|^2 + |S_{VV}|^2$ is the total scattering power.

2.2.2. Features from CP and DP Data

In this paper, to compare the performance among different polarization modes, the CP and DP data are simulated from the FP SAR data. We applied right-hand circular polarization (i.e., CTLR) [34,35] because circular transmission enables a better reconstruction of pseudo-FP information [36]. The CP data constructed from the FP data are shown as follows:

$$\begin{bmatrix} S_{RH} & S_{RV} \end{bmatrix}^T = \begin{bmatrix} S_{HH} - iS_{HV} & S_{VH} - iS_{VV} \end{bmatrix}^T \tag{7}$$

where S_{RH} and S_{RV} represent the scattering coefficients.

The second column of Table 3 ($c1$–$c9$) shows the features from the CP data. $c1$ and $c2$ are the amplitudes of S_{RH} and S_{RV}. $c3$ and $c4$ are the polarimetric coherence and phase difference between S_{RH} and S_{RV}, which we calculated using the same formulas as (3) and (4). Features $c5$ and $c6$ are the entropy and the alpha angle, respectively, extracted from the reconstructed coherency matrix proposed by Nord [36]. The formulas are

$$\begin{cases} H_{CTLR} = -\sum\limits_{i=1}^{2} p_i log_2 p_i \\ \alpha = \sum\limits_{i=1}^{2} p_i \alpha_i \end{cases} \tag{8}$$

respectively, where $p_i = \lambda_i / \sum_{i=1}^{2} \lambda_i$, α_i is an eigenvector corresponding to λ_i, and i = 1,2 . . . . $c7$–$c9$ are the components of a Cloude decomposition [37]. The formulas are

$$P_S = \frac{1}{2} g_0 m (1 + \cos 2\alpha_s) \tag{9}$$

$$P_d = \frac{1}{2} g_0 m (1 - \cos 2\alpha_s) \tag{10}$$

$$P_v = g_0 (1 - m) \tag{11}$$

respectively, where $\alpha_s = 1/2 \tan^{-1}(\sqrt{g_1^2 + g_2^2}/(-g_3))$. $c10$–$c15$ are components from Raney's decomposition using the Stokes parameters of the scattering matrix [34,38]. The formulas are

$$m = \frac{(g_1^2 + g_2^2 + g_3^2)^{1/2}}{g_0} \quad 0 < m < 1 \quad \tan\delta = -\frac{g_3}{g_2} \quad \sin 2\chi = -\frac{g_3}{m g_0} \tag{12}$$

$$\begin{bmatrix} P_d \\ P_v \\ P_s \end{bmatrix}_{m-\delta} = \begin{bmatrix} (m g_0 \frac{1-\sin\delta}{2})^{1/2} \\ (g_0(1-m))^{1/2} \\ (m g_0 \frac{1+\sin\delta}{2})^{1/2} \end{bmatrix} \tag{13}$$

The third column of Table 3 shows the features from the DP data, the first four of which are extracted from FP features: $d1 = f1$, $d2 = f2$, $d3 = f6$, and $d4 = f8$. $d5$ and $d6$ are the pseudo entropy and the alpha angle, respectively, and are calculated from the eigenvalue analysis of a 2 × 2 covariance matrix [39,40].

2.3. Sample Selection

To assess the performance of features for ship detection, we select samples with ships only, samples with sea only and samples with both ships and sea. The numbers of regions of interest in the samples are 75, 75 and 50, respectively.

3. Comprehensive Quantification and Evaluation of Features for Ship Detection

First, we introduce the Euclidean distance to quantitatively evaluate the capacity of features for detecting ships. Then, we identify good features for ship detection by measuring the redundancy via the mutual information. The results can provide a reference for feature selection in practical applications.

3.1. Evaluation of Different Features by Euclidean Distance

Based on the features' scattering differences, the Euclidean distance between ships and the surrounding sea area is used to evaluate the performance of 40 features extracted from FP, CP and DP decomposition. The distance is defined as in [41]:

$$D = \frac{|M_{SHIP} - M_{SEA}|}{\sqrt{\sigma_{SHIP}^2 + \sigma_{SEA}^2}} \tag{14}$$

where M_{SHIP} and M_{SEA} correspond to the statistical average of the samples of ships and the sea surface, respectively, and σ_{SHIP}^2 and σ_{SEA}^2 denote the variance in ships and sea surface, respectively. This equation implies that the larger the distance is, the better the performance of the features in distinguishing ships from the surrounding sea.

The five RADARSAT-2 images in Table 1 are used to calculate the Euclidean distance. For each image, we use 30 regions of interest: 15 ship samples and 15 surrounding sea samples. The final Euclidean distance is the average of the 15 calculated values.

Figure 4 shows the distances between ships and the sea area for $f1$–$f19$, $c1$–$c15$, and $d1$–$d6$. In general, the trend in the distance between different features is consistent in the five images. The distances between features in CP mode are generally larger than those in FP and DP mode; the next largest distances between features are in FP mode, and the distances between features in DP mode are the smallest, especially in scenes 02 and 05. Therefore, the features from CP mode are more suitable for ship detection than those from FP and DP mode. Moreover, with larger distances, $f9$, $f11$, $c4$, $c6$, $c11$ and $c12$ have good ship detection performance. In each scene, the values of these six features are several to more than ten times the values of the other polarization features, especially in scene 05. Then, $f7$, $c5$, $c15$ and $d5$ are smaller than $f9$, $f11$, $c4$, $c6$, $c11$ and $c12$ but larger than the other features. Note that the distances of $c4$, $c6$, $c11$ and $c12$ are larger than 10 in scene 05, which indicates the best ship detection performance among these features out of scenes 01–05.

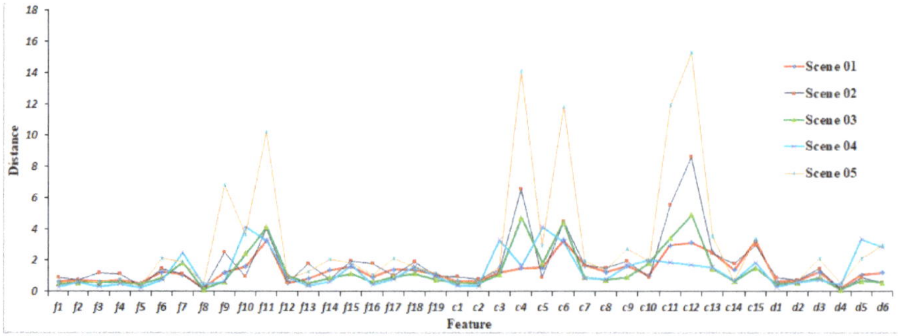

Figure 4. Euclidean distance between ships and the sea area in five images.

3.2. Mutual Information Analysis

Based on the selected features in Section 3.1, the information from the features ($f7$, $f9$, $f11$, $c4$, $c5$, $c6$, $c11$, $c12$, $c15$ and $d5$) should be quantified for better accuracy and efficiency in ship detection. The relevance between ships and features, and the redundancy among different features, should be further evaluated. Mutual information is a correlation measure based on the information-theoretical concept of entropy and has become an important measure in the analysis of informational content [42–44].

Given two random variables X and Y, the mutual information is defined as

$$I(X|Y) = H(X) - H(X|Y) \tag{15}$$

where $H(X)$ denotes the entropy of X and $H(X|Y)$ denotes the conditional entropy of X given Y. The formulas of $H(X)$ and $H(X|Y)$ are

$$H(X) = -\sum_i P(x_i) \log_2(P(x_i)) \tag{16}$$

$$H(X|Y) = -\sum_j P(y_j) \sum_i P(x_i|y_j) \log_2(P(x_i|y_j)) \tag{17}$$

where $P(x_i)$ are the prior probabilities for all values of X and $P(x_i|y_j)$ are the posterior probabilities of X given the values of Y.

The intuitive concept behind this definition of I describes the fraction of information that is shared mutually by both X and Y, called "information overlap." Moreover, the mutual information $I(X|Y)$ is symmetric in X and Y, which means that $I(X|Y) = I(Y|X)$ in a strictly mathematical sense. We normalize $I(X|Y)$ by dividing it by $H(X) + H(Y)$ to achieve increased comparability. The formula is as follows:

$$SU(X,Y) = \frac{I(X|Y)}{H(X) + H(Y)} \tag{18}$$

Twenty regions of interest are selected to calculate the normalized mutual information. The final mutual information value is the average of the calculated values. Table 4 shows the normalized mutual information of the ships and features. In this case, X is a ship and Y is a feature, and a high mutual information I implies a strong predictive value of feature Y for identifying ship X. The values of $f11$, $c4$, $c6$, $c11$ and $c12$ are greater than 0.6, which indicates a high relevance between the ship and the feature, and this is consistent with the conclusion mentioned in section A. The performance of the features selected above is thus further confirmed.

Table 4. Normalized mutual information of ship and feature.

Feature	$f7$	$f9$	$f11$	$c4$	$c5$	$c6$	$c11$	$c12$	$c15$	$d5$
$I(x\|y)$	0.383	0.522	0.603	0.610	0.519	0.625	0.658	0.631	0.315	0.359

Figure 5 shows the normalized mutual information of features. Figure 5a,b are symmetric about the main diagonal, which confirms the symmetry of the mutual information mentioned above. Furthermore, the normalized mutual information between a feature and itself is 1, while the values among different features are less than 1. In general, the trends in the normalized mutual information values in (a) and (b) are consistent with low information redundancy. The features in CP and FP mode have a higher relevance than the features in DP mode, which may be due to the loss of polarization information in DP mode. In detail, the feature pairs with relatively high information overlap are ($f11$, $c6$), ($f9$, $c4$) and ($c5$, $c6$). Among these, $f11$ and $c6$ represent the alphas extracted from the H/alpha decompositions; $f9$ and $c4$ represent the polarimetric coherence extracted from HH/VV and RH/RV, respectively; and $c5$ and $c6$ represent the entropy and alpha extracted from the H/alpha decomposition, respectively. This result indicates that features from the same polarimetric decomposition have higher redundancy. For the feature pairs, ($f7$, $c5$), ($f7$, $c6$), ($f9$, $c12$), ($c4$, $c12$) and ($c6$, $c11$) have a lower redundancy than ($f11$, $c6$), ($f9$, $c4$) and ($c5$, $c6$). There is little information overlap between both $c15$ and $d5$ and the other features. Hence, combined with the Euclidean distance and normalized mutual information, $c4$, $c6$, $c11$ and $c12$ are selected for further study.

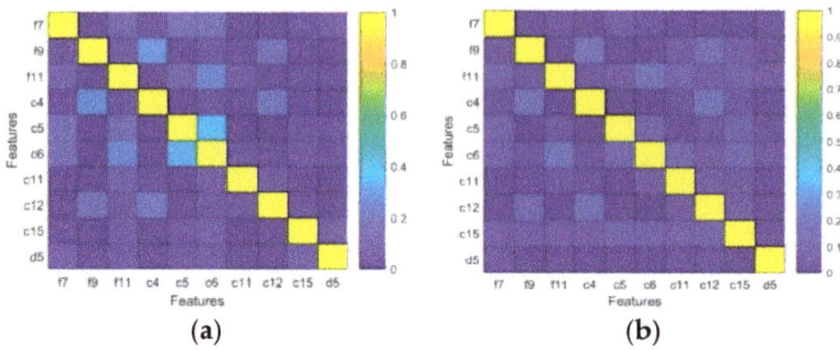

Figure 5. Normalized mutual information among different features, and samples are from scene 01 (**a**) and scene 03 (**b**).

4. A New Feature: Phase Factor

Section 3 concludes that the features in CP mode are more suitable than the DP and FP modes for ship target detection. Therefore, in this section, the features in CP mode are further studied. To analyze the theoretical ship detection performance of features, the relationship between the coherency matrix and the Stokes vector is established. Then, the X-Bragg scattering model is introduced to describe the Stokes vector. Finally, a new feature, which has a good ship detection performance, is proposed.

In CTLR mode, the radar antenna transmits a circular signal and simultaneously receives two orthogonal linear polarization signals. Consider a radar that transmits a right circular signal. The scattering vector [37,45] is

$$\vec{k}_l = [E_{RH}\ E_{RV}]^T = \frac{1}{\sqrt{2}}(S_{HH} - jS_{HV}, -jS_{VV} + S_{HV})^T \tag{19}$$

The coherency matrix T is defined by Huynen parameters [46]:

$$T = \vec{k}_p \vec{k}_p^H = \begin{bmatrix} T_{11} & T_{12} & T_{13} \\ T_{21} & T_{22} & T_{23} \\ T_{31} & T_{32} & T_{33} \end{bmatrix} = \begin{bmatrix} 2A_0 & C - jD & H + jG \\ C + jD & B_0 + B & E + jF \\ H - jG & E - jF & B_0 - B \end{bmatrix} \tag{20}$$

where $\vec{k}_p = 1/\sqrt{2}[S_{HH} + S_{VV}\ S_{HH} - S_{VV}\ 2S_{HV}]^T$, and A_0, B, B_0, C, D, E, F, G, and H are the Huynen parameters. Note that A_0, B_0, and F are rotation invariants.

Matrix T can be expressed by S_{HH}, S_{HV}, and S_{VV}, but it is extremely complicated [32]. In this case, a new idea is proposed by using the elements of the scattering vector:

$$\begin{cases} E_{RH} + jE_{RV} = S_{HH} + S_{VV} \\ E_{RH} - jE_{RV} = S_{HH} - S_{VV} - 2jS_{HV} \end{cases} \tag{21}$$

Then, a new matrix Y is given by

$$
\begin{aligned}
Y &= \begin{bmatrix} E_{RH}+jE_{RV} \\ E_{RH}-jE_{RV} \end{bmatrix} \begin{bmatrix} (E_{RH}+jE_{RV})^H & (E_{RH}-jE_{RV})^H \end{bmatrix} \\
&= \begin{bmatrix} S_{HH}+S_{VV} \\ S_{HH}-S_{VV}-2jS_{HV} \end{bmatrix} \begin{bmatrix} (S_{HH}+S_{VV})^H & (S_{HH}-S_{VV}-2jS_{HV})^H \end{bmatrix} \\
&= \begin{bmatrix} (S_{HH}+S_{VV})(S_{HH}+S_{VV})^* & (S_{HH}+S_{VV})((S_{HH}-S_{VV})^*+2jS^*_{HV}) \\ (S_{HH}-S_{VV}-2jS_{HV})(S_{HH}+S_{VV})^* & (S_{HH}-S_{VV}-2jS_{HV})((S_{HH}-S_{VV})^*+2jS^*_{HV}) \end{bmatrix}
\end{aligned} \tag{22}
$$

Combined with matrix T and the Huynen parameters, matrix Y can be obtained:

$$
Y = \begin{bmatrix} T_{11} & T_{12}+jT_{13} \\ T^*_{12}-jT^*_{13} & T_{22}+T_{33}-2\mathrm{Im}(T_{23}) \end{bmatrix} = \begin{bmatrix} 2A_0 & (C-G)+j(H-D) \\ (C-G)-j(H-D) & 2(B_0-F) \end{bmatrix} \tag{23}
$$

In [46], the Stokes vector of the scattered wave in CTLR mode is written as

$$
g = \begin{bmatrix} g_0 \\ g_1 \\ g_2 \\ g_3 \end{bmatrix} = \begin{bmatrix} A_0+B_0-F \\ C-G \\ H-D \\ -A_0+B_0-F \end{bmatrix} \tag{24}
$$

Therefore, Y can be derived from Equations (23) and (24):

$$
Y = \begin{bmatrix} T_{11} & T_{12}+jT_{13} \\ T^*_{12}-jT^*_{13} & T_{22}+T_{33}-2\mathrm{Im}(T_{23}) \end{bmatrix} = \begin{bmatrix} g_0-g_3 & g_1+jg_2 \\ g_1-jg_2 & g_0+g_3 \end{bmatrix} \tag{25}
$$

As a result, the Stokes vector is described by the coherency matrix:

$$
\begin{cases} g_0 = \frac{T_{11}+T_{22}+T_{33}}{2} - \mathrm{Im}(T_{23}) \\ g_1 = \mathrm{Re}(T_{12}) - \mathrm{Im}(T_{13}) \\ g_2 = \mathrm{Im}(T_{12}) + \mathrm{Re}(T_{13}) \\ g_3 = \frac{-T_{11}+T_{22}+T_{33}}{2} - \mathrm{Im}(T_{23}) \end{cases} \tag{26}
$$

Based on the theory mentioned above, the coherency matrix T is used to represent the Stokes vector through the constructed matrix Y. For a better description of the features, the X-Bragg scattering model is introduced below.

The X-Bragg scattering model was first introduced by Hajnsek to solve the case of nonzero cross-polarized backscattering and depolarization [23]. By assuming a roughness disturbance-induced random surface slope β, X-Bragg scattering is modeled as a reflection depolarizer by rotating the Bragg coherency matrix about an angle β and performing configurational averaging over a given distribution $P(\beta)$:

$$
[T] = \int_0^{2\pi} [T(\beta)]P(\beta)d\beta \tag{27}
$$

assuming that $P(\beta)$ is a uniform distribution of approximately zero with width β_1 ($\beta_1 < \pi/2$). The width β_1 describes the roughness component of the sea surface. The coherency matrix for the rough surface becomes Equation (28) with $\sin c(x) = \sin(x)/x$.

$$
T_{\text{X-Bragg}} = \begin{bmatrix} T_{11} & T_{12} & T_{13} \\ T_{21} & T_{22} & T_{23} \\ T_{31} & T_{32} & T_{33} \end{bmatrix} = \begin{bmatrix} C_1 & C_2 \sin c(2\beta_1) & 0 \\ C_2 \sin c(2\beta_1) & C_3(1+\sin c(4\beta_1)) & 0 \\ 0 & 0 & C_3(1-\sin c(4\beta_1)) \end{bmatrix} \tag{28}
$$

where

$$\begin{cases} C_1 = |R_S + R_P|^2 = |S_{HH} + S_{VV}|^2/2 \\ C_2 = (R_S + R_P)(R_S^* - R_P^*) = (S_{HH} + S_{VV})(S_{HH} - S_{VV})^*/2 \\ C_3 = \frac{1}{2}|R_S - R_P|^2 = |S_{HH} - S_{VV}|^2/4 \end{cases}$$

Substituting Equation (28) into Equation (26), the Stokes vector in CP SAR can be described by an X-Bragg scattering matrix

$$\begin{cases} g_0 = \frac{T_{11}+T_{22}+T_{33}}{2} - \mathrm{Im}(T_{23}) = \frac{C_1+2C_3}{2} \\ g_1 = \mathrm{Re}(T_{12}) - \mathrm{Im}(T_{13}) = \mathrm{Re}(C_2 \sin c(2\beta_1)) \\ g_2 = \mathrm{Im}(T_{12}) + \mathrm{Re}(T_{13}) = \mathrm{Im}(C_2 \sin c(2\beta_1)) \\ g_3 = \frac{-T_{11}+T_{22}+T_{33}}{2} - \mathrm{Im}(T_{23}) = \frac{-C_1+2C_3}{2} \end{cases} \tag{29}$$

Note that g_0 and g_3 are rotation invariants because they are independent of β_1, while g_1 and g_2 are related to the rotation angle β_1. Hence, the features described by g_0 and g_3 are stable for separating ships from sea, even in a high sea state.

For a better explanation of the features with strong ship detection abilities, the roundness ($c11$), delta ($c12$) and the *HESA* [16] are listed as examples. Combined with the model derived from Equation (29), the polarization features are derived by the X-Bragg scattering matrix, which shows the scattering difference between ships and the sea surface. On this basis, a new feature, the phase factor, is presented.

4.1. Roundness

The formula of roundness is

$$\sin 2\chi = -\frac{g_3}{\sqrt{g_1^2 + g_2^2 + g_3^2}} \tag{30}$$

According to Equations (29) and (30), roundness is given by

$$\sin 2\chi = \frac{C_1 - 2C_3}{\sqrt{4(C_2 \sin c(2\beta_1))^2 + (-C_1 + 2C_3)^2}} \tag{31}$$

In Equation (31), the sign of the roundness is consistent with that of $\sin 2\chi$. On the right side of Equation (31), the denominator is positive, so the sign of the roundness depends on the sign of the numerator. The numerator of Equation (31) can be derived as

$$C_1 - 2C_3 = 2\mathrm{Re}(S_{HH}S_{VV}^*)) \tag{32}$$

As shown in Equation (32), the value of C_1–$2C_3$ is depends on $\mathrm{Re}(S_{HH}S_{VV}^*)$. When single scattering is dominant, the sign of $\mathrm{Re}(S_{HH}S_{VV}^*)$ is positive, and when even scattering is dominant, the sign of $\mathrm{Re}(S_{HH}S_{VV}^*)$ is negative [47]. In fact, the sea surface is mainly characterized by single scattering, while ships are mainly characterized by even scattering. Consequently, the value of the sea surface should be positive, and the value of a ship should be negative. The areas shown in Figure 6a–c represent the red box insets shown in Figure 2a,b,e. The images are derived from RADARSAT-2 scenes 01, 02 and 05 respectively, which were each acquired at low sea state. The ships and the sea surface can be separated by a constant 0 in the feature roundness. Note that there exists a "ship" in the lower left corner of (a) without AIS information, so it is uncertain whether it is a ship or not.

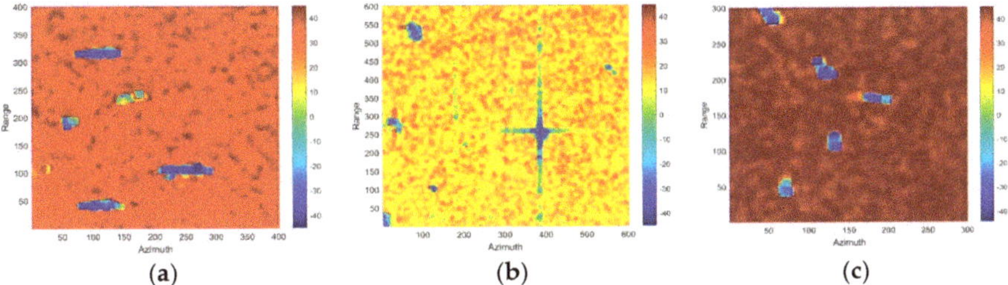

Figure 6. Three examples of the values of ships and the sea surface in the roundness. (**a**) Area in the red box from scene 01. (**b**) Area in the red box from scene 02. (**c**) Area in the red box from scene 05.

Combined with Equations (29) and (31), the roundness is related to angle β_1, which means that a higher sea state can lead to a decline of the roundness detector's performance. What's worse, small ships even cannot be distinguished from the sea clutter.

4.2. Delta

The formula for delta is

$$\delta = \tan^{-1}(\frac{g_3}{g_2}) \tag{33}$$

Then, delta is obtained by substituting Equation (29) into Equation (33):

$$\delta = -\frac{1}{\sin c(2\beta_1)} \frac{\mathrm{Re}(S_{HH}S_{VV}^*)}{\mathrm{Im}(S_{HH}S_{VV}^*)} \tag{34}$$

In Equation (34), for single scattering, the sign of delta is negative; for even scattering, the sign of delta is positive [47]. Due to the scattering differences, ships in the SAR image are mainly characterized by even scattering, while the sea is mainly characterized by single scattering. Therefore, the sign of delta for ships should be positive, and the sign of delta for the sea surface should be negative. The areas shown in Figure 7a–c represent the red box insets shown in Figure 2a,b,e. The images are derived from RADARSAT-2 scenes 01, 02 and 05 respectively, which were each acquired at low sea state. The constant 0 can be used to distinguish ships from the sea surface in the feature delta.

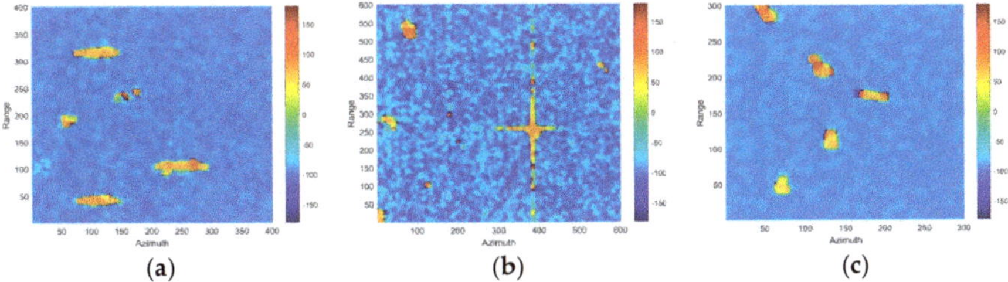

Figure 7. Three examples of the values of ships and the sea surface in the delta. (**a**) Area in the red box from scene 01. (**b**) Area in the red box from scene 02. (**c**) Area in the red box from scene 05.

Note that the value of delta is related to angle β_1 in Equation (34), and the surface roughness increases with the increasing sea state. Therefore, the value of delta is unstably influenced by β_1, making it difficult to use in distinguishing ships and the sea surface in a high sea state.

4.3. HESA

The formula of the *HESA* is

$$HESA = \sqrt{g_0 H_\omega} \tag{35}$$

where

$$H_{\omega} = -\sum_{i=1}^{2} p_i \log_2 p_i,\ p_{1,2} = \frac{g_0 \pm \sqrt{g_1^2 + g_2^2 + g_3^2}}{2g_0}$$

The areas shown in Figure 8a–c represent the red box insets shown in Figure 2a,b,e. The images are derived from RADARSAT-2 scenes 01, 02 and 05 respectively, which were each acquired at low sea state. In Equation (35), the value of the *HESA* is positive, as shown in Figure 8, and the outlines of ships are clear. However, the *HESA* is related not only to the dielectric constant and the incidence angle but also to the rotation angle β_1. β_1 represents the sea surface roughness, and the *HESA* may cause a severe decline when the sea state is high. The constant 1 can be selected to separate ships from the sea surface.

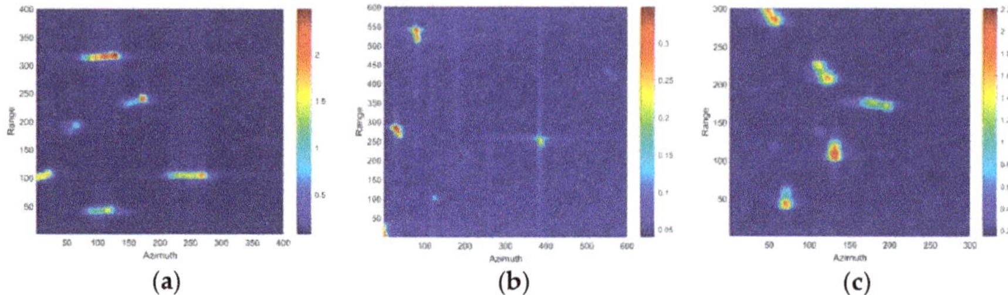

Figure 8. Three examples of the values of ships and the sea surface in the *HESA*. (**a**) Area in the red box from scene 01. (**b**) Area in the red box from scene 02. (**c**) Area in the red box from scene 05.

4.4. Phase Factor

Based on the analysis of the above features, a new feature ς, called the phase factor, is presented in this paper. The formula of the phase factor is

$$\varsigma = \tan^{-1}(\frac{g_0}{g_3}) \tag{36}$$

Combined with Equation (29), the phase factor can be derived by

$$\varsigma = \tan^{-1}(\frac{g_0}{g_3}) = \tan^{-1}(\frac{T_{11} + T_{22} + T_{33} - 2\mathrm{Im}(T_{23})}{-T_{11} + T_{22} + T_{33} - 2\mathrm{Im}(T_{23})}) = \tan^{-1}(\frac{C_1 + 2C_3}{-C_1 + 2C_3}) \tag{37}$$

Equivalently,

$$\begin{aligned} \tan\varsigma &= \frac{C_1+2C_3}{-C_1+2C_3} = \frac{|R_S-R_P|^2+|R_S+R_P|^2}{|R_S-R_P|^2-|R_S+R_P|^2} \\ &= \frac{|S_{HH}-S_{VV}|^2+|S_{HH}+S_{VV}|^2}{|S_{HH}-S_{VV}|^2-|S_{HH}+S_{VV}|^2} = \frac{|S_{HH}-S_{VV}|^2+|S_{HH}+S_{VV}|^2}{-4\mathrm{Re}(S_{HH}S_{VV}^*)} \end{aligned} \tag{38}$$

In Equation (38), the sign of the phase factor depends on the sign of $-\mathrm{Re}(S_{HH}S_{VV}^*)$. For single scattering, the value of $\mathrm{Re}(S_{HH}S_{VV}^*)$ is positive, so the value of the phase factor is negative; for even scattering, the value of the phase factor is positive [47]. Considering that ships are mainly characterized by even scattering, while sea surfaces are mainly characterized by single scattering, the sign of ships is positive, and the sign of the sea surface is negative. In other words, the phase factor is able to distinguish single scattering and even scattering to determine the dominant scattering mechanism. When the phase factor is positive, the even scattering is stronger than the surface scattering; when the phase factor is negative, the surface scattering is stronger than the even scattering. The areas shown in Figure 9a–c represent the red box insets shown in Figure 2a,b,e. The images are derived from RADARSAT-2 scenes 01, 02 and 05 respectively, which were each acquired at low sea state. In Figure 9, the sign of ships is positive, while the sign of the sea surface is negative, which means that the constant 0 can be used to distinguish ships and sea surface.

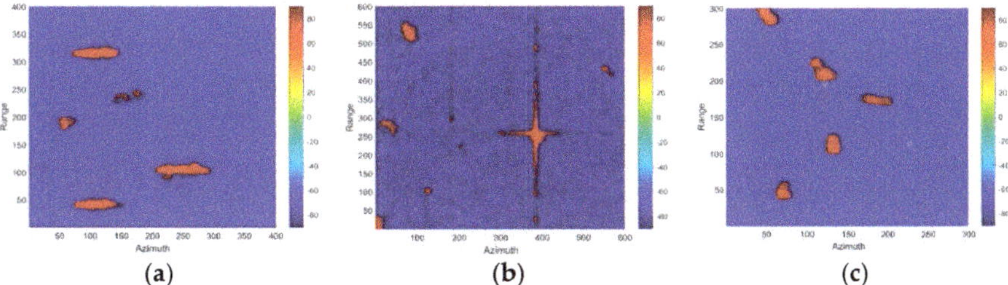

Figure 9. Three examples of the values of ships and the sea surface in the phase factor. (**a**) Area in the red box from scene 01. (**b**) Area in the red box from scene 02. (**c**) Area in the red box from scene 05.

Furthermore, Equation (38) shows that the value of the phase factor is related only to the dielectric constant and incident angle and is independent of the random surface slope β_1. This finding indicates that the phase factor is rotation invariant and stable to different sea states (especially the high sea state), which is of great benefit to ship detection. Therefore, the phase factor theoretically achieves a better detection performance than the other abovementioned polarization features.

5. Detection Results and Discussion

In this section, experiments were performed using CTLR mode emulated from C-band RADARSAT-2 FP SAR data to validate the superiority of the phase factor in ship detection. The phase factor is compared with the roundness, delta, *HESA* and CFAR detectors, respectively.

5.1. Comparisons Between Phase Factor and Roundness, Delta, HESA Detectors

In this section, comparisons are made among roundness, delta, *HESA* and phase factor detectors by analyzing the scattering difference between ships and the sea surface.

Two experiments comparing five detectors in ship detection are performed, as shown in Figure 10 (#1) and Figure 11 (#2). Figure 10 shows the detection results in a medium sea state. The roundness, delta, the *HESA* and the phase factor perform better than the amplitude in detection tasks because the detected ships in (b)–(f) all have clear outlines, and the ship pixels were very bright with respect to the surrounding sea clutter. The results indicate that the four features from CP decomposition can effectively distinguish ships from sea clutter. In (a), (d), (f) and (i), only one ship is detected by the amplitude and *HESA*, while three ships are detected by the other three features. For the roundness, delta and phase factor, the signs of the ship and sea clutter data are opposites, which facilitate distinguishing ships from the sea clutter by means of a constant 0. For the *HESA*, the signs of ships and sea clutter data are all positive, so it is hard to select a proper value to separate ships from sea surface. Note that the spans of the roundness, delta and the phase factor are dozens of times larger than that of the *HESA*.

For the sake of fairness, $P = |M_{SHIP} - M_{SEA}|/\max f$ is used to evaluate the ship detection performance, where M_{SHIP} and M_{SEA} correspond to the statistical average of the samples of ships and sea surface, respectively, and f represents features. Note that P ranges from 0 to 2, which can describe the scattering difference and can distinguish ships from the surrounding sea surface. This finding indicates that the higher the value P, the better the detection performance is. The results are listed in Table 5. Multiples represent the performance ratio of the roundness, delta, *HESA* or phase factor to the amplitude.

The performances of five detectors are listed in descending order: phase factor, roundness, delta, *HESA*, and amplitude of RV polarization. Note that the performances of the phase factor, roundness, delta, and the *HESA* are 65, 54, 41 and 9 times the amplitude of RV polarization, respectively. Thus, according to the value and the scattering difference between the ships and the sea surface, the phase factor can detect ships better than the other four detectors. In another respect, the phase factor is irrelevant to the sea surface roughness, and thus it is sufficiently stable with an increasing sea state, as shown in (e) and (j). In contrast, the roundness, delta and *HESA* are related to the sea surface roughness. The sea surface is very rough in a high sea state, and sea spikes can cause false alarms and an increased difficulty in the detection.

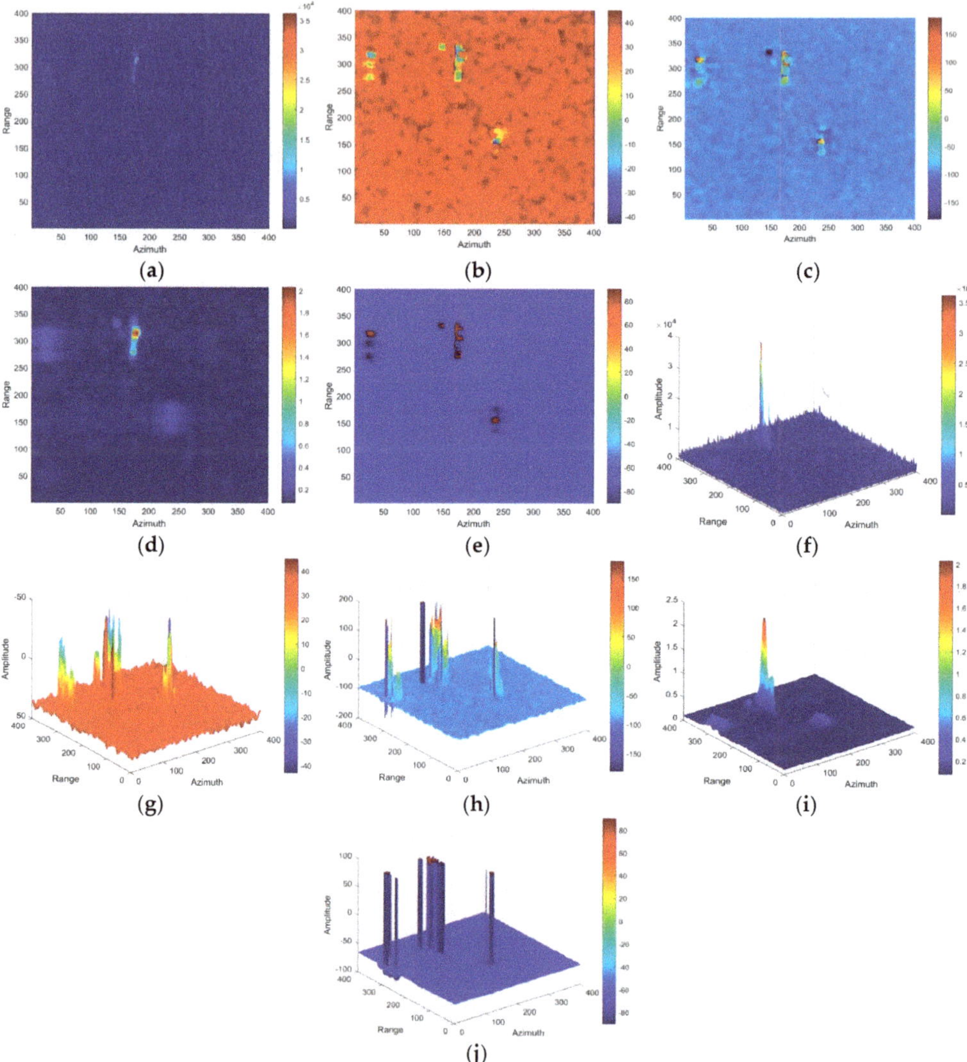

Figure 10. Comparison of the five detectors in a medium sea state. (**a**) Amplitude of RV polarization. (**b**) Roundness. (**c**) Delta. (**d**) *HESA*. (**e**) Phase factor. (**f**) 3-D display of amplitude. (**g**) 3-D display of roundness. (**h**) 3-D display of delta. (**i**) 3-D display of *HESA*. (**j**) 3-D display of phase factor.

Figure 11 is another comparison of the detectors in a high sea state (#2). Combined with the AIS, the image contains four small ships. All the ships can be detected with the five detectors. Influenced by strong winds, many false alarms appear in (b)–(d) and (g)–(i), resulting in the severe performance degradation of the roundness and *HESA* detectors.

Table 5. Performances comparison of the five detectors.

Feature	Experiment no.	Amplitude	Roundness	Delta	*HESA*	Phase Factor
P	#1	0.02	1.07	0.82	0.18	1.29
	#2	0.05	1.37	1.19	0.32	1.83
Multiples	#1	1	54	41	9	65
	#2	1	27	24	7	37

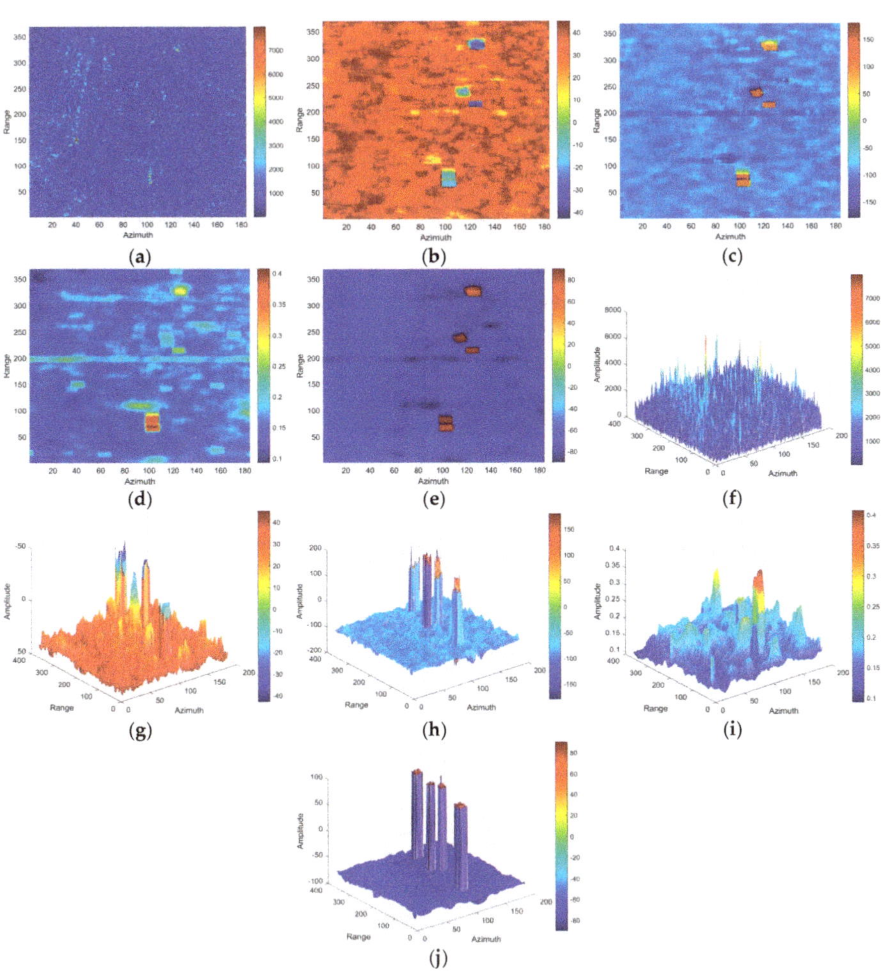

Figure 11. Comparison of the five detectors in a high sea state. (**a**) Amplitude of RV polarization; (**b**) Roundness; (**c**) Delta; (**d**) *HESA*; (**e**) Phase factor; (**f**) 3-D display of the amplitude; (**g**) 3-D display of roundness; (**h**) 3-D display of delta; (**i**) 3-D display of *HESA*; (**j**) 3-D display of phase factor.

The performances of the five detectors are shown in Table 5, which are consistent with results from #1. The performances of the phase factor, roundness, delta, and *HESA* are 37, 27, 24 and 7 times that of the amplitude of the RV polarization, respectively. According to the performance ratio, although the detectors in a high sea state are smaller than those in a medium state, the phase factor is always the best among the five detectors. The results demonstrate that the phase factor is an effective detector with strong robustness, especially in a high sea state, which is useful in practical applications.

5.2. Comparisons Between Phase Factor and CFAR Detectors

Comparisons between phase factor and CFAR detectors were made to verify the superiority of the phase factor detector for ship detection in low, medium and high sea states. The CFAR detector is based on the Weibull, Log-normal, G^0, K and generalized Gamma distribution (GΓD) of the sea clutter, and the method of log-cumulants (MoLC) based on the Mellin transform is used for the parameter estimation of the sea clutter model.

Considering the false alarm rate and detection rate, the *FOM* is used for the detection performance analysis [48]

$$FOM = \frac{N_{tt}}{N_{fa} + N_{gt}} \tag{39}$$

where N_{tt} and N_{fa} are the numbers of detected ships and false alarms, respectively. N_{gt} is the number of ships that matched with AIS. It is indicated from (39) that the larger the *FOM*, the better the detection performance.

The amplitude of RV (Radar transmit in right circular and receive in vertical) polarization emulated from the five RADARSAT-2 FP SAR images shown in Figure 2 is used for ship detection. 19 regions of interest, including 97, 40 and 28 ships in low, medium and high sea states, respectively, are extracted, and each area is 400*400 pixels. The false alarm rate is set to 0.001, which is the best after multiple tests for CFAR ship detection. The phase factor detector uses a constant 0 to distinguish ships and the surrounding sea surface. In low, medium and high sea states, Table 6 shows the detection results by the CFAR and phase factor detectors.

Table 6. Detection performance comparison of CFAR and phase factor detectors.

Model	Sea State	False Alarms	Correct Detections	*FOM*
Weibull-CFAR	Low	22	89	0.75
	medium	4	40	0.9
	high	16	28	0.64
Log-normal-CFAR	Low	1	83	0.85
	medium	1	36	0.88
	high	0	16	0.57
G^0-CFAR	Low	0	58	0.59
	medium	2	32	0.76
	high	0	12	0.43
K-CFAR	Low	74	94	0.55
	medium	14	40	0.74
	high	40	28	0.41
GΓD-CFAR	Low	6	74	0.72
	medium	1	36	0.88
	high	8	28	0.78
Phase factor	Low	5	96	0.94
	medium	0	40	1
	high	0	24	0.86

In low sea state, the *FOMs* of these detectors in descending order are phase factor, Log-normal-CFAR, Weibull-CFAR, GΓD-CFAR, G^0-CFAR and K-CFAR; in medium sea state, they are phase factor, Weibull-CFAR, Log-normal-CFAR, GΓD-CFAR, G^0-CFAR and K-CFAR; in high sea state, they are phase factor, GΓD-CFAR, Weibull-CFAR, Log-normal-CFAR, G^0-CFAR and K-CFAR. The results indicate that the phase factor detector has the best performance in low (*FOM*: 0.94), medium (*FOM:* 1) and high sea states (*FOM:* 0.86) for ship detection, followed by Weibull-CFAR, Log-normal-CFAR and GΓD-CFAR, while G^0-CFAR and K-CFAR are the worst, which is caused by high false alarms, low correct detection rates, or both. In contrast with the CFAR detector, the phase factor can discriminate ships and the sea easily by a constant 0 without complex calculation or false alarm rate setting. Moreover, the phase factor is independent of the sea surface roughness, and hence it can perform well in different sea states, even in high sea state.

Figures 12–14 show three examples of detection results in low, medium and high sea states respectively. In Figures 12–14, (b)–(g) are the ship detection results of the Weibull-CFAR, Log-normal-CFAR, G^0-CFAR, K-CFAR, GΓD-CFAR and phase factor detectors. The red boxes and red circles represent ships matched with AIS and false alarms respectively, and the red stars represent ships undetected. In Figure 12 (low sea state), the Weibull-CFAR, K-CFAR and phase factor detectors are the best without false alarms or missing ships, while a ship is missing in Log-normal-CFAR and GΓD-CFAR detection, what's worse, two ships are missing in G^0-CFAR detection.

In Figure 13 (medium sea state), the Log-normal-CFAR, GΓD-CFAR and phase factor detectors perform better than the other detectors. Two and three false alarms exist in Weibull-CFAR and K-CFAR respectively, and a ship in G^0-CFAR is failed to be detected.

In Figure 14 (high sea state), only the phase factor detector detects two ships without any false alarm. Weibull-CFAR, GΓD-CFAR, Log-normal-CFAR and G^0-CFAR missing one or two ships, and K-CFAR detected all ships but with too many false alarms. The results indicate that the CFAR method is not stable in different conditions, easily causing false alarms and missing detection. In general, the phase factor performs better than the other detectors even in high sea state, while the detection performance of the Weibull-CFAR, Log-normal-CFAR, G^0-CFAR, K-CFAR and GΓD-CFAR decrease with the increasing sea state. The results are in accordance with the theory presented in Section 4.4.

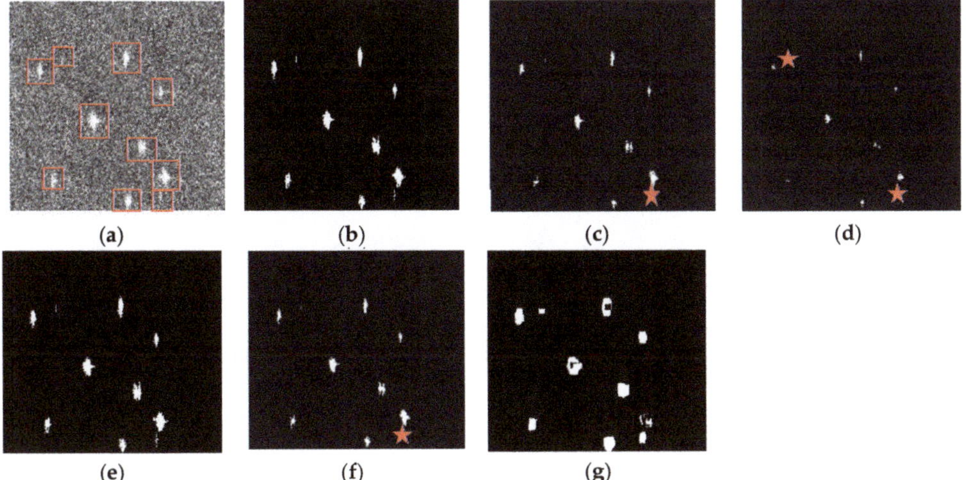

Figure 12. Detection performance comparison of CFAR and phase factor detectors in a low sea state. (**a**) Amplitude of RV polarization; (**b**) Weibull-CFAR; (**c**) Log-normal-CFAR; (**d**) G^0-CFAR; (**e**) K-CFAR; (**f**) GΓD-CFAR; (**g**) phase factor detector.

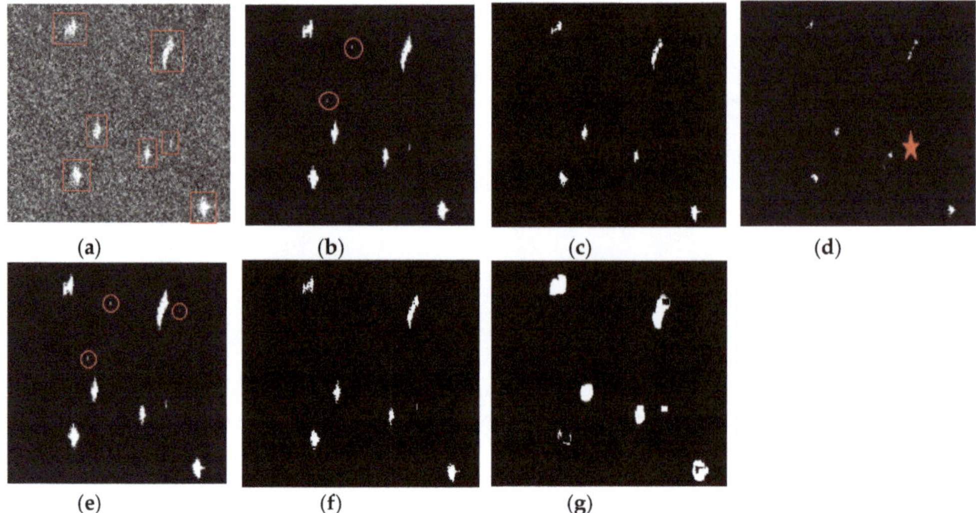

Figure 13. Detection performance comparison of CFAR and phase factor detectors in a medium sea state. (**a**) Amplitude of RV polarization; (**b**) Weibull-CFAR; (**c**) Log-normal-CFAR; (**d**) G^0-CFAR; (**e**) K-CFAR; (**f**) GΓD-CFAR; (**g**) phase factor detector.

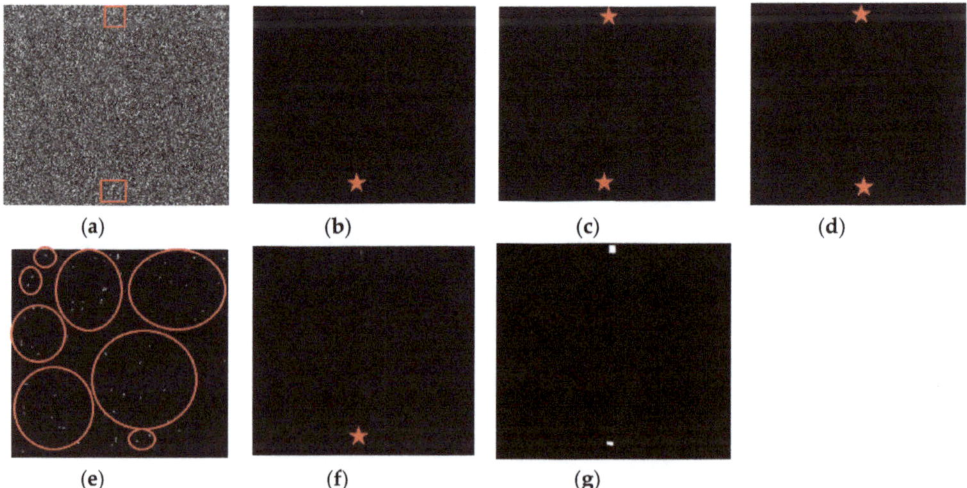

Figure 14. Detection performance comparison of CFAR and phase factor detectors in a high sea state. (**a**) Amplitude of RV polarization; (**b**) Weibull-CFAR; (**c**) Log-normal-CFAR; (**d**) G^0-CFAR; (**e**) K-CFAR; (**f**) GΓD-CFAR; (**g**) phase factor detector.

6. Conclusions

In this paper, in order to establish a reliable feature selection method for marine vessel monitoring purposes, CP and DP SAR data were simulated by five FP RADARSAT-2 images, and forty features were extracted from the FP, CP and DP decomposition. We comprehensively quantified and evaluated these features for ship detection by using the Euclidean distance. The result indicated that features $f7$, $f9$, $f11$, $c4$, $c5$, $c6$, $c11$, $c12$, $c15$ and $d5$ perform better than the other features. For features selected by the Euclidean distance, the relevance between ships and features, along with the redundancy among different features, are further analyzed. The ship detection performance of $f7$, $f9$, $f11$, $c4$, $c5$, $c6$, $c11$, *c*12, *c*15 and *d*5 from the mutual information are consistent with those from the Euclidean distance. Furthermore, the mutual information among the features *f*7, *f*9, *f*11, *c*4, *c*5, *c*6, *c*11, *c*12, *c*15 and *d*5 are low. In conclusion, *f*11, *c*4, *c*6, *c*11 and *c*12 are used for ship detection, which indicates that the features' performance in CP SAR mode is better than that in DP and FP SAR mode.

The features in CP SAR mode are further studied to present a new feature that is simple and suitable for use in complex sea states for ship detection. After a series of derivations and analyses by introducing the sea surface roughness, a new feature, named the phase factor, is proposed that can discriminate the ships and sea surface by a constant 0 and is simpler than the CFAR method without the need for false alarm setting and complex threshold calculations by using a segmentation algorithm. What's more, it is independent of the sea surface roughness and can achieve good performance even in a high sea state.

Experiments demonstrate that the phase factor is stable and better than the roundness, delta, *HESA* and CFAR detectors in low, medium and high sea states. The performances of the phase factor, roundness, delta, and the *HESA* are 65, 54, 41 and 9 times that of the amplitude of RV polarization, respectively. In comparison with CFAR method, the phase factor detector is best in low (*FOM*: 0.94), medium (*FOM*: 1) and high sea states (*FOM*: 0.86) for ship detection, followed by Weibull-CFAR, Log-normal-CFAR and GΓD-CFAR, while G^0-CFAR and K-CFAR are the worst, which is caused by high false alarms, low correct detection rates, or both. Therefore, the phase factor can be used in complex sea states for ship detection, especially for the detection of weak and small ship targets in a high sea state.

Author Contributions: X.Z. proposed the idea. C.C. and X.Z. developed the method. C.C., J.Z., J.M. and X.Z. analyzed the results and wrote the text. X.M. supervised the work. All authors commented the paper.

Funding: This research was supported by the National Key R and D Program of China (No. 2016YFC1401000) and the National Natural Science Foundation of China (No. 61971455).

Acknowledgments: The authors would like to thank the Canadian Space Agency for providing the RADARSAT-2 data.

Conflicts of Interest: The authors declare no conflicts of interest.

References

1. Nicolas, J.M.; Anfinsen, S.N. Introduction to second kind statistics: Application of log-moments and log-cumulants to the analysis of radar image distributions. *Trait. Signal* **2002**, *19*, 139–167.
2. Ai, J.; Qi, X.; Yu, W.; Yu, W.; Deng, Y.; Liu, F.; Shi, L. A new CFAR ship detection algorithm based on 2-D joint log-normal distribution in SAR images. *IEEE Geosci. Remote Sens. Lett.* **2010**, *7*, 806–810. [CrossRef]
3. Qin, X.; Zhou, S.; Zou, H.; Gao, G. A CFAR detection algorithm for generalized gamma distributed background in high-resolution SAR images. *IEEE Geosci. Remote Sens. Lett.* **2012**, *10*, 806–810.
4. Angelliaume, S.; Rosenberg, L.; Ritchie, M. Modeling the Amplitude Distribution of Radar Sea Clutter. *Remote Sens.* **2019**, *11*, 319. [CrossRef]
5. He, Z.; Zhou, X.G.; Lu, J.; Kuang, G.Y. A fast CFAR detection algorithm based on the G^0 distribution for SAR images. *J. Natl. Univ. Def. Technol.* **2009**, *31*, 47–51.
6. Mahdianpari, M.; Salehi, B.; Mohammadimanesh, F.; Brisco, B. An assessment of simulated compact polarimetric SAR data for wetland classification using random forest algorithm. *Can. J. Remote Sens.* **2017**, *43*, 468–484. [CrossRef]
7. Ouchi, K. Recent Trend and Advance of Synthetic Aperture Radar with Selected Topics. *Remote Sens.* **2013**, *5*, 716–807. [CrossRef]
8. Mazzarella, F.; Vespe, M.; Santamaria, C. SAR ship detection and self-reporting data fusion based on traffic knowledge. *IEEE Geosci. Remote Sens. Lett.* **2015**, *12*, 1685–1689. [CrossRef]
9. Sun, Y.; Wang, C.; Zhang, H.; Wu, F.; Zhang, B. A new PolSAR ship detector on RADARSAT-2 data. *Proc. SPIE* **2012**, *8525*. [CrossRef]
10. Sugimoto, M.; Ouchi, K.; Nakamura, Y. On the novel use of model-based decomposition in SAR polarimetry for target detection on the sea. *Remote Sens. Lett.* **2013**, *4*, 843–852. [CrossRef]

11. Shirvany, R.; Chabert, M.; Tourneret, J.Y. Ship and oil-spill detection using the degree of polarization in linear and hybrid/compact dual-pol SAR. *IEEE J. Sel. Top. Appl. Earth Observ. Remote Sens.* **2012**, *5*, 885–892. [CrossRef]
12. Touzi, R.; Vachon, P.W. RCM polarimetric SAR for enhanced ship detection and classification. *Can. J. Remote Sens.* **2015**, *41*, 473–484. [CrossRef]
13. Yin, J.; Yang, J. Ship detection by using the M-Chi and M-Delta decomposition. In Proceedings of the 2014 IEEE Geoscience and Remote Sensing Symposium, Quebec City, QC, Canada, 13–18 July 2014.
14. Yin, J.; Yang, J.; Zhou, Z.S.; Song, J. The Extended Bragg Scattering Model-Based Method for Ship and Oil-Spill Observation Using Compact Polarimetric SAR. *IEEE J. Sel. Top. Appl. Earth Observ.* **2015**, *8*, 3760–3772. [CrossRef]
15. Paes, R.L.; Nunziata, F.; Migliaccio, M. On the capability of hybrid-polarity features to observe metallic targets at sea. *IEEE J. Ocean. Eng.* **2016**, *41*, 346–361. [CrossRef]
16. Gao, G.; Gao, S.; He, J.; Li, G. Adaptive Ship Detection in Hybrid-Polarimetric SAR Images Based on the Power-Entropy Decomposition. *IEEE Trans. Geosci. Remote Sens.* **2018**, *56*, 5394–5407. [CrossRef]
17. Buono, A.; Nunziata, F.; Migliaccio, M. Analysis of full and compact polarimetric SAR features over the sea surface. *IEEE Geosci. Remote Sens. Lett.* **2016**, *13*, 1527–1531. [CrossRef]
18. Charbonneau, F.J.; Brisco, B.; Raney, R.K.; McNairn, H.; Liu, C.; Vachon, P.W.; Geldsetzer, T. Compact polarimetry overview and applications assessment. *Can. J. Remote Sens.* **2010**, *36*, S298–S315. [CrossRef]
19. Buono, A.; Nunziata, F.; Migliaccio, M. Polarimetric analysis of compact-polarimetry SAR architectures for sea oil slick observation. *IEEE Trans. Geosci. Remote Sens.* **2016**, *54*, 5862–5874. [CrossRef]
20. Souyris, J.C.; Imbo, P.; Fjortoft, R. Compact polarimetry based on symmetry properties of geophysical media: The/spl pi//4 mode. *IEEE Trans. Geosci. Remote Sens.* **2005**, *43*, 634–646. [CrossRef]
21. Raney, R.K. Comparing compact and quadrature polarimetric SAR performance. *IEEE Geosci. Remote Sens. Lett.* **2016**, *13*, 861–864. [CrossRef]
22. Gao, G.; Gao, S.; He, J. Ship Detection Using Compact Polarimetric SAR Based on the Notch Filter. *IEEE Trans. Geosci. Remote Sens.* **2018**, *56*, 5380–5393. [CrossRef]
23. Hajnsek, I.; Pottier, E.; Cloude, S.R. Inversion of surface parameters from polarimetric SAR. *IEEE Trans. Geosci. Remote Sens.* **2003**, *41*, 727–744. [CrossRef]
24. Hersbach, H. *CMOD5: An Improved Geophysical Model Function for ERS C-Band Scatterometry*; European Centre for Medium-Range Weather Forecasts: Reading, UK, 2003.
25. Hersbach, H.; Stoffelen, A.; deHaan, S. An improved C-band scatterometer ocean geophysical model function: CMOD5. *J. Geophys. Res. Oceans* **2007**, *112*. [CrossRef]
26. Beaufort_Scale. Available online: https://en.wikipedia.org/wiki/Beaufort_scale (accessed on 20 May 2018).
27. Harati-Mokhtari, A.; Wall, A.; Brooks, P.; Wang, J. Automatic Identification System (AIS): Data reliability and human error implications. *J. Navig.* **2007**, *60*, 373–389. [CrossRef]
28. Tetreault, B.J. Use of the Automatic Identification System (AIS) for maritime domain awareness (MDA). In Proceedings of the OCEANS 2005 MTS/IEEE, Washington, DC, USA, 18–23 Septmber 2005.
29. Cloude, S.R.; Pottier, E. An entropy based classification scheme for land applications of polarimetric SAR. *IEEE Trans. Geosci. Remote Sens.* **1997**, *35*, 68–78. [CrossRef]
30. Sabry, R.; Ainsworth, T.L. SAR Compact Polarimetry for Change Detection and Characterization. *IEEE J. Sel. Top. Appl. Earth Observ. Remote Sens.* **2019**, *12*, 898–909. [CrossRef]
31. Freeman, A.; Durden, S.L. A three-component scattering model for polarimetric SAR data. *IEEE Trans. Geosci. Remote Sens.* **1998**, *36*, 963–973. [CrossRef]
32. Yamaguchi, Y.; Sato, A.; Boerner, W.M.; Sato, R.; Yamada, H. Four-component scattering power decomposition with rotation of coherency matrix. *IEEE Trans. Geosci. Remote Sens.* **2011**, *49*, 2251–2258. [CrossRef]
33. Yamaguchi, Y.; Moriyama, T.; Ishido, M.; Yamada, H. Four-component scattering model for polarimetric SAR image decomposition. *IEEE Trans. Geosci. Remote Sens.* **2005**, *43*, 1699–1706. [CrossRef]
34. Raney, R.K. Hybrid-polarity SAR architecture. *IEEE Trans. Geosci. Remote Sens.* **2007**, *45*, 3397–3404. [CrossRef]
35. Espeseth, M.M.; Skrunes, S.; Jones, C.E.; Brekke, C.; Holt, B.; Doulgeris, A.P. Analysis of evolving oil spills in full-polarimetric and hybrid-polarity SAR. *IEEE Trans. Geosci. Remote Sens.* **2017**, *55*, 4190–4210. [CrossRef]
36. Nord, M.E.; Ainsworth, T.L.; Lee, J.S.; Stacy, N.J. Comparison of compact polarimetric synthetic aperture radar modes. *IEEE Trans. Geosci. Remote Sens.* **2009**, *47*, 174–188. [CrossRef]

37. Cloude, S.R.; Goodenough, D.G.; Chen, H. Compact decomposition theory. *IEEE Geosci. Remote Sens. Lett.* **2011**, *9*, 28–32. [CrossRef]
38. Raney, R.K.; Cahill, J.T.; Patterson, G.W.; Bussey, D.B.J. The m-chi decomposition of hybrid dual-polarimetric radar data with application to lunar craters. *J. Geophys. Res. Planets* **2012**, *117*, E12. [CrossRef]
39. Shan, Z.; Wang, C.; Zhang, H.; Chen, J. H-Alpha Decomposition and Alternative Parameters for Dual Polarization SAR Data. In Proceedings of the PIERS, SuZhou, China, 12–16 September 2011.
40. Cloude, S.R. Dual-versus quad-pol: A new test statistic for radar polarimetry. In Proceedings of the PolInSAR Conference—ESAESRIN, Frascati, Italy, 28 January–1 February 2009.
41. Cao, C.H.; Zhang, J.; Zhang, X. The analysis of ship target detection performance with C band compact polarimetric SAR. *Period. Ocean Univ. China* **2017**, *47*, 85–93.
42. Peng, H.; Long, F.; Ding, C. Feature selection based on mutual information: Criteria of max-dependency, max-relevance, and min-redundancy. *IEEE Trans. Pattern Anal. Mach. Intell.* **2005**, *27*, 1226–1238. [CrossRef]
43. Estévez, P.A.; Tesmer, M.; Perez, C.A.; Zurada, J.M. Normalized mutual information feature selection. *IEEE Trans. Neural Netw.* **2009**, *20*, 189–201. [CrossRef]
44. Bi, N.; Tan, J.; Lai, J.H.; Suen, C.Y. High-dimensional supervised feature selection via optimized kernel mutual information. *Expert Syst. Appl.* **2018**, *108*, 81–95. [CrossRef]
45. Dabboor, M.; Geldsetzer, T. Towards sea ice classification using simulated RADARSAT Constellation Mission compact polarimetric SAR imagery. *Remote Sens. Environ.* **2014**, *140*, 189–195. [CrossRef]
46. Wang, H.; Zhou, Z.; Turnbull, J. Three-Component Decomposition Based on Stokes Vector for Compact Polarimetric SAR. *Sensors* **2015**, *15*, 24087–24108. [CrossRef]
47. Van Zyl, J.J. Unsupervised classification of scattering behavior using radar polarimetry data. *IEEE Trans. Geosci. Remote Sens.* **1989**, *27*, 36–45. [CrossRef]
48. Zhang, X.; Zhang, J.; Meng, J.M.; Chen, L.M. A novel polarimetric SAR ship detection filter. In Proceedings of the IET International Radar Conference 2013, Xi'an, China, 14–16 April 2013.

Letter

Sea Ice Extent Detection in the Bohai Sea Using Sentinel-3 OLCI Data

Hua Su [1,*], Bowen Ji [1] and Yunpeng Wang [2]

[1] Key Laboratory of Spatial Data Mining and Information Sharing of Ministry of Education, National & Local Joint Engineering Research Centre of Satellite Geospatial Information Technology, Fuzhou University, Fuzhou 350108, China; n175527011@fzu.edu.cn

[2] State Key Laboratory of Organic Geochemistry, Guangzhou Institute of Geochemistry, Chinese Academy of Sciences, Guangzhou 510640, China; wangyp@gig.ac.cn

* Correspondence: suhua@fzu.edu.cn; Tel.: +86-591-6317-9863

Received: 2 September 2019; Accepted: 17 October 2019; Published: 20 October 2019

Abstract: Sea ice distribution is an important indicator of ice conditions and regional climate change in the Bohai Sea (China). In this study, we monitored the spatiotemporal distribution of the Bohai Sea ice in the winter of 2017–2018 by developing sea ice information indexes using 300 m resolution Sentinel-3 Ocean and Land Color Instrument (OLCI) images. We assessed and validated the index performance using Sentinel-2 MultiSpectral Instrument (MSI) images with higher spatial resolution. The results indicate that the proposed Normalized Difference Sea Ice Information Index ($NDSIII_{OLCI}$), which is based on OLCI Bands 20 and 21, can be used to rapidly and effectively detect sea ice but is somewhat affected by the turbidity of the seawater in the southern Bohai Sea. The novel Enhanced Normalized Difference Sea Ice Information Index ($ENDSIII_{OLCI}$), which builds on $NDSIII_{OLCI}$ by also considering OLCI Bands 12 and 16, can monitor sea ice more accurately and effectively than $NDSIII_{OLCI}$ and suffers less from interference from turbidity. The spatiotemporal evolution of the Bohai Sea ice in the winter of 2017–2018 was successfully monitored by $ENDSIII_{OLCI}$. The results show that this sea ice information index based on OLCI data can effectively extract sea ice extent for sediment-laden water and is well suited for monitoring the evolution of Bohai Sea ice in winter.

Keywords: Bohai sea ice; sea ice extent; OLCI imagery; sea ice information index

1. Introduction

The Bohai Sea is a semi-enclosed sea in China and is the southernmost area of the frozen sea in the Northern Hemisphere. Seasonal sea ice occurs there every winter from December to March and severely influences maritime activities and the marine economy of the surrounding areas when accumulated sea ice blockades ports and obstructs sea routes. In a particularly cold winter, sea ice can destroy marine facilities, coastal ports, and mariculture and lead to substantial property damage [1–4]. Therefore, monitoring the distribution and spatiotemporal pattern of sea ice is crucial for disaster prevention and maritime management [5]. The distribution of the sea ice is also a key climatic indicator as it can reflect regional climate change and is essential for studying long-term climatic changes in response to recent global warming.

Large-scale monitoring and evaluation of sea ice in high-latitude frozen zones have been carried out by means of remote-sensing technology, including microwave and optical remote sensing. Passive microwave and synthetic aperture radar (SAR) imagery are the main data sources for ice detection as they have all-day and almost all-weather imaging capability [6–8]. The operational sea ice concentration products, such as OSI-450, SICCI-25km, and so forth, were provided by passive microwave data, and can help us better understand the evolution of the Earth's ice cover [9]. Ice extent is most commonly estimated on the basis of the sea ice concentration retrieved from passive microwave data. For example, the contour corresponding to an ice concentration of 15% is commonly used to define the ice extent [10]. However, owing to the coarse spatial resolution of ice concentration data derived from passive microwave imagery (5–25 km), the sea ice concentration is underestimated when the floe size is small or ice cover is sparse. Near coastlines, passive microwave datasets with a large footprint are subject to land contamination, resulting in a mixed land–sea signal being received. This land contamination can cause the extent of sea ice to be overestimated [11]. High-spatial-resolution space-borne SAR datasets such as Radarsat-2 and Sentinel-1A/B can be used to monitor sea ice more subtly. The sea ice concentration can be estimated from single-band SAR data either directly or via a classification scheme [7,12,13]. Space-borne SAR can provide all-weather observations with a much higher spatial resolution (5–100 m) than the passive microwave, but it is challenging to obtain because of the high cost and the long revisit period for most of them.

Optical remote-sensing data have also been widely used to estimate the extent and concentration of sea ice. Although the use of optical remote sensors is constrained by the weather conditions, they have the merits of finer spatial resolution, low cost, and a short revisit period (one day or less). Thus, optical remote sensors such as the Advanced Very High-Resolution Radiometer (AVHRR) [14], Moderate Resolution Imaging Spectroradiometer (MODIS) [15–18], Geostationary Ocean Color Imager (GOCI) [19–21], and FengYun-3 Medium Resolution Spectral Imager (MERSI) [22] have been effectively employed to extract sea ice distribution information via a variety of methods. For example, rapid and effective sea ice extraction has been achieved with a ratio-threshold segmentation method based on the red and infrared bands of MODIS images [2,23]. Sea ice detectability in coastal regions has been improved using texture features derived from MODIS images to accurately detect sea ice in sediment-laden water [24]. The identification of sea ice and the accuracy of image interpretation have also been improved by processing, respectively, optical and microwave images by hue–intensity–saturation (HIS) adjustment and wavelet transformation and further fusing these through principal component analysis (PCA) [5]. Different classifiers such as a decision tree and a support vector machine have been used to directly distinguish sea ice on the basis of multispectral remote-sensing imagery [25,26], in some cases combining multiple features like image texture and surface temperature to improve the accuracy of sea ice extent estimation [27,28].

Data are now available from a new-generation sensor called the Ocean and Land Color Instrument (OLCI), which is carried on the Sentinel-3 satellite. This sensor has relatively high spectral resolution and spatiotemporal resolution in the visible and near-infrared spectra and thus is well suited to the requirements of large-scale coastal environmental monitoring. OLCI data have already been used to monitor and evaluate water quality [29–31] but have as yet rarely been used to study sea ice. In this study, sea ice information indexes based on OLCI multispectral imagery are developed to detect the extent of sea ice and then employed to monitor the spatial and temporal variation of sea ice in the Bohai Sea in the winter of 2017–2018.

2. Study Area and Data

The Bohai Sea (37°07′–41°0′N, 117°35′–121°10′E), located on the northeast of China, borders three land areas and one sea (Figure 1). It covers a total area of 73,686 km^2 and has an average depth of 18 m. It comprises three bays: the Liaodong Bay in the north, the Bohai Bay in the west, and the Laizhou Bay in the south. Over 40 tributaries flow into the sea, the largest four of which are the Yellow River, Haihe River, Luanhe River, and Liaohe River, which carry large quantities of freshwater and sediment into the sea from the land. The salinity of the seawater is only about 30 PSU, making it the least saline of China's coastal waters. Seasonal sea ice usually first occurs at the coast in late December then accumulates along the shoreline and gradually expands into the central basin. Ice coverage finally comes to an end in March of the next year. The thickness of the ice can reach up to 40 cm in extremely cold winters [23], and it usually reaches its maximum extent at the midpoint of the sea ice evolutionary process in late January to early February.

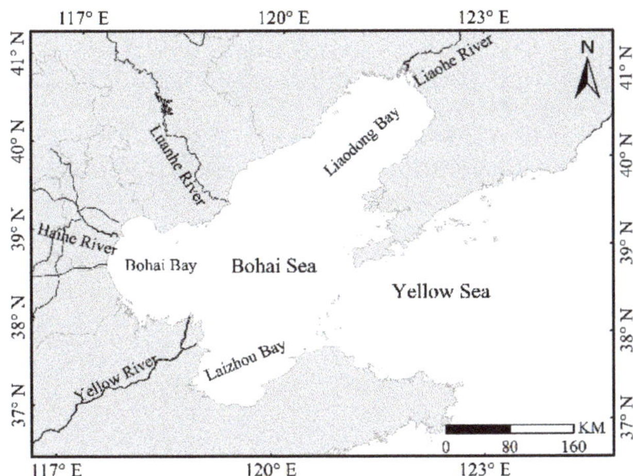

Figure 1. The study area of the Bohai Sea, including the Liaodong Bay, Bohai Bay, and Laizhou Bay.

The new-generation optical sensor OLCI is the successor of ENVISAT's MERIS, having higher spectral resolution and more spectral channels. The OLCI dataset is composed of 21 distinctive spectral bands spanning the spectral range 400–1020 nm across the visible and near-infrared spectra. These multispectral data are very well suited to studying coastal sea ice. An overview of the OLCI bands is given in Table 1. Full-resolution (300 m) OLCI images (OLCI level 1b) acquired from the European Space Agency (ESA) data hub (https://scihub.copernicus.eu/) are employed for Bohai Sea ice detection in this study. Image preprocessing, including subsetting, reprojecting, and radiance-to-reflectance transformation, is conducted using the SNAP 6.0 toolbox (Sentinel Application Platform, http://step.esa.int/main/toolboxes/snap/), which was designed for processing and analyzing Sentinel satellite products.

Table 1. OLCI band characteristics.

Band Number	Central Wavelength (nm)	Full Width at Half Maximum (nm)	Signal-to-Noise Ratio
Band 1	400	15	2188
Band 2	412.5	10	2061
Band 3	442.5	10	1811
Band 4	490	10	1541
Band 5	510	10	1488
Band 6	560	10	1280
Band 7	620	10	997
Band 8	665	10	883
Band 9	673.75	7.5	707
Band 10	681.25	7.5	745
Band 11	708.75	10	785
Band 12	753.75	7.5	605
Band 13	761.25	2.5	232
Band 14	764.375	3.75	305
Band 15	767.5	2.5	330
Band 16	778.75	15	812
Band 17	865	20	666
Band 18	885	10	395
Band 19	900	10	308
Band 20	940	20	203
Band 21	1020	40	152

The Sentinel-2 MultiSpectral Instrument (MSI) provides multispectral, high-resolution imagery in 13 spectral bands: four bands at 10 m, six bands at 20 m, and three bands at 60 m spatial resolution. The instrument's imaging bands cover visible, near-infrared (NIR), and short-wave infrared (SWIR) spectra. In this study, six MSI images (MSI level 1C) acquired by Sentinel-2B over the Bohai Sea on February 1, 2018, were obtained from the ESA data hub (https://scihub.copernicus.eu/) and were processed as comparison and validation data. These images were first preprocessed by the atmospheric correction software [32] (Sen2Cor-02.05.05-win64, http://step.esa.int/main/third-party-plugins-2/sen2cor/sen2cor_v2-5-5/) and were resampled at 60 m resolution to obtain L2A products with all bands of imagery. In addition, MSI data were employed to derive the Normalized Difference Snow Index (NDSI), which was useful for sea ice detection [26,33], as a comparison.

3. Methods

3.1. Normalized Difference Sea Ice Information Index

A total of 10,570 pixels were manually selected as samples and classified as sea ice, seawater, turbid seawater, land, snow and cloud by visual interpretation. The samples were distributed across four different OLCI images in the Bohai Sea on 24 January, 28 January, 1 February, and 12 February, 2018. Descriptive statistics were computed for these samples for characteristic bands to obtain the mean and standard deviation of the top of the atmosphere (TOA) reflectance values.

The TOA reflectance of sea ice in Band 20 (930–950 nm) is higher than that in Band 21 (1000–1040 nm) in OLCI imagery; the opposite is true for all other objects, such as land and cloud cover (Figure 2a). Significant differences such as this in the spectral characteristics of land cover types are the basis for remote-sensing detection, and this particular characteristic is utilized to detect sea ice using the band ratio strategy. The Normalized Difference Sea Ice Information Index ($NDSIII_{OLCI}$) is the normalization of this band ratio so that its value ranges between −1 and 1. The $NDSIII_{OLCI}$ feature was extracted using the following equation:

$$NDSIII_{OLCI} = (B20 - B21)/(B20 + B21), \tag{1}$$

where B20 and B21 are the TOA reflectances of Band 20 and Band 21 in OLCI images, respectively.

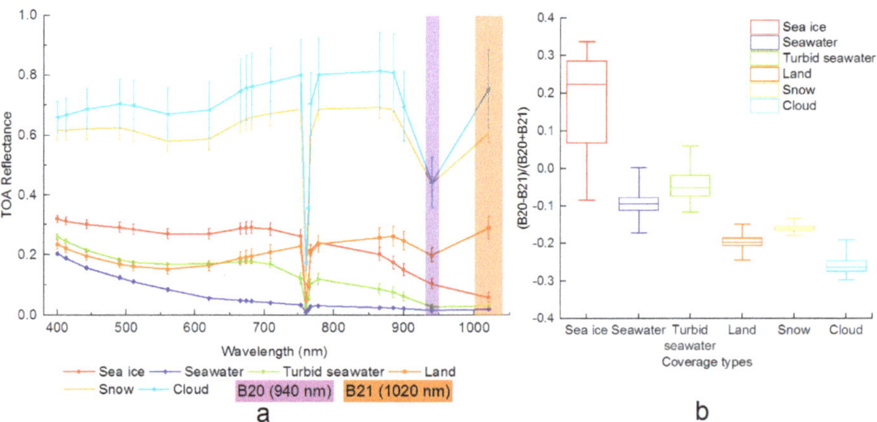

Figure 2. TOA reflectance values in OLCI all bands (**a**) and $NDSIII_{OLCI}$ values (**b**) for sea ice, seawater, turbid seawater, land, snow, and cloud cover in the Bohai Sea. The whiskers in (**a**) depict the standard deviations of the TOA reflectance samples. The whiskers in (**b**) indicate the maximum and minimum ratio values of the sample. The box is determined by the 25th and 75th percentiles of the ratio values of the sample. The median value is marked as the line within the box.

The box plot in Figure 2b indicates that sea ice information is emphasized in the $NDSIII_{OLCI}$ feature, and other cover types present in the OLCI images are de-emphasized. Sea ice has the most significant feature with the highest index value among all types of coverage. This enhancement of sea ice information and suppression of other surfaces can effectively reduce the interferences in sea ice information extraction in the Bohai Sea. Particularly, the values for sea ice are positive because its numerator is greater than zero, and the values of other cover types are negative (Figure 2b). This figure also indicates that over 75% percent of the ratio value of sea ice was greater than 0. Therefore, sea ice, which has the brightest pixels, can be directly segmented out from an $NDSIII_{OLCI}$ feature image with a single threshold value of 0. According to the distribution of the box plot, turbid seawater is most likely to interfere with sea ice detection, as it might not be easy to separate from sea ice in $NDSIII_{OLCI}$.

3.2. Enhanced Normalized Difference Sea Ice Information Index

The complex water environment of the Bohai Sea makes it challenging to extract sea ice precisely using traditional remote-sensing technology. The main reason for the incomplete separation of seawater and sea ice in remote-sensing images is spectral confusion between sea ice and the suspended sediment in turbid seawater [24,27,28]. To distinguish them better, we have developed the Enhanced Normalized Difference Sea Ice Information Index ($ENDSIII_{OLCI}$) by adding consideration of Band 12 (750–757.5 nm) and Band 16 (771.25–786.25 nm) to the $NDSIII_{OLCI}$.

The TOA reflectance characteristics of sea ice and turbid seawater in Bands 12, 16, 20, and 21 in OLCI imagery are shown in Figure 3. It shows subtle differences in TOA reflectance between Band 12 and Band 16 and between Band 20 and Band 21 for turbid seawater but a more visible reduction in TOA reflectance between these bands for sea ice. These spectral characteristics indicate that sea ice and turbid seawater can be separated using a spectral feature that combines these band ratios. Therefore, the discriminant for identifying sea ice in turbid seawater is expressed as follows:

$$\begin{cases} B12 - B16 > 0 \\ B20 - B21 > 0 \end{cases} \tag{2}$$

Linear summation was utilized to combine these two discriminants. The discrimination condition is expressed as follows:

$$B12 - B16 + B20 - B21 > 0 \tag{3}$$

The difference between sea ice and turbid seawater is further emphasized by summing the terms in the discrimination condition (3) to construct the Enhanced Normalized Difference Sea Ice Information Index ($ENDSIII_{OLCI}$) as follows:

$$ENDSIII_{OLCI} = \frac{B12 - B16 + B20 - B21}{B12 + B16 + B20 + B21} \tag{4}$$

where B12, B16, B20, and B21 correspond to the TOA reflectance values of Bands 12, 16, 20, and 21 in OLCI images, respectively.

$ENDSIII_{OLCI}$, which considers Band 12 and Band 16, is a further extension of $NDSIII_{OLCI}$. Sea ice can be distinguished from turbid seawater in $ENDSIII_{OLCI}$ by combining the two-criterion equations (2). After normalization, the index performed stably in sea ice detection from OLCI images.

3.3. Determinaton of Threshold Values

To obtain optimal threshold values for sea ice separation, the segmentation thresholds were identified through sampling of index values for $NDSIII_{OLCI}$ and $ENDSIII_{OLCI}$ during the three main stages of the Bohai Sea ice: the freezing stage (early January), the stable stage (late January to early February), and the melting stage (late February to early March) in the winter of 2017–2018 (Figure 4). While the background coverage types, such as land, snow, and cloud, were significantly suppressed in our index and can be easily distinguished from sea ice, this was not the case for seawater, particularly turbid seawater. To address this, a total of 389 points were manually selected from nine images that included the freezing stage (1, 5 and 20 January 2018), the stable stage (28 and 31 January and 4 February 2018) and the melting stage (16 and 20 February and 21 March 2018), and classified as either sea ice or seawater by visual interpretation. Thresholds were determined from the sampling histogram using the Jenks natural break method [34], which maximizes interclass variance while minimizing intraclass variance by iteratively comparing clusters of data.

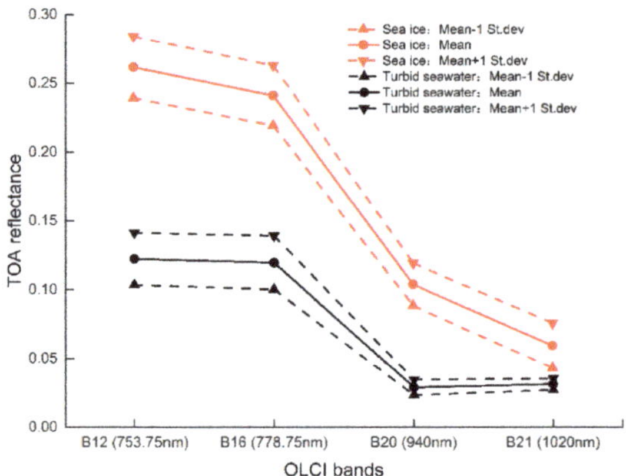

Figure 3. Sample TOA reflectance values for sea ice and turbid seawater in OLCI Bands 12, 16, 20, and 21 in the Bohai Sea.

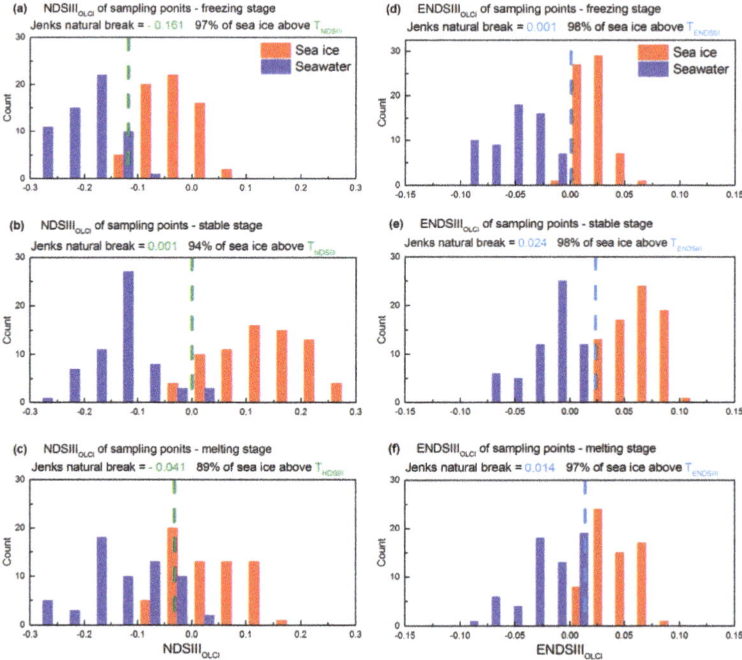

Figure 4. The histogram distributions of sampling points for the three sea ice development stages (freezing, stable and melting stages) with $NDSIII_{OLCI}$ **(a–c)** and $ENDSIII_{OLCI}$ **(d–f)**. The T_{NDSIII} and $T_{ENDSIII}$ threshold values (dashed vertical lines) were determined for sea ice separation by the Jenks natural break method.

Figure 4 shows the threshold values, T_{NDSIII} for $NDSIII_{OLCI}$ and $T_{ENDSIII}$ for $ENDSIII_{OLCI}$, obtained from the sampling dataset for sea ice extraction. According to the sampling, the threshold values of $NDSIII_{OLCI}$ and $ENDSIII_{OLCI}$ are not fixed and vary somewhat depending on the samples and ice stages. The thresholds obtained using the Jenks method performed fine in the freezing stage (Figure 4a,d), where 97% and 98% of sea ice values were above T_{NDSIII} and $T_{ENDSIII}$ respectively, and also acceptably performs in the stable stage (Figure 4b,e) and the melting stage (Figure 4c,f). Exceedance was still 94% and 98% of sea ice in the stable stage, and 89% and 97% of sea ice in the melting stage for the T_{NDSIII} and $T_{ENDSIII}$ thresholds, respectively. The sea ice can be extracted more completely using $T_{ENDSIII}$ instead of T_{NDSIII}.

The T_{NDSIII} and $T_{ENDSIII}$ were determined using the samples from the winter of 2017–2018. They were suitable for the ice detection during the 2017–2018 winter, but they may vary year by year depending on the ice conditions, such as ice developing stages, ice thickness and snow-covered situations. It is better to reset the threshold values when applying the indexes for sea ice detection in other years because the ice conditions vary with the years. The threshold values varied a little for sea ice detection in different ice stages in the 2017–2018 winter, however, the relatively stable values can provide a valuable reference for the threshold determinations of the ice extraction in other years.

3.4. Normalized Difference Snow Index

In polar and high-latitude regions, snow detection is intimately linked to sea ice detection, as the sea ice cover is mostly covered by snow. The Normalized Difference Snow Index (NDSI) has been used by many studies to detect the presence of sea ice in open water (Equation (5)) [35].

$$NDSI = (Green - SWIR)/(Green + SWIR) \tag{5}$$

The NDSI takes advantage of the contrasting spectral behaviors of snow and sea ice cover in the visible and short-wave infrared parts of the spectrum. Snow and sea ice will have a high NDSI value because they exhibit a large contrast in reflectance between the shot-wave infrared band (SWIR Band 11: 1.613 μm) and the visible band (Green Band 3: 0.56 μm). However, the OLCI instrument lacks the short-wave infrared bands required to derive NDSI. In this study, we used the MSI images (resampled from 60 to 300 m spatial resolution) to extract the NDSI feature, and we compared this with our efforts to detect sea ice in the Bohai Sea.

3.5. Support Vector Machine Classifier

The support vector machine (SVM) is a machine learning method based on statistical learning theory. Supervised classification using the SVM method has been widely used in image analysis to identify the class affiliated with each pixel. The basic idea of SVM classification is to use the kernel function to map linearly indivisible points in a low-dimensional space into linearly separable points in a high-dimensional space [36–38]. The goal of SVM classification is to find the optimal separating hyperplane that maximizes the margins between different classes. The output of SVM classification is a decision value of each pixel for each class, and it can extract good classification results from complex and noisy data.

We chose a radial basis function (RBF) to build the SVM classifier because it performs well in most cases [39]. The parameter of the Gamma (G) and penalty (C) in the kernel function were quantitatively analyzed and set to the following empirically optimized values: G = 1/feature number and C = 100 [27].

4. Results

4.1. Sea Ice Detection and Validation

Finally, the feature images based on $NDSIII_{OLCI}$ and $ENDSIII_{OLCI}$ were obtained from OLCI data, which significantly enhanced the sea ice information. We also added a feature image which considered the normalized ratio of Bands 12 and 16 as an important transition factor from $NDSIII_{OLCI}$ to $ENDSIII_{OLCI}$. Samples from the feature images were classified into the types of land cover as sea ice, seawater, turbid seawater, land, snow, and cloud through visual interpretation of OLCI true-color composite imagery. In Figure 5, sea ice is the brightest feature among the different types of land covers, suggesting its extent can be easily extracted from the feature images via threshold segmentation. The feature distribution histogram, which is a statistical representation of pixel values in feature images, was also considered for determining an appropriate threshold. In principle, sea ice information (shown in red) has the highest value of the two feature types in both normalized ratio histograms.

An additional feature of the normalized band ratio of B12 and B16 was given in response to the important role that Bands 12 and 16 play in the reduction of interference of sea ice detection in turbid seawater areas. Sea ice with bright pixels can be visually distinguished from turbid seawater with its darker pixels in the southern Bohai Sea (Figure 5c). The statistical histograms (Figure 5f) of sea ice (shown in red) and turbid seawater (shown in green) which were sampled from (B12−B16)/(B12+B16) feature image distribute separately with a boundary. This significant difference between sea ice and turbid seawater in the feature can enable us to separate them easily, but this feature could not be used to distinguish sea ice from seawater when they have approximated brightness in the feature image.

Land and cloud cover, which will be masked with great care when using other approaches, were not major sources of contamination error for sea ice identification using these OLCI imagery-based sea ice information indexes. The spectral characteristics of land and cloud enable them to be clearly identified from optical remote-sensing datasets containing rich spectral information. In the sea ice information indexes, their signals were attenuated by considering the normalized ratio of characteristic bands and were centered around −0.2 (land) and −0.3 (cloud) in $NDSIII_{OLCI}$ and −0.125 (land and cloud) in $ENDSIII_{OLCI}$ (Figure 5b,d). The obvious visible separation between the normalized ratios of these two types of cover and that of sea ice meets the condition of sea ice extraction using optical images without masking by land or cloud.

Another cover type that will impact the accuracy of sea ice mapping with optical images is snow, which has high reflectance at visible and near-infrared wavelengths. Snow-covered ice will be confused with snow-covered land when using optical data to detect sea ice. Little snow-covered ice occurs in the Bohai Sea region in winter [40]. Furthermore, the region covered by snow on land has a low normalized ratio in the sea ice information index, generally well below the value for sea ice.

The most difficult step in sea ice extraction is to divide sea ice cover from turbid seawater. The high concentration of suspended sediment in turbid seawater leads to spectral confusion and affects sea ice identification. In the feature histogram of $NDSIII_{OLCI}$ in Figure 5e, the normalized ratio of Band 20 to Band 21 for the area covered by sea ice is greater than 0, and those for seawater, land, snow, and cloud are less than T_{NDSIII} which is 0.001 in the stable stage. It is noteworthy that the normalized ratio for some turbid seawater areas is also greater than T_{NDSIII}, giving $NDSIII_{OLCI}$ insufficient ability to distinguish sea ice from turbid seawater with a high sediment concentration. Seawater and turbid seawater may be extracted with sea ice in $NDSIII_{OLCI}$ when a lower threshold value is used for segmentation. The misclassification caused by spectral confusion did not appear with $ENDSIII_{OLCI}$, which also considers OLCI Bands 12 and 16. The normalized ratio for the area covered by sea ice is greater than $T_{ENDSIII}$ which is 0.024 in the stable stage, and that of other land cover types is less than this value, including seawater and turbid seawater. Sea ice information can therefore be extracted accurately from sediment-laden water using threshold segmentation in $ENDSIII_{OLCI}$ feature images. On the basis of these results, regions with sea ice were extracted in this study by threshold segmentation of $NDSIII_{OLCI}$ and $ENDSIII_{OLCI}$ feature images using certain thresholds.

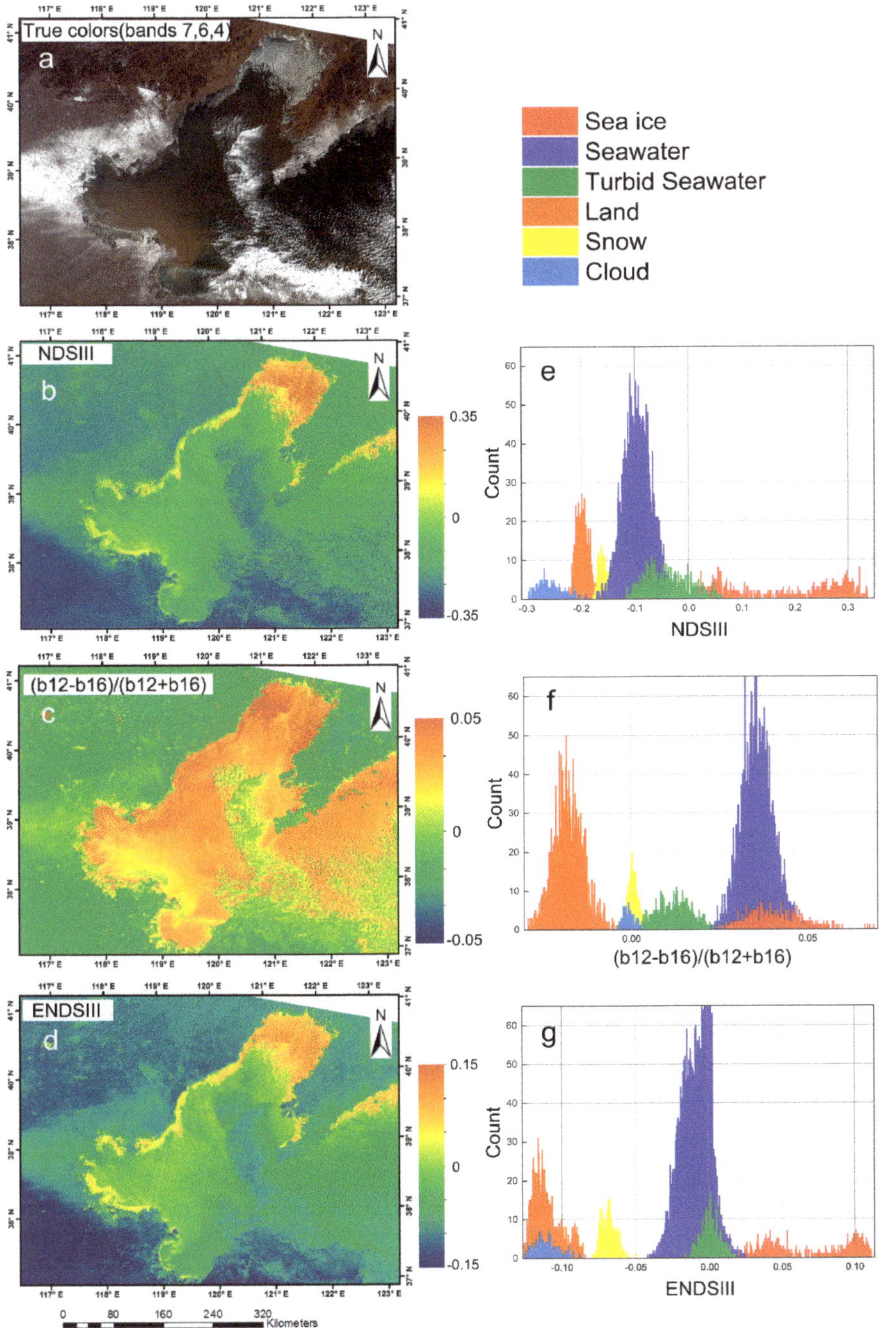

Figure 5. An example of true-color (**a**), $NDSIII_{OLCI}$ (**b**), (b12−b16)/(b12+b16) (**c**), and $ENDSIII_{OLCI}$ (**d**) feature extraction from OLCI imagery on 24 January 2018. A statistical histogram of the main types of surface cover (sea ice, seawater, turbid seawater, land, snow, and cloud) in each image is displayed alongside the corresponding image (**e**–**g**).

After sea ice extent extraction from OLCI imagery on the basis of threshold segmentation which was established using the Jenks natural break method from different stages of sea ice, the extraction results were compared with NDSI and SVM methods and validated using a simultaneously acquired high-resolution Sentinel-2 MSI image with a spatial resolution of 60 m after preprocessing (Figure 6).

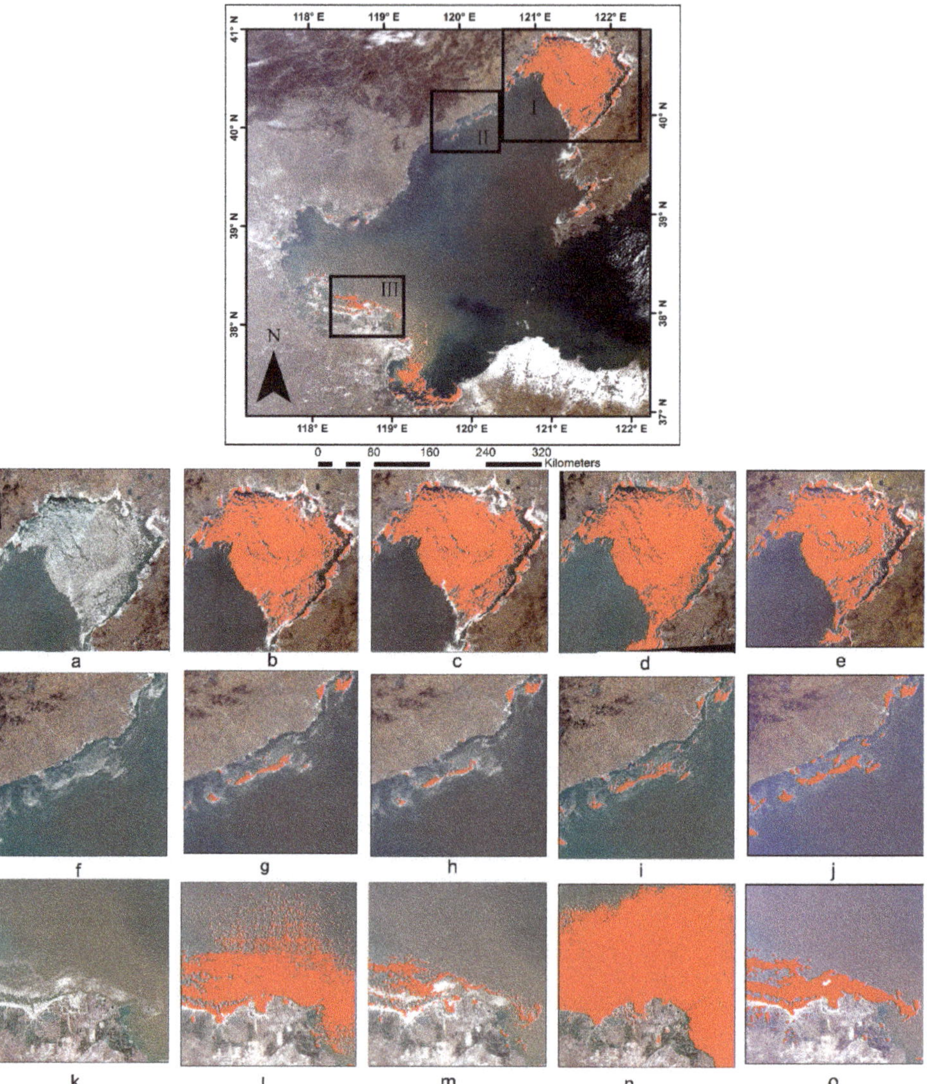

Figure 6. An image of sea ice extraction result for the entire Bohai Sea (top) on 1 February 2018 using threshold segmentation from ENDSIII$_{OLCI}$. Three true-color images in the first column (**a**,**f**,**k**) are the enlarged MSI validation images indicated by boxes I, II, and III in the top image, respectively. The next four columns present the results of sea ice extent extraction using NDSIII$_{OLCI}$ (**b**,**g**,**l**), ENDSIII$_{OLCI}$ (**c**,**h**,**m**), NDSI (**d**,**i**,**n**), and SVM (**e**,**j**,**o**).

Figure 6 shows a comparison among the different methods of sea ice extraction from satellite imagery in the Bohai Sea on February 1, 2018. Three representative scenes, including high concentration of ice (Figure 6a), low concentration of ice (Figure 6f), and sea ice in turbid water (Figure 6k), were selected from the MSI images for detection validation. The MSI image was also employed in sea ice detection via NDSI using threshold segmentation, so as to compensate for the deficiency in NDSI extraction from OLCI images (Figure 6d,i,n). Sea ice extent was also extracted from OLCI images using the SVM classification method as a comparison (Figure 6e,j,o).

Generally, the spectral-characteristic-based sea ice detection method was well capable of identifying the Bohai Sea ice in OLCI images and enabled the details of its extent, such as the ice edges and ice lanes, to be rapidly and precisely determined. The largest critical sea ice hazard, in Liaodong Bay (Figure 6a), was extracted from OLCI images using $NDSIII_{OLCI}$ (Figure 6b), $ENDSIII_{OLCI}$ (Figure 6c), NDSI (Figure 6d), and SVM (Figure 6e), and the first three distribution maps give similar depictions of the sea ice, but omission of sea ice detection occurred when performing the SVM classifier. The three indexes had consistently good performance in critical regions with thick, extensive sea ice cover in the northern Bohai Sea. A comparison of the results shows that the sea ice area extracted by the $NDSIII_{OLCI}$ (Figure 6l) and NDSI (Figure 6n) are larger than that extracted by the $ENDSIII_{OLCI}$ (Figure 6l) and SVM (Figure 6o). This is mainly attributed to the complex seawater environment and different sea ice features near the Yellow River estuary in Laizhou Bay (Figure 6k) where the concentration of suspended sediment reaches 100 mg l^{-1} in winter. The $NDSIII_{OLCI}$ and NDSI are likely to overestimate the extent of sea ice in coastal waters where the sediment concentration is high. However, the results of the $ENDSIII_{OLCI}$ and SVM are not affected by turbid seawater contamination, and comparison with the reference images indicate that they can accurately depict the outer edge of sea ice in areas of turbid sea. In addition, omitted extraction in the extent of sea ice was observed in both indexes at the western coast of the Bohai Sea where the floe size is small or the ice cover is sparse (Figure 6f). Validation against the MSI image indicates that only thicker sea ice with higher brightness in the remote-sensing image was well identified using these approaches. Thus, thin ice was not effectually detected when extracting the Bohai Sea ice from OLCI imagery using the multispectral-bands ratio indexes employed in this study.

Comparison of the results confirms that land and cloud do not contribute to the sea ice signal. Threshold segmentation based on sea ice information indexes is efficiently capable of extracting sea ice extent without masking by land and cloud. Additionally, snow-covered land cannot influence the sea ice detection using our indexes, even though the snow was perceived to exist in the shore side region beside sea-ice-covered areas.

The validation results clearly show that the different methods achieved different sea ice detection accuracies. The accuracy of our $ENDSIII_{OLCI}$ was high, with an overall accuracy of 94.83% and a Kappa coefficient of 76.54%, close to the accuracy of SVM, and higher than $NDSIII_{OLCI}$ or NDSI (Table 2). The results indicate that the main source of error was the mislabeling of turbid seawater as sea ice. Given the spectral confusion between sea ice and turbid seawater, the error was relatively significant in double-bands ratio methods, such as NDSI and $NDSIII_{OLCI}$. The SVM method reached high detection accuracy through image classification, but it needs sample training in the complex classifier, which is relatively time-consuming and inefficient. However, the $ENDSIII_{OLCI}$ has the advantage of rapid and effective detection of sea ice while outperforming the other methods. These results suggest that our $ENDSIII_{OLCI}$ method is well suited for sea ice monitoring in the Bohai Sea, even with its complex seawater environment during winter.

The sea ice extraction results via $ENDSIII_{OLCI}$ were also validated using another two simultaneous MSI images on different dates (29 January 2018 (Figure 7a) at ice stable stage and 16 February 2018 (Figure 7c) at ice melting stage) which are available. The sea ice extraction results from $ENDSIII_{OLCI}$ (Figure 7) show that the method can effectively extract sea ice extent at different ice stages. A few areas with high reflectance in the image were not extracted as sea ice, which may be caused by the snow cover. The snow cover area was small and had little effect on the sea ice extraction.

Table 2. Contingency table for accuracy validation of sea ice detection based on different methods on 1 February 2018.

			Ground Truth			
			Sea Ice	**Other**	**Total**	**Commission Error**
ENDSIII$_{OLCI}$	**Map**	Sea Ice	89	11	100	11.00%
		Other	35	754	789	4.44%
		Total	124	765	889	
		Omission Error	28.23%	1.44%		Overall Accuracy
		Kappa		76.54%		94.83%
ENDSIII$_{OLCI}$	**Map**	Sea Ice	94	35	129	27.13%
		Other	30	730	760	3.95%
		Total	124	765	889	
		Omission Error	24.19%	4.58%		Overall Accuracy
		Kappa		70.05%		92.69%
NDSI	**Map**	Sea Ice	107	77	184	41.85%
		Other	18	798	816	2.21%
		Total	125	875	1000	
		Omission Error	14.40%	8.80%		Overall Accuracy
		Kappa		63.88%		90.50%
SVM	**Map**	Sea Ice	97	19	116	16.38%
		Other	28	762	790	3.54%
		Total	125	781	906	
		Omission Error	22.40%	2.43%		Overall Accuracy
		Kappa		77.51%		94.81%

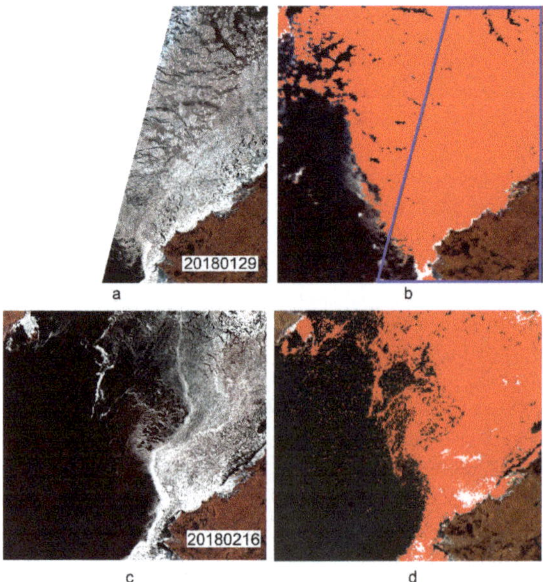

Figure 7. Sea ice extraction results based on ENDSIII$_{OLCI}$ from OLCI images with 300 m spatial resolution on 29 January 2018 (**b**) and February 16, 2018 (**d**). Two true-color images in the first column (**a**,**c**) are the MSI validation images with 60 m spatial resolution. The blue box in (**b**) represents the boundary of the validation image (**a**).

4.2. Spatiotemporal Evolution of the Bohai Sea Ice in the 2017–2018 Winter

Sea ice coverage significantly expanded in the Bohai Sea from December 2017 to January 2018. The $ENDSIII_{OLCI}$, which efficiently reduces the interference of turbid seawater in the southern Bohai Sea, was further applied to monitoring the variability in sea ice extent with 300 m spatial resolution in the Bohai Sea during the winter of 2017–2018. Owing to the limits of cloud coverage and the revisit cycle of the satellite, only 18 images were acquired by the Sentinel-3 OLCI instrument in the Bohai Sea region from 1 January 2018 to 8 March 2018. All of the images were utilized for sea ice extent extraction and determination of spatiotemporal change in sea ice coverage during the 2017–2018 winter season. Several clear-sky scenes were acquired by Sentinel-3 OLCI prior to January 1, 2018, in December 2017. At that time, only sporadic sea ice coverage could be identified near the coastal region in the Liaodong Bay (results not shown). In early January 2018, most of the sea ice was confined to the northern part of the Liaodong Bay region. The average sea ice coverage from January 1 to January 20 (Figure 8a–e) was less than 1,400 km^2; because of cloud contamination over the sea ice area, there was some underestimation of the extent of sea ice on January 13 and 20.

A significant increase in sea ice coverage occurred in the Bohai Sea in mid-January 2018 (Figure 9), with a particularly pronounced expansion occurring between 20 January 2018 (Figure 8e) and 24 January 2018 (Figure 8f). In those four days, the sea ice expanded to cover a large offshore area in Liaodong Bay, as well as some areas in Bohai Bay and Laizhou Bay. On 24 January 2018, the total sea ice coverage was 10,827 km^2 (Figure 8f). By 28 January 2018, it had further expanded to both northern and western Liaodong Bay, causing the total sea ice coverage in the Bohai Sea to jump to its peak value for the entire winter season, 13,060 km^2 (Figure 8g). The sea ice began its first retreat in late January and early February. The sea ice coverages on January 29, January 31, and February 1 were 7,457 km^2 (Figure 8h), 6,489 km^2 (Figure 8i), and 5,963 km^2 (Figure 8j), respectively.

The sea ice showed a notable resurgence in early February (Figure 9). Three days after the first retreat, on February 4, 2018, the sea ice had again covered half of Liaodong Bay and had reached a coverage of 10,497 km^2 (Figure 8k). On 5 February 2018, the ice coverage was insistent, at 9,935 km^2 (Figure 8l). After 12 February 2018, when the total coverage was 12,954 km^2 (Figure 8m), into late February, the sea ice coverage rapidly reduced and became more fragmented. The sea ice melted from the south to the north, the opposite direction to its growth, with a gradual downward trend in the ice coverage from 16 February 2018 to 8 March 2018, during which period successive images showed coverages of 6,337, 4,820, 1,932, 3,063, and 1,470 km^2 (Figure 8n–r, respectively). The remaining sea ice was mainly concentrated in the north of Liaodong Bay and had drifted to and accumulated in the east of Liaodong Bay under the action of external forces such as wind and waves. The sea ice had finally completely melted away in mid-March.

Figure 8. Spatiotemporal evolution of the extent of Bohai Sea ice during the winter of 2017–2018 from OLCI images using the ENDSIII$_{OLCI}$ method (**a**–**r**). Red areas depict sea ice coverage.

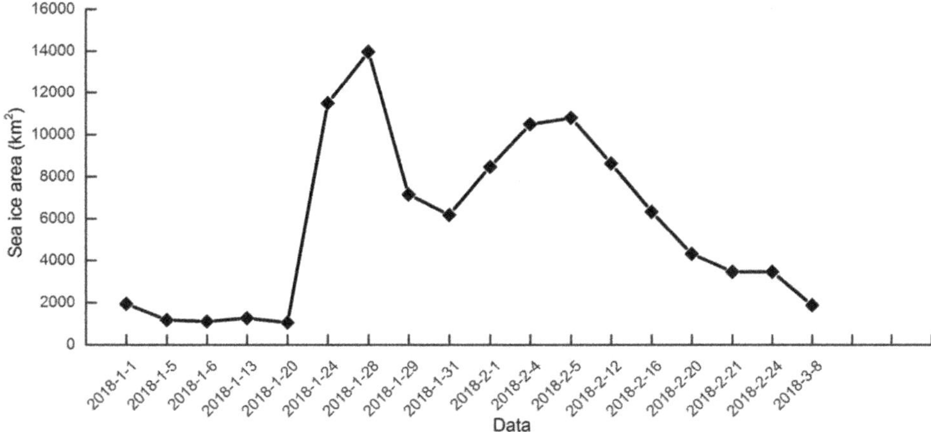

Figure 9. The evolution of the Bohai Sea ice area extracted from OLCI images during the winter of 2017–2018.

5. Conclusions

Two sea ice information indexes have been developed to quickly and accurately extract the extent of sea ice from OLCI remote-sensing data. Comparison of the extraction results with higher-resolution Sentinel-2 MSI imagery verifies that these indexes enable sea ice mapping with OLCI data in the Bohai Sea. The Normalized Difference Sea Ice Information Index ($NDSIII_{OLCI}$), which is the normalized ratio of Band 20 to Band 21 in OLCI TOA reflectance images, and the Enhanced Normalized Difference Sea Ice Information Index ($ENDSIII_{OLCI}$), which is a modification of $NDSIII_{OLCI}$ in which Bands 12 and 16 are also incorporated, can effectively detect sea ice information in the Bohai Sea and suppress most background information, such as coverage by land, cloud, and snow. Comparison between the results from our indexes, famous NDSI, and SVM methods indicates that sediment-laden water can interfere with sea ice extraction with the $NDSIII_{OLCI}$ and NDSI but that the $ENDSIII_{OLCI}$ and SVM suffer from less such interference. However, these four methods have poorer performance in detecting thin sea ice in the western Bohai Sea than they do in detecting thick sea ice. The accuracy evaluation suggests that our $ENDSIII_{OLCI}$ index can rapidly and accurately detect and map the sea ice extent in the Bohai Sea during winter. The results also show our approach can extract most of the sea ice (including nilas ice, gray ice, and gray-white ice) in OLCI images, but the new ice which is small and thin is hard to interpret and detect from the medium-resolution OLCI images due to the limitation of the spatial resolution. Moreover, it would be better to reset the threshold values when employing our indexes to detect sea ice extent in other years because the ice conditions vary with the years.

The spatiotemporal evolution of the Bohai Sea ice in the winter of 2017–2018 was monitored by applying the $ENDSIII_{OLCI}$ to OLCI images. Two major increases were detected in the sea ice extent in mid-January and early February. The largest extent of the sea ice was 13,060 km^2 on January 28. After reaching its peak in late January 2018, sea ice coverage remained high until early February, and the sea ice then gradually melted from south to north in mid-February. The whole period when there was ice coverage lasted for about four months, within which there was a significant expansion in mid-January and a final fading away in early March. Overall, our proposed method provides a convenient and effective technique for sea ice detection and evolution study in the Bohai Sea, which can help monitor the recent impacts of global warming.

Author Contributions: H.S. and B.J. conceived and designed the experiments; B.J. performed the experiments; H.S. and B.J. analyzed the results; H.S. and B.J. wrote the paper; H.S. and Y.W. revised the paper.

Funding: This research was funded by National Natural Science Foundation of China (41971384, 41601444, 41630963), Natural Science Foundation of Fujian Province, China (2017J01657), Outstanding Young Scientists Program in Universities of Fujian Province (KJ2017-17), and Central Guide Local Science and Technology Development Projects (2017L3012).

Acknowledgments: We thank the European Space Agency (ESA) data hub for the Sentinel-3 OLCI data and Sentinel-2 MSI imagery (https://scihub.copernicus.eu/), which are freely accessible to the public.

Conflicts of Interest: The authors declare no conflict of interest.

References

1. Ning, L.; Xie, F.; Gu, W.; Xu, Y.; Huang, S.; Yuan, S.; Cui, W.; Levy, J. Using remote sensing to estimate sea ice thickness in the Bohai Sea, China based on ice type. *Int. J. Remote Sens.* **2009**, *30*, 4539–4552. [CrossRef]
2. Su, H.; Wang, Y.; Yang, J. Monitoring the Spatiotemporal Evolution of Sea Ice in the Bohai Sea in the 2009–2010 Winter Combining MODIS and Meteorological Data. *Estuaries Coasts* **2012**, *35*, 281–291. [CrossRef]
3. Shi, W.; Wang, M. Sea ice properties in the Bohai Sea measured by MODIS-Aqua: 1. Satellite algorithm development. *J. Mar. Syst.* **2012**, *95*, 32–40. [CrossRef]
4. Ouyang, L.; Hui, F.; Zhu, L.; Cheng, X.; Cheng, B.; Shokr, M.; Zhao, J.; Ding, M.; Zeng, T. The spatiotemporal patterns of sea ice in the Bohai Sea during the winter seasons of 2000–2016. *Int. J. Digit. Earth* **2017**, 1–17. [CrossRef]
5. Liu, M.; Dai, Y.; Zhang, J.; Zhang, X.; Meng, J.; Xie, Q. PCA-based sea-ice image fusion of optical data by HIS transform and SAR data by wavelet transform. *Acta Oceanol. Sin.* **2015**, *34*, 59–67. [CrossRef]
6. Teleti, P.R.; Luis, A.J. Sea ice observations in polar regions: Evolution of technologies in remote sensing. *Int. J. Geosci.* **2013**, *4*, 1031–1050. [CrossRef]
7. Karvonen, J. Baltic sea ice concentration estimation based on C-band dual-polarized SAR data. *IEEE Trans. Geosci. Remote Sens.* **2014**, *52*, 5558–5566. [CrossRef]
8. Ivanova, N.; Pedersen, L.; Tonboe, R.; Kern, S.; Heygster, G.; Lavergne, T.; Sørensen, A.; Saldo, R.; Dybkjær, G.; Brucker, L. Inter-comparison and evaluation of sea ice algorithms: Towards further identification of challenges and optimal approach using passive microwave observations. *Cryosphere* **2015**, *9*, 1797–1817. [CrossRef]
9. Lavergne, T.; Macdonald Sørensen, A.; Kern, S.; Tonboe, R.; Notz, D.; Aaboe, S.; Bell, L.; Dybkjær, G.; Eastwood, S.; Gabarro, C.; et al. Version 2 of the EUMETSAT OSI SAF and ESA CCI sea-ice concentration climate data records. *Cryosphere* **2019**, *13*, 49–78. [CrossRef]
10. Meier, W.N.; Fetterer, F.; Stewart, J.S.; Helfrich, S. How do sea-ice concentrations from operational data compare with passive microwave estimates? Implications for improved model evaluations and forecasting. *Ann. Glaciol.* **2015**, *56*, 332–340. [CrossRef]
11. Agnew, T.; Howell, S. The use of operational ice charts for evaluating passive microwave ice concentration data. *Atmosphere-Ocean* **2003**, *41*, 317–331. [CrossRef]
12. Berg, A.; Eriksson, L.E.B. SAR algorithm for sea ice concentration—Evaluation for the Baltic Sea. *IEEE Geosci. Remote Sens. Lett.* **2012**, *9*, 938–942. [CrossRef]
13. Deng, H.; Clausi, D.A. Unsupervised segmentation of synthetic aperture radar sea ice imagery using a novel Markov random field model. *IEEE Trans. Geosci. Remote Sens.* **2005**, *43*, 528–538. [CrossRef]
14. Huck, P.; Light, B.; Eicken, H.; Haller, M. Mapping sediment-laden sea ice in the Arctic using AVHRR remote-sensing data: Atmospheric correction and determination of reflectances as a function of ice type and sediment load. *Remote Sens. Environ.* **2007**, *107*, 484–495. [CrossRef]
15. Shi, W.; Wang, M. Sea ice properties in the Bohai Sea measured by MODIS-Aqua: 2. Study of sea ice seasonal and interannual variability. *J. Mar. Syst.* **2012**, *95*, 41–49. [CrossRef]
16. Yuan, S.; Liu, C.; Liu, X. Practical Model of Sea Ice Thickness of Bohai Sea Based on MODIS Data. *Chin. Geogr. Sci.* **2018**, *28*, 863–872. [CrossRef]
17. Drüe, C.; Heinemann, G. High-resolution maps of the sea-ice concentration from MODIS satellite data. *Geophys. Res. Lett.* **2004**, *31*. [CrossRef]
18. Zhang, D.; Ke, C.; Sun, B.; Lei, R.; Tang, X. Extraction of sea ice concentration based on spectral unmixing method. *J. Appl. Remote Sens.* **2011**, *5*, 053552. [CrossRef]
19. Liu, W.; Sheng, H.; Zhang, X. Sea ice thickness estimation in the Bohai Sea using geostationary ocean color imager data. *Acta Oceanol. Sin.* **2016**, *35*, 105–112. [CrossRef]

20. Lang, W.; Wu, Q.; Zhang, X.; Meng, J.; Wang, N.; Cao, Y. Sea ice drift tracking in the Bohai Sea using geostationary ocean color imagery. *J. Appl. Remote Sens.* **2014**, *8*, 083650. [CrossRef]
21. Yan, Y.; Huang, K.; Shao, D.; Xu, Y.; Gu, W. Monitoring the Characteristics of the Bohai Sea Ice Using High-Resolution Geostationary Ocean Color Imager (GOCI) Data. *Sustainability* **2019**, *11*, 777. [CrossRef]
22. Wang, X.; Wu, Z.; Wang, C.; Li, X.; Li, X.; Qiu, Y. Reducing the impact of thin clouds on Arctic Ocean sea ice concentration from FengYun-3 MERSI data single cavity. *IEEE Access* **2017**, *5*, 16341–16348. [CrossRef]
23. Su, H.; Wang, Y. Using MODIS data to estimate sea ice thickness in the Bohai Sea (China) in the 2009–2010 winter. *J. Geophys. Res. Oceans* **2012**, *117*. [CrossRef]
24. Su, H.; Wang, Y.; Xiao, J.; Li, L. Improving MODIS sea ice detectability using gray level co-occurrence matrix texture analysis method: A case study in the Bohai Sea. *ISPRS J. Photogramm. Remote Sens.* **2013**, *85*, 13–20. [CrossRef]
25. Han, Y.; Li, J.; Zhang, Y.; Hong, Z.; Wang, J. Sea ice detection based on an improved similarity measurement method using hyperspectral data. *Sensors* **2017**, *17*, 1124. [CrossRef]
26. Gignac, C.; Bernier, M.; Chokmani, K.; Poulin, J. IceMap250—Automatic 250 m sea ice extent mapping using MODIS data. *Remote Sens.* **2017**, *9*, 70. [CrossRef]
27. Su, H.; Wang, Y.; Xiao, J.; Yan, X. Classification of MODIS images combining surface temperature and texture features using the Support Vector Machine method for estimation of the extent of sea ice in the frozen Bohai Bay, China. *Int. J. Remote Sens.* **2015**, *36*, 2734–2750. [CrossRef]
28. Zhang, N.; Wu, Y.; Zhang, Q. Detection of sea ice in sediment laden water using MODIS in the Bohai Sea: A CART decision tree method. *Int. J. Remote Sens.* **2015**, *36*, 1661–1674. [CrossRef]
29. Toming, K.; Kutser, T.; Uiboupin, R.; Arikas, A.; Vahter, K.; Paavel, B. Mapping water quality parameters with sentinel-3 ocean and land colour instrument imagery in the Baltic Sea. *Remote Sens.* **2017**, *9*, 1070. [CrossRef]
30. Bi, S.; Li, Y.; Wang, Q.; Lyu, H.; Liu, G.; Zheng, Z.; Du, C.; Mu, M.; Xu, J.; Lei, S.; et al. Inland Water Atmospheric Correction Based on Turbidity Classification Using OLCI and SLSTR Synergistic Observations. *Remote Sens.* **2018**, *10*, 1002. [CrossRef]
31. Lin, J.; Lyu, H.; Miao, S.; Pan, Y.; Wu, Z.; Li, Y.; Wang, Q. A two-step approach to mapping particulate organic carbon (POC) in inland water using OLCI images. *Ecol. Indic.* **2018**, *90*, 502–512. [CrossRef]
32. Vuolo, F.; Zóltak, M.; Pipitone, C.; Zappa, L.; Wenng, H.; Immitzer, M.; Weiss, M.; Baret, F.; Atzberger, C. Data service platform for Sentinel-2 surface reflectance and value-added products: System use and examples. *Remote Sens.* **2016**, *8*, 938. [CrossRef]
33. Riggs, G.A.; Hall, D.K.; Ackerman, S.A. Sea ice extent and classification mapping with the moderate resolution imaging spectroradiometer airborne simulator. *Remote Sens. Environ.* **1999**, *68*, 152–163. [CrossRef]
34. Jenks, G.F. The data model concept in statistical mapping. *Int. Yearb. Cartogr.* **1967**, *7*, 186–190.
35. Dozier, J. Spectral signature of alpine snow cover from the landsat thematic mapper. *Remote Sens. Environ.* **1989**, *28*, 9–22. [CrossRef]
36. Burges, C.J.C. A tutorial on support vector machines for pattern recognition. *Data Min. Knowl. Discov.* **1998**, *2*, 121–167. [CrossRef]
37. Vapnik, V.N. An overview of statistical learning theory. *IEEE Trans. Neural Netw.* **1999**, *10*, 988–999. [CrossRef]
38. Weston, J.; Watkins, C. Support vector machines for multi-class pattern recognition. In Proceedings of the 7th European Symposium on Artificial Neural Networks (ESANN-99), Bruges, Belgium, 21–24 April 1999; Volume 99, pp. 219–224.
39. Zhang, H.; Zhang, Y.; Lin, H. Compare different levels of fusion between optical and SAR data for impervious surfaces estimation. In Proceedings of the 2nd International Workshop on Earth Observation and Remote Sensing Applications, EORSA 2012, Shanghai, China, 8–11 June 2012; pp. 26–30. [CrossRef]
40. Yuan, S.; Gu, W.; Liu, C.; Xie, F. Towards a semi-empirical model of the sea ice thickness based on hyperspectral remote sensing in the Bohai Sea. *Acta Oceanol. Sin.* **2017**, *36*, 80–89. [CrossRef]

MDPI
St. Alban-Anlage 66
4052 Basel
Switzerland
Tel. +41 61 683 77 34
Fax +41 61 302 89 18
www.mdpi.com

Remote Sensing Editorial Office
E-mail: remotesensing@mdpi.com
www.mdpi.com/journal/remotesensing

www.ingramcontent.com/pod-product-compliance
Lightning Source LLC
LaVergne TN
LVHW070200170726
843515LV00007B/1965

* 9 7 8 3 0 3 9 2 8 6 5 8 4 *